THE ETERNAL TRAIL

A Tracker Looks at Evolution

OTHER BOOKS BY MARTIN LOCKLEY

Dinosaur Tracks and Traces (with David Gillette)

Tracking Dinosaurs

Dinosaur Tracks and Other Fossil Footprints of the Western United States (with Adrian Hunt)

Dinosaur Tracks and Other Fossil Footprints of Europe (with Christian Meyer)

THE ETERNAL TRAIL

A Tracker Looks at Evolution

MARTIN LOCKLEY

PERSEUS BOOKS

Cambridge, Massachusetts

Many of the designations used by manufacturers and sellers to distinguish their products are claimed as trademarks. Where those designations appear in this book and Perseus Publishing was aware of a trademark claim, the designations have been printed in initial capital letters.

A CIP record for this book is available from the Library of Congress
ISBN: 0-7382-0362-9

Copyright © 1999 by Martin Lockley

Perseus Publishing is a member of the Perseus Books Group

All illustrations by Martin Lockley except figures 1.3, 2.1, 10.4

Text design by Heather Hutchison
Set in 11-point Minion

1 2 3 4 5 6 7 8 9 10—03 02 01 00
First paperback printing, July 2000

Perseus Publishing books are available at special discounts for bulk purchases in the U.S. by corporations, institutions, and other organizations. For more information, please contact the Special Markets Department at HarperCollins Publishers, 10 East 53rd Street, New York, NY 10022, or call 1-212-207-7528.

Find us on the World Wide Web at
http://www.perseuspublishing.com

To all fellow travelers on the eternal trail, especially family, friends, colleagues, and students:

My parents, Jill and Ronald; my children, Peter and Katie; my brother Stephen and sister Ann; my stepchildren, Linda and Luis; my tracker co-conspirators Christian "Jacques Cousteau" Meyer and Adrian "Hidden Genius" Hunt; my sometimes spiritual mentors Gretchen and Giuseppe; colleagues Karen, Masaki, Marvin, Vanda, and Vicki; practitioners in the healing arts Jacob, Jean, Rowan, and Vega; and last but not least, to a potential new generation of unruly dinosaur trackers: Anne, Beth, Emma, Fabbio, Joanna, John, Mike, Rebecca, and Richard. May the road be smooth and may you see farther down the trail than I ever will.

I hope my mind will not grow tall
to look down on things,
but wide to embrace all sorts of things.
                    –Evelyn Underhill

CONTENTS

Preface   xi
Acknowledgments   xvii

Introduction: The Trail from There to Here   1

1 **First Impressions**   9

Act I: Cast of Characters, 9
Love Is the Aboriginal Tracker, 11
Our Long Track Record, 15
Tracking Extinct Animals, 16
Elite Feet, 17
Memories, Impressions, and Reflections, 19
Individual Signatures, 21
Binary Bipeds and Digital Digressions, 23
The Beat of the Feet, 25
The Trail Through Time, 26
Pioneers Along the Trail, 29
Leading with the Legs, 31
Male Feet and Female Footprints, 39

2 **Paleozoic Prelude**   43

Act II: Cast of Characters, 43
The Deep Structure of Sediments and Strata, 44
Cruising the Cambrian: Fossil Art in the Making, 47
Walking in Circles: The First Footprints on Land, 53
Fish Out of Water: Giant Steps for Vertebrates, 54
Monster Millipedes, 56
The First Reptile Tracks? 58
Giant Swamp Dwellers, 60
Dimetropus: Our Earliest Mammal Ancestors, 61
Permian Murders and Dirty Devils, 63

The Mammal Underground: Of Burrows and Bush Pigs, 64
Running from Floods in the Desert, 67
Shape Shifting, 71
Spiders in the Dunes, 74
The Italian Bigfoot, 76

3 Hand Signals   79

Act III: Cast of Characters, 79
Worldwide Wanderings, 80
The Hand Animal, 82
A Message in the Lady's Hand, 86
Now You See 'Em, Now You Don't, 89
Tracking the Crimson Crocs, 90
The First Biped, 93
A Protomammal with Hairy Feet?, 94
Lizards Galore, 95
The First Giants, 96
Now We See 'Em: A New Look at Dinosaur
    National Monument, 98
Gateway to the World of Miniature Mammals, 99
Dinosauroids and the First Dinosaurs, 100
The Stress of Missing and Broken Pieces, 102

4 Noah's Raven   105

Act IV: Cast of Characters, 105
Fibonacci's Fingers: The Sacred Geometry of
    Hands and Feet, 106
The First Dinosaur Tracker, 108
Big Blue Bananas in the Great Rift Valley, 110
Grallator: The "Early Bird" That Went on Stilts, 112
True Thunder, 113
The Giant Animal, 115
Odd Foot, 118
The Diminutive Dune Runners, 120
Hop to It, 122
Left Limp, 125
The World's Longest Brontosaur Tracks, 126
Glimpse of a Dinosaur from the Dark Ages, 127

5 Lost Souls   129

Act V: Cast of Characters, 129
Megalosaurs and Megatracksites, 129

Holy Father of the Elephants: The World's Longest Dinosaur
     Trackways, 134
Padded Out: Of Flesh and Bones, 137
Pterosaur Runways, 138
Dinehichnus: The People's Track, 141
Social Commentary, 143
The Rise and Fall of Thunder Beings and Thunder Lizards, 144
Dinosaur Trails in Purgatory, 147
Trampled to Death, 150
Crocodiles in the Channel, 151
Sacred Mule Tracks From Lobster Bay, 153
Teil hard's Enigmatic Chinese Footprint, 155

6 **With God on Our Side**   **159**

Act VI: Cast of Characters, 159
Tracking Dinosaurs in *The Lost World,* 160
A Pterosaurian Bigfoot or Two, 162
A Rare Commodity: Tracks of Armored Dinosaurs, 163
The Early Birds: New Perspectives on Bird Evolution, 165
Ships That Pass in the Night: A Mesozoic Crossroads, 166
Baby Brontosaurs, 169
Wading in Deeper, 170
Ghost Prints, Phantom Tracks, and Phantom Limbs, 172
Bird's Herd, 173
Trackers and Attackers, 176
Sauropod Serenade, 180
Creative Midnight Chisel Work: Evolution Versus Creation, 181
Evolution as Creation: Science and Spirituality, 185

7 **Dancing Dinosaurologists**   **191**

Act VII: Cast of Characters, 191
Migrating Along the Dinosaur Freeway, 192
A Promenade Along Dinosaur Ridge, 194
More Flying Crocodiles, 195
Dinosaurs and Dynamite, 197
The Battle of Carenque, 199
A Dinosaur Stampede, 201
The Kangaroo Factor, 203
A Korean Duck Pond, 204
Pterosaur Psynchronicity, 205
The Mystery Dinosaur, 207
T. Rex Tracks and Dancing Dinosaurologists, 209
The Wall: Getting High in the Andes, 212

Of Hooves and Horned Dinosaurs, 215
Celestial Messages on the Iridium Band, 221
The Last Dinosaur Tracks, 224

8 Out of Africa   **225**

Act VIII: Cast of Characters, 225
Tracking The Rise of the Himalayas: Mammals
   Masquerading as Dinosaurs, 226
Horse Tracks at Dawn, 227
Of Frogs and Flamingos, 229
Tracking the Symbolism of the Cloven Hoof, 231
Pliocene Dressage: Early Equestrianism, 235
Leakey Tracks Lucy, 239
Blessed Trail of Our Ancestors, 245

9 Spirit Trails   **249**

Act IX: Cast of Characters, 249
Mammoths and Mastodons, 249
A Sloth with Sandals, 252
Spoor of the Carnivore, 254
Giant Wombats Dead in Their Tracks, 256
Moa Tracks but No Moa Track Maker, 257
Tracking Bigfoot, 261
A Convergence of Trails, 268

10 The Signature of Humanity   **275**

Act X: Cast of Characters, 275
Mirror, Mirror on the Wall, 276
Fingerprints of Our Genes, 279
Electrostatic Footprints, 282
Over the Moon: The Trail Leaves Our Planet, 283
Tetrapods on Mars, 287
This Land Is Your Land, This Land Is My Land, 290
Sacred Ground, 292
Save the Last Dance for Me, 296

Notes   **301**
Index   **325**

PREFACE

Although some scientists and philosophers may argue that time does not exist, most of us perceive life as a journey with a beginning and an end. For most of us it is also an epic journey that encompasses the greater part of our experience, the hardships of tough expeditions and the exhilaration of new discoveries. But our mortal journeys do not make up all of our experience, for we are inextricably linked with the journeys of our ancestors and our descendants; we better understand the purpose of our travels through the cosmos if we know where we came from and where we are heading. As we become aware of the universal chain of being, we realize that evolution began long before the first creature set foot on land, and will continue long after future generations of astronauts set foot on faraway planets. For the philosophical, holistic or cosmic tracker, the first footprints on the shores of ancient Paleozoic sea ways and the random scatter of hominid trackways on the moon are merely obvious, physical manifestations of the eternal trail, the evolution of the cosmos. On a deeper level such bits of track evidence are part of a huge, seamless network of trails left by living organisms, broken only where time and erosion have removed or hidden pieces of the puzzle. But the essential fabric of the web remains as a coherent system representing the shadow and spirit of all creatures that have walked this earth.

Just as our daily journeys and excursions blend into each other without discrete beginnings and endings, so we find that the trails of our ancestors, fellow travelers, and those who take the baton to run the next leg are all woven together into a web of endless trails. It is our awareness of past roads and future orientations that gives us direction through the eternal now. When we look back over our shoulders at the tracks of dinosaurs and the ruins of Babylon, we can revisit the trails made by previous travelers in the sands of time. These signatures of the past are not mute markings of a lost world, but a record of the past that is full of meaning. More than this, such excerpts from the eternal trail can be read, like books, and so bring about a convergence of past and present experience. In this sense, then, it is true that time does not exist, for like the psychic archaeologist who actually experiences the past when standing on an ancient site, we can all feel linked to the past by touching a dinosaur track, or discovering an old family photograph.

This book is about the eternal trail that connects past, present, and future, and about the connectedness of all beings that have left their signatures in the soils and substrates of planet Earth and its environs. As a paleontologist who has specialized in the study of fossil footprints, I have access to the ancient track record, comprising literally millions of fossil footprints, as a tangible record of where we have been. But in what language is this fossil footprint literature written? The simple answer is that it is written in the language of the eternal trail. It is a language with which no one is fully conversant, and for which no magic Rosetta Stone has been unearthed, yet we all speak and write a few words of it, enough, at least, to leave our signatures and footprints somewhere along the trail. More than this I would say that this eternal language is in part the foundation of our own spoken and written language, for what did humankind read before we composed written words and symbols of our own making? We read the bountiful book of nature, to be sure, but specifically we understood tracks as personalized signatures of other beings. They were inscribed on substrates that now, a few thousand years later, are incorporated into the pages of our earth science journals and textbooks. Among vowels alone, *E*, *O*, and *U* are dead ringers for bird, elephant, and horse tracks!

Trackers such as paleontologists and geologists have tried to decipher this ancient language, and have begun to put together a glossary of terms. As this vocabulary emerges we are beginning to create the first draft of a more comprehensive dictionary. Individual words, like steps, can be pieced together into phrases, which like trackways begin to reveal a rhythm and cadence appropriate to the measure and beat of the creatures we are tracking. Like words in our dictionaries, tracks on the eternal trail may have many meanings, some obvious and some subtle. Ultimately no meaning is entirely right or entirely wrong: as a result of misinterpreting a track or signature, we often pause, look deeper, and so arrive at a more profound understanding.

It is my hope that as a tracker I can discover something of the deeper meaning of the eternal trail. Perhaps the best approach is to begin at the beginning and piece together an understanding of trackways and the animals that made them with simple observations and descriptions. My research group, the University of Colorado at Denver Dinosaur Trackers, has spent considerable time beside the trail patiently observing and recording the trackways of thousands of extinct animals. Such endeavors have allowed us to speak with some confidence about former travelers on the eternal trail, and draw conclusions that confirm, embellish, challenge, and even refute conventional wisdom in paleontology. How far along we are in this apprenticeship is hard to judge objectively, but we should always be ready to take the next step. Allow us, then, to ask what meaning might be found in these trackways beyond the physical manifestations of footprint evidence.

Scientists are trained to record what they see, seek new explanations, and fathom meaning. Such quests are more than a mere by-product of intellectual training. They are part of an inner exploration, an innate characteristic of an inquisitive consciousness that yearns to know itself. Our awareness of the ancient eternal trails that brought us from deep time to the present has made us want to revisit these ancient byways and travel them again; to relive our lost past and collapse a billion years of Earth history into a few generations of paleontological literature. Why are we remapping the eternal trail? And what will we find as a result of this endeavor? I can not claim that this book will provide all the answers, but I can at least pose the questions. I offer it only as my signature—another footprint on the eternal trail, a booklike image projected through the holographic negative of Earth's trampled skin. It is an image that will soon be fossilized, but yet may endure as have other footprints. If it reminds future trackers to look over their shoulders and reread the ancient trail, it will have served as a footprint that links our last step with the one we are about to make.

•   •   •

As a professional geologist, it is my natural inclination to want to stress the scientific underpinnings that lead me to draw conclusions about fossil footprints. Science is traditionally regarded as a rational and objective pursuit that should remain uncluttered by too much conjecture or speculation. But men and women of science often embark on scientific quests because of subjective callings—by the "belief" that new discoveries lie outside the known realm. Such discoveries are often anticipated intuitively—perhaps driven by metaphysical currents we do not understand or are afraid to admit for fear of appearing unscientific. (But these too are potentially accessible to scientific inquiry.) Of course, any discoveries must ultimately be verified by scientific means if they are to be widely accepted, but many appealing ideas enjoy popular and scientific approval without being rigorously tested until long after they have become common currency.

The word "science" comes from the Latin word *scientia,* meaning "having knowledge," and "philosophy" comes from the Greeks *philosophos,* "lover of wisdom." Science's realm of study is the entire cosmos, which includes both inner and outer space. If science seeks answers to questions, those who ask such questions, like wide-eyed children, have faith that the potential exists to discover answers. This conscious quest for enlightenment exists initially outside the realm of knowledge. Most scientists, whether they acknowledge it or not, are driven by a highly subjective zeal or love of this quest.

This same subjective zeal often dictates our choice of field or area of exploration. I have faith that by studying the unexplored realm of fossil footprints I will ultimately shed some new light on Earth's history. My faith has been rewarded by discoveries that exceeded my expectations. Herein lies the rub. Perhaps there is a correlation between faith and the magnitude of discovery that lies ahead. If you don't believe it, you often won't see it. Lack of faith in the utility and meaning of fossil footprints has lead many to conclude that they reveal little of significance. But the power of positive thinking does pay off in hard scientific currency.

Throughout this book my conclusions are based on the scientific literature that our research group has helped create, and on the broader literature of available knowledge. Any departure, however tangential, from the scientific underpinnings of research into fossil footprints is proposed simply as an exploratory next step. Having reported what we think we know, for us it is logical to ask "Where do we go from here?" The old adage holds that there is no such thing as a stupid question. To this end I have perhaps raised too many unanswered questions, or naively asked whether certain tenuous connections have any meaning. My ruminations about possible connections are just that; they are not statements that these connections are known to exist. But as almost everything is connected in some way, perhaps I have probability on my side.

With enough study, luck, and faith in the utility of footprints, we may begin to find answers we never dreamed possible. It is in this spirit that science and human endeavor proceeds best. Taoist science requires "a path of virtuous conduct"—a thoughtful search for enlightenment, not a pursuit of knowledge for knowledge's sake, or for selfish gain. If we equate the evolution of increasing complexity in the history of life with a more sophisticated consciousness and ever wider exploration of environment, Earth, and universe, it is perhaps reasonable to view the quest for and love of knowledge as being as old as the eternal trail we all tread. If this is so, we must surely consider that the trail leads somewhere, possibly to a better, or at least more challenging, complex and multidimensional future, and not, as some may believe, in a purely random direction. If the Tao, the creative principle that "orders" the universe, has order, then it also has direction—for what is a path without direction?

We know something of the past, however, and can step back onto the eternal trail with the maps that paleontology has provided for us. By following in the footsteps of our ancestors, perhaps we shall learn whether the trail is a random walk or a path with direction. At some point we shall meet our own species and our very selves on the trail and shall be able to pose the same perennial question: "Quo vadis?"; "Whither are you going?"; "Where do we go from here?" No matter how we phrase it, the question begins from the same fundamental premise, that we are all travelers on the eternal trail. There is no way to step off the trail, so it is only natural that we should periodically ask for direction and ponder what lies ahead.

Martin Lockley
Golden, Colorado

ACKNOWLEDGMENTS

The research on which much of this book is based would not have been possible without the sponsorship and support of many institutions, agencies, and individuals. For more than a decade the National Science Foundation provided research grant sponsorship to study fossil footprints in the western United States, and South Korea. National Geographic provided support for research conducted in East Africa and Central Asia, and I have also benefited from the support of the National Park Service, and the Swiss National Science Foundation, among many other agencies. My colleagues Christian Meyer (Switzerland), Adrian Hunt and Spencer Lucas (New Mexico), Masaki Matsukawa (Japan), Vanda Faria dos Santos (Portugal), Seong Young Yang and Seong Kyu Lim (South Korea) have been long-time friends and collaborators on dozens of research papers and research initiatives. I also acknowledge my colleagues among the faculty and students at the University of Colorado at Denver, where we base our Dinosaur Trackers Research Group. In this regard I particularly thank my colleagues Karen Houck, Marvin Loflin, and Vicki Spencer, and my good friends at Dinosaur Ridge, especially Bob Raynolds, Joe Tempel, and Karen Hester for ongoing assistance and moral support. I am also grateful to Paul Koroshetz for artwork, and to Rowan Jackson for reading the manuscript and offering helpful suggestions. In this regard, I have also benefited from several days of stimulating discussion with Wolfgang Schad, and I also thank Jim Farlow for his critical reading of my first draft of Chapter 6. I also thank Gretchen Minney for reading the manuscript with a copy editor's eye. Last but not least, thanks go to Regula Noetzli, my literary agent, who gave me encouragement and advice, and Amanda Cook, my editor at Perseus Books.

THE TRAIL FROM THERE TO HERE

We lie in the lap of immense intelligence, which makes us receivers of its truth and organs of its activity.

Ralph Waldo Emerson

The eternal trail is nothing less than the evolution of the universe. In this book we will take the role of scientists following this trail like trackers following an animal trail, to try to understand where it has led, and where it will lead. If we are to follow a trail, it is best to try to begin at the beginning. The universe started with a Big Bang, or so we are told. Space-time expanded outward and matter began to condense from the void, first giving us subatomic particles, then elements, gas clouds, nebulae, stars, and solar systems like our own. Rocky planets such as ours began to accumulate water and create the suitable cradle for life (*bios*), ultimately leading to the emergence of sentient hominids who evolved a language and such concepts as awareness, knowledge, and consciousness. We got here from there, a journey of between 10 billion and 20 billion years, as a result of a long series of events that many believe resulted largely from chance; if "someone" were to replay the tape it would all be different. But I do not believe, however, that it's all chance, that there is no pattern or design.

And I am not alone. Others, too, view this scientific, "accepted," explanation of how the universe came into being as speculation, an incomplete hypothesis based on incomplete information. The physicist Werner Karl Heisenberg noted that "we can say almost nothing clearly," and the philosopher Karl Popper noted that the principle of scientific verification is untenable.[1] Popper's classic assertion is a kind of philosophic uncertainty principle. This recognition should not be seen as a cop-out, or a license to dream up any old idea. Rather it is a recognition that at best we can agree on hypotheses and theories that reflect current awareness of the universe. But for knowledge to advance, we must be prepared to reinterpret our old ideas and integrate them with new knowledge. Our awareness of the universe continues to expand, leaving many older and previously cherished ideas lacking the fullness and power they held for previous generations of trackers on the eternal trail. We cannot rely only on "conventional scientific wisdom," for science is, or should be, the birthplace of philosophy, and should reflect the wisdom of the sages. Imagine the scientist who has "scaled the mountains of ignorance; he is about to conquer the highest peak; as he pulls himself over the highest rock he is greeted by a band of theologians who have been sitting there for centuries."[2]

I agree with the contemporary philosopher Ken Wilber, when he asserts that this Big Bang business is "very weird"—a theory in need of serious scrutiny.[3] I will make only a few observations. One can hardly follow the entire eternal trail of evolution by picking it up near the "end." But if it is an eternal trail how can we explain or even conceptualize a beginning ("back there") or an end (beyond the "here" and now). There can be no talk of "age" in the face of infinite duration. Even if we postulate finite beginnings and endings, we still face the intractable problem of trying to conceive what went before and after. The anthropologist Bruce Holbrook noted that the idea of the Big Bang invokes an unscientific "supernatural" cause. He states that "there is . . . no reason to suppose that the universe had a beginning, that it is not eternal. . . . [T]here is reason to suppose that it is cyclic. Among all known phenomena there is not one which is not cyclic."[4] This does not mean that there was not a Big Bang—simply that it was not the "beginning" of eternity. The psychoanalyst Carl Jung pointed out that "fixed" concepts of time such as beginnings, endings, and measurements are mental constructs that form "indispensable coordinates" to help us understand a universe that is largely beyond our comprehension.[5] They are the means by which we construct order for ourselves and make sense of nature. According to this notion, the Big Bang was only the starting gun for this particular lap around the arena of eternity. Our known universe is merely a leg in an eternal relay run from an unknown before to an unknown after. The eternal trail is open-ended.

The British physicist David Bohm, and other like-minded intellectuals, have been widely cited in popular cosmological literature because they help "explain" the manifestation or emergence through time of progressively more complex matter, from wispy gases to rocky planets and complex life forms. The universe is "driven" or "drawn toward" some goal by intelligent or at least purposeful forces. Hidden potentialities—also known as deep structure, or implicate (Bohm's term) order—manifest themselves periodically as atoms, stars, life, and human consciousness. We can regard such potentialities as the energetic fabric, spirit, or consciousness of the universe. Some people say that such concepts converge with their definition of God (nondenominational, I hasten to add). In their worldview, God represents the as yet unfathomable unknown or beyond, which by definition has not yet been explained by science. We shall return to this topic in Chapter 6, when the eternal trail runs into the evolution/creation controversy and broader issues of science and spirituality.

Some people consider talk of God and spirituality unscientific. But a mystical or intuitive approach to understanding existence is not incompatible with the scientific one. Many leading scientists freely admit that they have allowed themselves to be guided by intuition for insights that led to scientific breakthroughs. In *The End of Science,* John Horgan charges that many of our most revered scientists have breached the boundaries of empirical science, and are engaged in what he calls "ironic science": they no longer have their feet on the ground because their high-falutin' ideas and models have taken the place of observation and explanation of the real physical world, which is traditionally the raw material of empirical science.[6] By examining ancient and modern animal tracks we shall try to keep our feet on the ground as much as possible, but we should acknowledge that our ultimate goal is to reconstruct the story of creatures that are already extinct or long departed from the scene of the crime.

For now I shall postulate that time's arrow and the eternal trail have direction. We can keep in mind that in the view of many philosophers, evolutionary progression toward complexity leads to greater depth of understanding of the universe and greater consciousness. The theories of evolution help explain "how" complexity emerges, but the real mystery is not only "*how* evolution occurs—but *that* it occurs at all."[7]

The hidden, implicate forces that bind and form the structure of the universe are not just "out there" in the wide cosmological exterior, but are also "within" all animate and inanimate objects. This implies that all entities have deep structure, spirit, or consciousness. We may not wish to endow a rock with all these characteristics, though it can respond to heat and cold, but we surely accept that our language is steeped with awareness of the spirit and consciousness of all life forms, particularly of animals. As scientist-trackers we cannot appreciate the worldview of the world's great aboriginal trackers, such as the Bushmen of southern Africa, without ourselves being infused with a little of their consciousness and spirit. The same goes for the track makers: we must ultimately try to get under their skins a little if we are to understand their spirit and behavior. It is no coincidence that the rise of mechanized and industrialized society, which has generated exploitation and abuse of animals, has also seen a marked decline in our native tracking skills, which our hunter-gatherer forebears honed so well as part of their repertoire of survival skills. Now, however, a resurgence of interest in animal rights and in the feelings and emotional lives of animals is under way, which has coincided with renewed interest in ecology, wilderness conservation, and tracking.[8] (It is even possible to participate in tracking workshops, to relearn wilderness survival skills and the ways of nature.) On a fundamental level, we are examining and exercising our own emotions and internal ecology, whether we know it or not.

We shall follow the evolution of the universe from there to here through three well-defined phases. First we shall look at the physiosphere—the world of rocks and atmospheres; then the biosphere—the integrated web of plant and animal life; and finally, the "noosphere"—the name given to the phenomenon of superorganic mind by such luminaries as the great paleontologist-philosopher-priest Pierre Teilhard de Chardin, widely regarded as one of the great thinkers of the twentieth century, and the biologist-philosopher Julian Huxley.[9] Along the way we shall not only be looking at the most tangible manifestations of matter, from atoms to organisms—what Wilber, referring to what he terms the mechanistic side of science (i.e., Newtonian mechanics), calls "moving rocks"[10]—but shall also investigate their shadows, tracks, and traces that reverberate just beyond the materially "visible" realm.

In our vibrant universe, matter and energy are different expressions of the same whole, for as matter is energetically propelled through the universe it leaves a trace, however ephemeral. The potential therefore exists quite literally to find the tracks of subatomic particles that were in motion billions of years ago. Physicists have reconstructed the first moments of "creation" in sufficient detail to write entire books on the first few minutes.[11] Indeed one of the cornerstones of the theory of "the" Big Bang (or was it one of many?) is that with the right radio receivers we can still hear the echo of the big footstep in the low-frequency radio signal that reverberates consistently all through the observable universe.

As we move from sound tracks to light tracks, it is extraordinarily intriguing that light from the more distant stars is, in many cases, a signal or light trail constituting all that remains of a sun that has long since burned out. When we look at the light from such stars we are literally "looking back in time," seeing light actually emitted during the age of the dinosaurs. Like the tracks of long-lost dinosaurs, these trails of starlight record the former existence of material entities that have long since transformed or recycled themselves into ethereal energy and stardust. Yet we think nothing of employing these ephemeral, yet reliable, sidereal signals for navigation, in both the high-tech world of space flight and military aviation and the alternative world of astrology. Are we really so different from moths that navigate by moonbeams? At any time one or more of these light trails might end, when, from our Earthbound perspective, the star beam shuts itself off like a burned-out light bulb.

Much of our knowledge of the exterior cosmos is obtained from spectroscopic analysis of light and nonvisible electromagnetic waves. The Nobel laureate biologist Christian de Duve subscribes to the idea that all matter is "vital." In his book *Vital Dust,* he says: "Substances present in outer space act as filters [in such a way that] the substance causing the absorbtion or emission can be identified from the spectral patterns, which serve as fingerprints of the substances involved."[12] The principle is the same as that of a shadow cast by sunlight shining through a lace curtain, or rainbow colors generated by a prism. Every shadow or reflection is a fingerprint or signature of both the substance and the light that shines on it. So subtle is this dance of energy and matter that in summertime an ancient dinosaur footprint may be preserved because it was baked by the sun, while the next footprint in the sequence remained in a moist, muddy, shadow and was soon to be washed away or obliterated. In winter the process might be reversed as the sun thaws one frozen footprint and allows water to wash it away, while the next remains rock hard in the frozen shade. Such extraordinary interplay of matter and energy are manifest wherever we go on our terrestrial and extraterrestrial journeys. Wherever celestial light or electromagnetism illuminates and energizes earthly matter, we establish Jung's "indispensable coordinates," mental constructs to center ourselves in time, space, and meaning. As the new biophysics now shows, we do this because we are literally vessels of light, an idea we will explore in Chapter 1.

Talk of stars and planets gets us part of the way from there to here. By the time our Earth took shape, the Big Bang was already an echo some 10 billion years old. When we pick up the eternal trail on terra firma we find that the oldest tracks left here on Earth that are known to geologists are *fission tracks,* which are the signatures of subatomic particles that break loose from unstable elements such as uranium and tear through the lattice of surrounding crystals, leaving an actual trail of destruction that is preserved as a pattern visible under the microscope. Physicists in pursuit of elusive subatomic particles that lack substance must also find a way to identify these particles' "tracks" as they whiz by; they do this by using "cloud chambers." Just as the reading of an animal's tracks tells us something of the track maker's energy and evolutionary history, so the reading of light spectra and fission tracks reveals the history of stars and planets.

Geologists can date the age of rocks by computing the rate of decay of unstable elements and using the number of fission tracks in the rock's crystals. In fact the whole concept of chance and probability is based on measuring these decay rates: the tracks of subatomic particles, often preserved in the dense medium of rock matrix after the decay of unstable elements, and read by geologists billions of years later, give us such concepts as chance, statistical probability, random mutation, and gambling. Our faith in the rate of decay of unstable elements is such that we accept the statistical validity of insurance premiums, the even odds of winning a coin toss and the improbability of breaking the bank in the casinos of Monte Carlo or Las Vegas.

From the physiosphere, I shall hasten toward my main theme—animal tracks in the living realm of the biosphere. Between the earliest, prelife chapters in Earth history, 4 billion years ago, and the first days of multicellular animal mobility on the late Precambrian sea floor, 600 million years ago, not much biotic track-making activity was registered in the fossil record. Yet the rock record is full of intricate sedimentary structures that record the deep structure of the cosmos in exquisite detail. These geological traces, written like a script on the sands of time, record the mechanical and energetic structure of a whole range of intra– and extra–Solar System dynamics, ranging from storms, tides, and seasons here on Earth to orbital periodicity of the planet Earth to the impact of extraterrestrial objects. On this inanimate but nonetheless dynamic parchment the dance of life began tracing its steps.

And what a dance ensued as the biosphere evolved! Once Precambrian organisms became emancipated from their primitive sedentary existence they expressed their mobility by leaving highly ordered tracks and trails. These configurations of footprints, from those of the first Cambrian trilobites through Mesozoic dinosaurs to recent hominids, are the sheet music of the biosphere, and indeed are the record of life's eternal dance. The rhythmic peristaltic expansions and contractions of worms, the heaving of clams and snails in their burrows, and the rhythmic beat of feet has marked time through half a billion years of vital evolution.

Then in the twentieth century the biosphere gathers momentum for the quantum leap to noosphere, and the eternal trail leaves the planet. Mind adds another layer of complexity to the yet unfathomed intricacies of the biosphere. A few tens of thousands of years ago *Homo sapiens* set foot in the deepest cave recesses of the Paleolithic underground, and left a message that reads: "Here begins an attempt to rewrite the history of creation in our own language." It is little different from the footprint language of trilobite and dinosaur consciousness spelled out in much older scripts. However, because it tells our story, it is more familiar, more complex, and more rich in meaning than the language and spoor of animals. But herein lies a great paradox. For all that we might view the signs, spoor, and lan-

The Standard Geologic Column

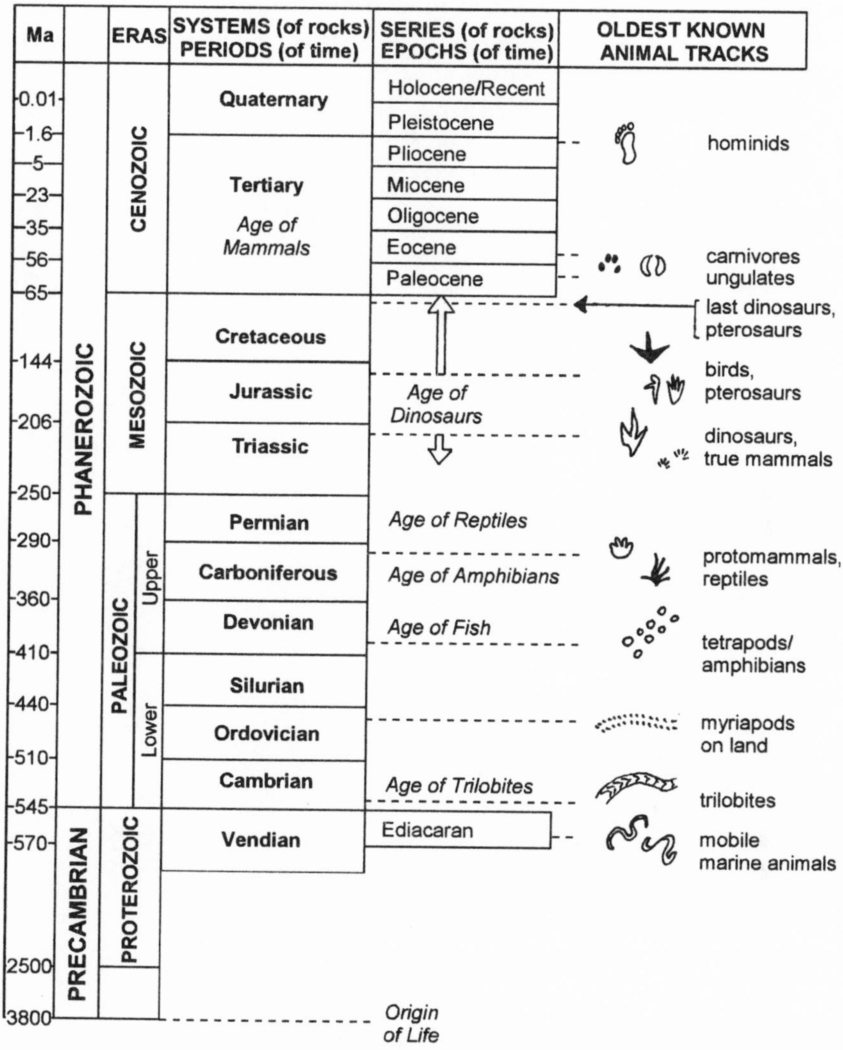

Ma	ERAS	SYSTEMS (of rocks) PERIODS (of time)	SERIES (of rocks) EPOCHS (of time)	OLDEST KNOWN ANIMAL TRACKS	
-0.01-	CENOZOIC	Quaternary	Holocene/Recent		
-1.6-			Pleistocene		hominids
-5-			Pliocene		
-23-		Tertiary	Miocene		
-35-			Oligocene		
-56-		*Age of Mammals*	Eocene		carnivores
-65-			Paleocene		ungulates
	MESOZOIC	Cretaceous			last dinosaurs, pterosaurs
-144-		Jurassic	*Age of Dinosaurs*		birds, pterosaurs
-206-		Triassic			dinosaurs, true mammals
-250-		Permian	*Age of Reptiles*		
-290-	PALEOZOIC (Upper)	Carboniferous	*Age of Amphibians*		protomammals, reptiles
-360-		Devonian	*Age of Fish*		tetrapods/ amphibians
-410-		Silurian			
-440-	PALEOZOIC (Lower)	Ordovician			myriapods on land
-510-		Cambrian	*Age of Trilobites*		trilobites
-545-	PROTEROZOIC	Vendian	Ediacaran		mobile marine animals
-570-					
2500					
3800			*Origin of Life*		

(Left margin: PHANEROZOIC spans Cenozoic–Paleozoic; PRECAMBRIAN spans Proterozoic.)

FIGURE 0.1 Standard geological time scale showing time of first appearance of major track-making groups.

guage of animals as simple in comparison to our own, we barely understand even them. In fact, we need higher consciousness to fathom the deeper mystery.

In a single century, modern psychology, anthropology, archaeology, and paleontology have simultaneously delved deep into the complex web of relationships that make up our personal family histories, our 3-million-year-old hominid phylogeny, and the entire prehistory of the biosphere. The web grows to critical levels of interconnectedness, and the pathways of the eternal trail become ever more diversified. The protoplasm of the biosphere builds its neural network, as happened a half billion years before the appearance of hominids, when animals added nervous systems to simple tissue. This, I believe, is where the eternal trail leads: from the simple unselfconscious beginnings of the biosphere to its fully conscious maturation to the birth of the noosphere. In our cyclic universe we come full circle, so that we may meet ourselves again and be renewed in deeper understanding. We are emerging noosphere, offspring of the biosphere. The eternal trail indeed leads precisely from there to here.

FIRST IMPRESSIONS

ACT I: CAST OF CHARACTERS

This book is ostensibly about fossil footprints, the spoor left by extinct animals. But there is more to it than that. When we walk by a mirror we see our reflection, and it leaves an imprint in our mind. When we walk by a lake, again we see our reflection and we also leave imprints along the shoreline. There is the material realm and there is the less tangible realm of shadows and reflections—the world of known reality, conscious and tangible, and the world of the unknown, of the unconscious and of intangible dreams. But both worlds are real. The artist and scientist who convert images and concepts from the mind's eye into paintings and scientific treatises are merely transporting intangibles from the immaterial ether into the manifest, material realm of canvas and paper. One day a melody in the mind of Mozart, the next a brilliant musical score.

In paleontology, tangible material objects are very important. The science could not exist without the bare bones that are its stock in trade. Fossil footprints, while being themselves tangible, are like scripts that tell of the less tangible dimensions of existence of extinct animals. They tell of behavior and give us subtle insights into the spirits of animals and how they interacted with contemporary species and the environments that they called home. They allow us to view the tangible world of fossil bones from alternate perspectives, and revisit conventional wisdom.

Our cast of characters in this tracking adventure is all the animals that ever left a footprint in the geological record, and a few rather exceptional human beings. Some, such as the Bushmen of southern Africa are themselves master trackers, or readers of the trail. Others are trail blazers, authors of scientific and philosophical works that cannot be ignored if we hope to read the important message left by the wayside. In this chapter we shall meet a few of them, philosophical trail blazers, who have given us insights into biology and the biosphere. We are at a turning point in the history of biology. After several generations of emphasis by scientists on molecules, genes, and other bits of organisms, and the "mechanisms" that help them to function, it is becoming apparent that the story of bits is just a bit of the story. Organisms are coherent wholes, as shown very elegantly by researchers such as Mae-Wan Ho, of the bioelectrodynamics lab at Britain's Open University, one of the most brilliant holistic biologists of the present generation, and the biologist Wolfgang Schad, whose extraordinary book *Man and Mammals: Towards a Biology of Form,* casts new and highly compelling light on the biology of mammals.[1]

Why is this new holistic approach to organisms important to trackers such as ourselves and all who seek information, truth, and understanding? At first glance, tracks tell us so little that if we view them as isolated bits of evidence we will gain correspondingly limited insight into the animals that made them. If we change our perception, our way of seeing, however, we begin to notice relationships between tracks and the whole organism. There is a relationship, for example, between the shapes of hooves and horns. It was an intuitive, holistic understanding of such relationships, born of lifetimes of experience, that made Bushmen such great trackers. Having explored and benefited from the bio-"mechanical" approach to animal locomotion and track patterns, we can begin to explore the benefits of the holistic approach, which tell us we can always read something of the spirit of the animal from its spoor.

We must put on our tracker's caps. Ultimately it is all about the perceptions we bring to bear on the subject of our attention. We should try to look deeply, not superficially, and make what the anthropologist William Irwin Thompson calls an archaeological excavation of our culture and species consciousness.[2] In doing so we begin to understand Paleolithic (Stone Age) cultures, the very ones that spawned the best trackers, in a radically new light. We then realize that reevaluating our perception applies to all our endeavors, including science and the stories we tell in journals and in books such as these. We cannot be good trackers if we are not keeping track of our own perceptions. If we do this we can retain a modicum of humility and a sense of humor that will stand us in good stead when the mysteries of the universe again transcend our feeble perceptions and we again stand before nature and humbly ask that she may remind us anew that we are her offspring.

LOVE IS THE ABORIGINAL TRACKER

Heaven is under our feet as well as over our heads.
Henry David Thoreau

Increasingly we live in a world of concrete, tarmac, television, and computer screens. These are great mechanical and electronic constructs of the human intellect—but they blot out our view of the rising and setting sun and keep our feet from being grounded in the earth, which in many cultures is a sacred act of communion. Our exterior technological constructs come between us and our inner hearts and souls, and certainly block out tracks and signs of our fellow creatures. As one of the greatest of all modern trackers, Laurens van der Post, has written: "[T]here is a great lost world to be rediscovered and rebuilt . . . in the wasteland of our spirit."[3]

It is easy for us as twentieth-century paleontologists to talk in convenient and self-congratulatory sound bites about our efforts to rediscover a "lost world" of the type reconstructed in Sir Alfred Conan Doyle's story of the same name[4] and bring it to the attention of an inquisitive public eager for an improved understanding of our ancient origins. Or, as Michael Crichton has done, we can convert the lost world into a fictitious domain of ferocious raptors run amok with an insatiable desire for blood and Hollywood stardom.[5] But when van der Post wrote, in *The Lost World of the Kalahari*, "Love is the aboriginal tracker, the Bushman on the faded desert spoor of our lost selves," he was clearly speaking of a much greater human quest, the search for our very souls, and our search for grounding as an integral part of all nature. He was talking not of a mythical lost world, but of a real world, albeit one we can lose sight of all too easily.

Van der Post truly loved the sensitive African Bushmen, whom he studied closely in their natural habitat. No one who had ever seen their tracks could forget them, for they had little feet. They touched the earth very gently. Though of this world, they were really of another world, an interior, collective spirit world, that we have lost touch with in our relentless quest to rebuild the earth into the artificial construct in which we often find so little meaning and peace. The Bushmen believed that "there is a dream dreaming us"—so perhaps they intuitively knew of the implicate inner order of things. We on the other hand are always trying to "make our dreams come true" by seeking their manifestation in the material, exterior realm.

The Bushmen were great trackers, they could follow anything anywhere and read the very spirit of the animals they loved. We on the other hand are generally lousy trackers, and very often cannot see things that are staring us in the face. Certainly, in our urban environments we feel very little of the pulse of nature around us, and so have to go to the wilderness to be revitalized and reborn. If we have the ability to revive our instinctive desire to commune with nature, and so love it, then we cannot have lost touch completely with our natural souls and the spirits of life around us. We are just out of practice. So it is with tracking. Few of us can identify the spoor of animals around us, much less the trails of insects and a myriad of more subtle signs. But it doesn't take much time to begin to learn. The anthropologist Elizabeth Marshall Thomas, who lived with the Bushmen, described in her classic, *The Harmless People,* how at an early age a Bushman child can pick out its mother's footprints from among those of the rest of the tribe.[6] These accomplishments by so called "primitive" or "aboriginal" peoples may seem remarkable to us, but they pale into insignificance beside the tracking skills of animals that follow trails by scent alone. When crossing the path of a rabbit, a dog can quickly tell which way the animal was heading. By making a regulated series of short sniffs of precisely the same duration, it can determine if the scent is increasing or decreasing in intensity in a given direction, and will immediately turn around if it is not heading down the trail in the same direction as its quarry. A dog can literally sniff out time's arrow.

When we lose touch with nature, we lose touch with our own natures, and so begin to think of rocks, plants, animals, and even people as somehow less important than the stressful job of fitting into the artificial world we are so busy creating. We lose our sense of perspective on history, and time becomes something we save or make, despite our awareness that there are no time banks or time storage units. We are so befuddled that we have to make time for people, and even make time to eat! All the while the Bushman's dream is dreaming us, and the inconceivably vast space-time continuum rolls on as it has from time immemorial.

Aboriginal trackers believed in a powerful sympathetic or spiritual connection between a creature and its tracks. In his classic anthropological work *The Golden Bough,* Sir James George Frazer cites many examples of peoples who believed that an adversary could be injured, sickened, or caused to limp by deliberately tampering with his or her footprint, or removing earth from the track to perform the appropriate ritual or "voodoo" ceremony. Are our modern views so different when fans visit Hollywood to feel closer to their filmstar idols by standing in their footprints? Don't we have closer connections to friends or idols by possessing their autographs and letters? Even the most skeptical paleontologist would feel a warm glow or special sensation if he or she were in possession of, or "in touch" with, an original manuscript penned by Charles Darwin. Surely such a sensation of connectedness is no less superstitious than that of the aboriginal tracker. And what of the analytical forensic tracker who must explore every creature-track connection to solve a crime or paleontological puzzle—possibly even "feeling" a warped spirit in the course of following a criminal's trail?

It is the premise of this book that footprints have great meaning—for like the hand of Charles Darwin, each is the energetic signature of the species and consciousness that made it. It is well known in literature and language that "hand" is synonymous with the actions of the head, mind, or spirit—that hands are the instruments of the soul. The guiding and surprisingly scientific principle behind palmistry (or *chieromancy*) is just this: the hand is a microcosm of the whole makeup of the individual. As I shall note later, geneticists make the same claim for genes. The same principle is now programmed into computers to expose the "hand" of authors who attempt to write anonymous, politically motivated best-sellers. If the universe is conscious, so is every living creature and substance. This view comes naturally to many aboriginal people, theologians, and mind-body physicians. Here each species, whether animal vegetable or mineral, manifests its own unique vital force or consciousness—what Einstein would call the field, and what homeopaths call the energetic "footprint" of the species.

It is my belief that the footprint of each species, or life form, is distinctive. Modern studies support this conclusion in most cases, though in the fossil record it is not always easy to distinguish the tracks of two or more closely related species. But this does not mean that it cannot be done with practice, or will not be done in the future. Current research shows that we are moving in the direction of being able to distinguish different track types with greater confidence and read characteristics of pace and stride as manifestations of the behavior and very spirit of animals. Trackers also pay attention to where and when tracks are found, for each species had a habitat or range, and its own time slot or season in Earth's history.

The eternal trail is well symbolized by the continuous trail of footprints of terrestrial animals that began almost a half billion years ago. This trail continues uninterrupted today as paleontologists map out the footprints of centipedes, dinosaurs, and hominid ancestors, to which they add their own paper trail of words, paragraphs, and text-file stories, some at least as fanciful as any told by aboriginal tribes. We are converting real track fossils into our own symbolic messages scribbled on paper and stored in our sacred academic vaults. Let us also remember with humility that the Bushman is a serious, full-time tracker, not just of animals but of the very elements of air, wind, and water. According to Elizabeth Marshall Thomas, the Bushmen could tell from fallen leaves whether they had been disarranged by wind or by passing animals. "Bushmen" she wrote "are always right when it comes to tracks."[7] I would rather have a Bushman's guidance through any physical or spiritual desert than be compelled to rely on most of what has been written "scientifically" on the subject of tracks.

Like the Bushman tracking the spirits of all life, we also transform our record of tangible measured trails into ephemeral words and interpretations uttered in scientific meetings and media sound bites. Until, like a trail growing old, our electronic messages become spirit trails and whispers in the ether, with all the components of myth and uncertainty that we regard as fanciful when someone else tells a tall story. My last e-mail message on tracks has vaporized amid the same starlight in which the Bushman's tracker mythology is constructed. How is this circumstance different from the songlines of the Australian Aborigines, described by Bruce Chatwin in his book *The Songlines.* The Aborigines' ancestors "while traveling through the country [were] thought to have scattered a trail of words and musical notes along the line of [their] footprints," leaving a trail to be followed. Half the world gets through daily life humming along with their favorite musical "tracks" and the spirit messages they contain. "It would seem," Chatwin noted, "that there exists, at some deep level of the human psyche, a connection between path finding and law."

Scientifically we cannot say where the eternal trail begins, or where it will end. Before the first cousin of a common "rolypoly" wood louse walked on land, marine worms and other invertebrates were already leaving countless trails as they nosed their way along the Precambrian seafloor. These creatures incidentally were our own billionth cousins. Then came tracks of amphibians, reptiles, dinosaurs, birds, and mammals, eventually leading to the footprints of Lucy striding with her kin across the African savannah. (Lucy represents our archetypal hominid Eve, though some have suggested she was in fact a "he.")[8] When we gaze on these famous tracks, the spoor of our own kin, only a few thousand generations removed from the tracks on the moon, can we avoid pausing to think of the billions of miles traveled?

These countless miles cannot be measured simply on the terrestrial odometer, for with every step our little planet has been hurtling through space at the breakneck speed of 18.5 miles per second, racking up the ultimate tally of frequent flier miles through the compounding of Earth-speed with footspeed. And with every epoch track makers have been evolving their own paths, behaviors, and spirits. As twentieth-century paleontologists we are just the latest to step on the stage, to discover where millions have passed before. Let us revere our ancestors, and the tracking skills we can learn from them, not just in the narrow physical realms of geological strata and zoological biomechanics, but in the profound insights they give into the great chain of being written as an eternal trail throughout space and time.

OUR LONG TRACK RECORD

That turnpike earth!—that common highway all over dented with the marks of . . . heels and hoofs.

Herman Melville, *Moby Dick*

About 400 million years ago a new type of backboned animal with fishy origins began to develop legs and walk on land. These animals and their descendants have been leaving footprints ever since. The central theme of this book is the global "track record" of land-based animals (mainly vertebrates) that begins in the Ordovician Period, about 450 million years ago, and continues into the present era, when in our generation NASA's twelve disciples made footprints on the moon, and space probes made trails on Mars.

It would take a considerable effort in arithmetic to estimate how many four-footed vertebrate species (*tetrapods*) have walked on land, and even greater powers of extrapolation and speculation to calculate how many individuals of each species have walked on Earth at one time or another. Thus the reader will forgive the omission of a realistic estimate of the astronomical number of footprints made by all individuals of all species since the first creatures set foot on land. If Earth's 5 billion mobile human inhabitants took an average of 1,000 steps each day (the equivalent of strolling around for about half a mile), they would produce the equivalent of 5,000 billion (5 trillion) seen or unseen footprints in a day. This is to say nothing of the tracks made by elephants in Africa, polar bears in the Arctic, and mice in the attic. And these are just a few of several thousand mammal species. Let us not forget ducks by the pond, lizards in the desert, and frogs on the bayou. Our busy human population could make almost 2,000 trillion footprints in a year (2 quadrillion) and 200 quadrillion in a century. If only one in 100,000 of these human tracks were made on soft ground where a clear footprint would register, and if only one in a million of those were to be preserved in the fossil record, there would still be an accumulation of 2 million tracks per century.

Without even multiplying these numbers to allow for other vertebrates, these conservative estimates for humans and their ancestors still add up to 20 billion footprints being preserved after a million years, or 8 trillion preserved since the first one was made on land. However conservatively or generously we do these calculations, the results come out the same–lots of fossil footprints. If only one in a million of these preserved footprints is accessible among the ragged edges of strata exposed at the earth's surface today, there are still millions available for study.

TRACKING EXTINCT ANIMALS

"Hallo" said Piglet, "What are you doing?"
"Tracking something" said Winnie-the-Pooh very mysteriously.
"Tracking what?" said Piglet, coming closer.
"That's exactly what I ask myself. I ask myself, What?"

A. A. Milne, *Winnie-the-Pooh*

The investigation of fossil footprints is not quite the same as tracking modern animals. It is quite easy to become proficient at identifying modern tracks because sooner or later one can see the animal that is making them. Remarkable tracking skills were common among hunter-gatherer societies, as recounted by Laurens van der Post in *The Lost World of the Kalahari*. When tracking extinct species, however, there is always some uncertainty as to "whodunnit." We must draw also on our knowledge of fossil skeletal remains to reconstruct the identity of the possible track makers. This does not mean we cannot study tracks if we don't have the skeletons of track makers available because, in many cases skeletal remains are unknown. Ultimately, however, it helps if we can put the track and bone records together in an integrated picture.

This exercise in fitting track makers into tracks is one of the obvious objectives of a good tracker. Ten years ago it would have been hard to get trackers to agree on what track types correspond to particular extinct groups, but much progress has been made in recent years. While writing this book I was struck by how many track types have been matched with track makers with reasonable confidence. Such progress is encouraging in the light of previous sometimes hilarious errors. For example, large millipedes have been confused with amphibians, and pterosaurs mistaken for crocodiles. This progress comes from careful comparison of tracks with foot skeletons, and from looking more closely at the age of the rocks in which the tracks occur. This way we place the correct culprits at the scene of the crime and do not attempt the paleontological equivalent of trying to fit our australopithecine ancestor Lucy into the footprints on the moon.

Finding the foot that fits the footprint is what I have dubbed "Cinderella syndrome."[9] Ultimately there was only one individual responsible for making any particular footprint. In this context it is helpful to remind ourselves of the first principles of detective work. A track maker cannot be held responsible for making a track if it did not have the means, opportunity, and motive. Means corresponds most closely with having the right foot anatomy. Only elephants have the means to make elephant tracks. Opportunity relates to time and place. Elephants have the opportunity to make tracks today, in certain areas, but did not have the opportunity to make tracks during the age of dinosaurs, and extinct *Australopithecus* could not make tracks on the moon. Motive is quite another matter, since it involves interpretation of the behavior and cannot be established as easily as means and opportunity.

Fitting fossil feet into footprints is not always easy because we are tracking animals in the flesh and not tracking foot skeletons. An excellent illustration of this difficulty is provided by our own feet. Let us imagine for a moment that we are anthropologists from a future millennium, and that we have found one or two human foot skeletons. How would we reconstruct the footprints? Would we get all the pads and flesh in the right position and proportion? Flesh is almost invariably lost from the fossil record. We must remind students of skeletons to look at tracks as the imprint of the *living* part of the foot.

Ultimately a deeper question arises: not simply what are we tracking, but how and why are we tracking? Are we seeking knowledge for some "useful" purpose? If we do not know what we are tracking, then we are like Winnie-the-Pooh, that bear of very little brain. Tracking extinct animals could easily be considered a futile exercise, leading to no results other than a cluttering of the scientific literature with obscure verbiage. If the exercise results in the rewriting of textbooks, and captures the imagination of children who get a better education, then perhaps the endeavor shows promise. If we recognize that fossil footprints are part of the present landscape, we may be able to preserve them as national parks and not destroy irreplaceable natural heritage in the name of development and technological progress. We cannot live in the past, but we cannot forget our past altogether. For a rewarding future we must integrate ourselves with past and present.

ELITE FEET

How beautiful are the feet of them that . . . bring glad tidings of good things.

Romans, 10:15 (King James)

Just as some feet are more beautiful than others, so some footprints are more attractive than others, at least from a scientific viewpoint. Vertebrate

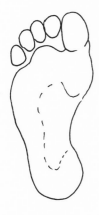
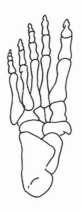

FIGURE 1.1   Fleshing out feet: A comparison of a human footprint with a human foot skeleton reveals that the impression of the bones in mud would be nothing like a track.

paleontologists naturally look to whole skeletons for a complete picture, but such skeletons are much rarer than most people might infer from a casual survey of paleontology books. Complete foot skeletons are even rarer, because they consist of small bones that are the first to wash away after an animal dies. Even so, tracks represent only feet, and only give a partial picture of the track maker. Some have even suggested that only lunatics would study tracks, and in one infamous case a scientific manuscript on footprints was rejected without being read on the grounds that tracks cannot provide any information of value. The British tracker Bill Sarjeant told me that a paper of his had been rejected by a leading journal of earth sciences simply because the subject was footprints. Ironically, a more accommodating journal later published the article without revisions.

Fortunately, such extreme views are rare. Tracks obviously reveal evidence of the living animal in the flesh. A brontosaur track is more than a mere foot skeleton with individual toe bones touching the ground. They consist of large elephantine footprints that show very large fleshy areas in contact with the substrate. Such experience tells us that we must treat tracks and foot skeletons as two different but related types of evidence. Paleontological tradition recognizes such differences and so we use different scientific terms to describe tracks and bones. The naming of plants and animals falls in the science of *taxonomy* (literally, the arrangement or classification of names) or *systematics*. The naming of fossil footprints is *ichnotaxonomy* (from the Greek *ichnos* meaning "trace"), and so trackers can be referred to as *ichnologists* (not to be confused with *ichthyologists*, who study fish).

Taxonomy labels different classes of animals, including species and genera, on the basis of anatomy; in addition, we have *ichnogenera* and *ichnospecies*, based on tracks. For example, if the skeleton is genus

Tyrannosaurus, the track ichnogenus is *Tyrannosauripus* (meaning "tyran-nosaur foot"). Such a system may seem wordy and duplicative, but it has some distinct advantages. It allows paleontologists to recognize their lim-itations. We would be less than honest if we claimed we could match every track with a track maker. Worse, we would make too many mistakes. Yet there are many examples of tracks that match with a particular genus, as in the case of the tyrannosaur track (discussed in detail in Chapter 7). Often, however, we are not quite so lucky, so we typically identify the track maker at the more general level of a family, or other larger grouping. This means that we have not yet found Cinderella. The shoe is known to fit a lady who was at the ball but we don't yet know her identity.

The first and most important tracker's rule is "Don't name crummy tracks." Put another way, "Be elitist!" and work primarily with pretty tracks in which the details of anatomy are well preserved. I have coined the term "elite tracks" to describe just such tracks of "museum quality."[10] A complete set of these guidelines has actually been presented in the form of "Ten Commandments" of tracking.[11] Unfortunately these command-ments were not passed down from on high until 1989, so many crummy tracks have been named in the past. Little by little, however, trackers are learning to use common sense and track down and name only the foot-prints that are useful to science.

MEMORIES, IMPRESSIONS, AND REFLECTIONS

When orthodox conceptions prove unfruitful . . . a little scientific
heterodoxy may not be amiss.
   Edmund W. Sinnott, *The Biology of the Spirit*

Earth has recorded the passage of multitudes of creatures. Some registered their sojourns decisively with obvious, elite signatures, others left only un-intelligible smudges, mere ghost signatures. (We will talk more about these ghost signatures in Chapter 6.) The Earth therefore has a memory of its former inhabitants locked in the track record. But how much is re-membered, and how are such memories recorded? For every particle emitted by an unstable element, only a few are recorded as fission tracks in the matrix of a receptive crystal. When a creature walks across the bare rock of a granite massif no trace is registered. Similarly, a thief leaves no fingerprint when touching the rough texture of fabric or upholstery. To leave an impression or signature, the earth must be receptive and absorb, or "register," the signal sent it. We can read only the tracks that are acces-sible to us with our present level of scientific knowledge and tracking skills. But, like tracks yet to be discovered, more information awaits our discovery as our tracking skills improve.

The ideal condition to record a clear fingerprint is to prepare a smooth surface of ink on a plate of glass. The ink is rolled out flat, pristine and dark, like the emulsion on a fresh roll of film, and like film emulsion it is receptive, designed to register signals sent to it from elsewhere. Like a mirror, it is smooth and shiny enough to reflect the image of the approaching finger, even before the finger makes an impression. As a fingertip touches the plate, only the raised ridges make a full impression in the inky layer. As the finger is removed white furrows remain and the light pattern of dermal loops, whorls, or arches is registered like light etching the emulsion of a negative. As the inky fingertip is transferred to white paper, the zebra pattern is recorded in reverse, making a positive print: the black ink on the ridges is transferred to, "printed" on, the paper, and the inkless furrows register as white lines between the sticky black trails.

Our fingerprint patterns resemble those on a zebra skin or the ripple-marked surface of a beach or riverbed. Explanations for such patterns point to certain inherent properties of matter under what scientists call *boundary conditions*. Hence, sand forms into ripples at the boundary between the sea bed and the sea, and color patterns and fingerprint skin texture reflect the dynamics of growth at the boundary between organism and environment, a growth pattern that begins in the embryo.

The sand of a beach can be compared to our crime-lab inked glass plate, in that it is an ideal surface to register and record the activity of passing animals. Without such sensitive surfaces, capable of registering the impression of nomads on the eternal trail, we would have no track record. These surfaces have the ability to record and so "remember" the morphology of the feet that walk all over them. This process of recording the history of the Earth and the Solar System has been going on since the beginning of time. Record keeping is not something invented by our species, and memory is not just a phenomenon of the human mind. Animals "remember" to migrate, plants "remember" to sprout in the spring, and the planets "remember" to orbit, all with the utmost precision. Somehow, without our intervention, the universe remembers to stay organized and do all kinds of things, many of which are still beyond our comprehension. So I should interject a sincere note of thanks to whatever forces are responsible for doing most of the legwork in recording all the data on which this book is based. We have merely learned how to read some of the eternal trail long after most of it was written.

Can we talk of sandflats and mudflats having memory? Are they more than just inert layers of sludge that absorb the physical impact of footprints? Where do we draw the line in attributing memories to creatures or entities other than humans? Does a crystal have memory if it records the decay of unstable elements? Does the earth have memory if it records the impact of meteorites and the fallout of iridium? Does photographic film have memory if it records our most memorable moments by responding to split-second light signals? Does our skin have memory if it records the scars of battle and rugby matches? Does the vertebrate body of a lizard or a frog have memory if it can regrow a tail or a leg? Could we have reconstructed anything of Earth's history if, before the advent of our species, our rocky little planet had not accumulated a rich storehouse of memories, written in a multiplicity of geological and paleontological languages?

INDIVIDUAL SIGNATURES

The art of signature teaches us to give each being its true name in accordance with its innate nature. . . . There is nothing that Nature has not signed in such a way that man may discover its essence.

Paracelsus, from *Paracelsus: Essential Readings*

We are all aware that individual fingerprints are so distinctive that they are routinely used to identify individuals in criminal cases. Despite a global population of 6 billion, governments, immigration officials, and law enforcement agencies keep track of us through our fingerprints. This puts fingerprint experts into the category of master trackers, ultimately capable of tracking down any one of us who leaves our individual signatures on a glass or door knob.

If it is hard to consistently identify different species of track makers from the fossil footprint record, it would therefore seem even harder to identify individuals. Paradoxically, however, using size and shape we can distinguish one individual from the next, more easily than we can distinguish sexes, races, or species. Within a sample of fossil footprints some may tend to be slender while others are stout. Phyllis Jackson, a now-retired English podiatrist, has recounted how during World War II her practice was suddenly flooded with refugee patients of Scottish, Irish, and Welsh descent, whose feet appeared to be different from those of typical Anglo-Saxons. She observed that Anglo-Saxons had broad feet and that Celts had longer and slimmer feet. Although she did not use statistical tests to establish her thesis, the simple observation that the Saxon shoes did not fit proved her point. Ms. Jackson has since attracted the attention of professional archaeologists by demonstrating that she could tell ancient Celts from Anglo-Saxons by their foot skeletons, and her most recent work reveals a huge range of variation in foot shape all across modern and ancient Europe.[12]

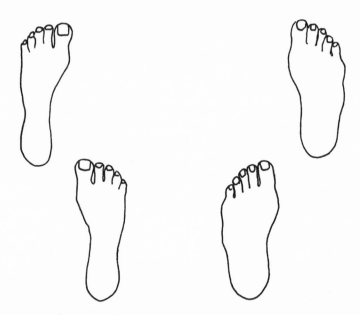

FIGURE 1.2   A comparison of long Celtic feet with those of wide-footed Saxons.

If we can distinguish different races or ethnic groups by their feet and foot skeletons, then presumably we should be able to use footprints to distinguish different hominid species, assuming we can find sufficient tracks. Among the various species of the genera *Homo* and *Australopithecus* that make up our own taxonomic family, the Hominidae, anthropologists readily identify distinctive robust (heavyset) and gracile (lightweight) varieties. As their footprints are known, we must ask ourselves how easy it is to track our own immediate ancestors? Would we recognize a *Homo erectus* or *Gigantopithecus* track as anomalous if we found one on the beach among those of our own species?

The study of individual signatures is an aspect of the broader study of variation among individuals in a population or species; Darwin made this variation among individuals the cornerstone of his theory of evolution. We do not really know why individual fingerprints or footprints differ, but it is a subject of great interest to scientists and forensic experts. There is evidence that palms and hand shapes fall into various categories that reveal much of the character of the person. In palmistry an elongate hand is regarded as a sign of the gift of high intelligence, sensitivity, intuition, and psychic ability, whereas a stout hand, of the type sometimes called the square or useful hand, is considered a sign of a "salt of the Earth," commonsense personality. Differences in hand shape that we can verify for ourselves by simple observations may also reflect gender to some degree; usually the female hand is less stout than a man's. Anthropologists recognize broad and narrow heads and body types expressed within all major racial groups. So broadness or narrowness is seen in the whole body, in the head, in the hand, in the foot. A pattern, surely. As we shall see, inherent qualities are also associated with narrowness and breadth, so we might infer from their foot shape that ancient Celts were more intuitive and mystical, whereas Saxons more practical and down to earth.

There is substantial scientific evidence that fingerprints and palm crease patterns fall into distinct categories.[13] For example, susceptibility to Alzheimer's, and other diseases has been correlated with certain distinctive fingerprint patterns. There is also strong evidence that left-handedness correlates with certain fingerprint types and personalities.[14] Lefties live shorter lives and are more likely to end up in jail or in mental institutions, but we don't know why. If genes for such diseases and psychological profiles are linked to skin or hand anatomy, then surely such correlations should be studied seriously. At present forensics studies fingerprints but not palms. Why? Perhaps because it is easier and less messy to collect and store fingerprints than palm prints. Or is there a taboo against studying palms, based on scepticism toward the ancient art of palmistry? This book would be sketchy and incomplete if trackers studied only the impressions of toe tips and claws. In the absence of any efforts to explore the scientific underpinnings of palmistry, we shall remain ignorant of its potential, and not appreciate the adage that hands and feet are the instruments of the soul. Within the broad realm of medical science there are many correlations between disease and physical attributes and less tangible characteristics such as behavior and personality. On a more practical level, our immense global fingerprint database could be used to identify individuals with susceptibilities to particular diseases. But a palm-print database would provide considerably more information. Such prospects lend a humanitarian perspective to a data set otherwise used primarily to facilitate bureaucracy and law enforcement.

BINARY BIPEDS AND DIGITAL DIGRESSIONS

I shot an arrow into the air,
It fell to earth, I knew not where.
**Henry Wadsworth Longfellow,
"The Arrow and the Song"**

As humans we share many common traits with all other tetrapods and vertebrates. We have bilaterally symmetrical bodies. Externally our right and left sides are mirror images in all but the smallest details. The mirror plane is called the sagittal plane (from the Latin *sagitta,* meaning "arrow"). This bilaterality is reiterated in sutures between the symmetrical pairs of bones that make up the skull and the division between the right and left brain, and is reflected in our language and culture. We speak of there being two sides to a discussion or the need to see the other side of the story. Our simplest method of counting is the base–2 binary system of mathematics, which is used in computers because it can be expressed as a simple on-or-off electrical pulse. We use light switches that are generally either on or off and legal arguments where the answer is either yes or no.

The Desana Indians of Colombia view the sky as a giant brain, divided into two hemispheres by the Milky Way.[15] This view of the plane of the galaxy makes it analogous to the sagittal plane that bisects our brains and bodies. To stand in sagittal alignment with the night sky is to integrate our entire beings with the cosmos. In fact, the Milky Way has great significance to many cultures as a spirit trail or river aligned in the vastness of eternity.

In Western cultures the existence of the sagittal plain influences many of our images and social and cultural constructs. We tend to be preoccupied with two sides. We turn either to the right, or dextral, side (which by extension also means correct, fitting, decent, or honorable) or to the left, or sinistral, side (meaning unlucky, unfavorable, or evil). Such entrenched notions of right and left, right and wrong, and what is *right*eous or what should be *left* out, has made it hard on left-handers, who have in some cultures become literal outcasts. What is the larger meaning of this preoccupation with sides? Is there an evolutionary message in our bicameral world of right and wrong, light and dark, day and night?

As a tracker I note that we leave symmetrical trackways as we walk endlessly, moving astride the sagittal plane. We are perhaps too preoccupied with duality and think too much in terms of polarity between opposing right or wrong theories. On this subject it has been suggested that our ancestors were ambidextrous (neither hand dominant) and so perhaps more holistic than we in their view of the world.[16] It has also been suggested that "sidedness" arose from the habit of tool using where one hand (the left) holds an object, while the other (the right) actively strikes it to manufacture an artifact. Right makes might?

On either side of the sagittal plane we plant five toe impressions with every footfall and wave five fingers at the world. So the number ten registers constantly in the unconscious abacus of our minds. Is it any wonder that we devised the decimal system? Things might have been different had we evolved with more or less than five digits. Indeed, the very term *digit*, which we commonly use to refer to numbers, is the Latin word for fingers and toes, which paleontologists still number, Roman style, I through V, beginning on the inside, with the thumb or big toe. Books on evolution traditionally talk of the "rise of animals" from lowly beginnings to higher status. This is seen quite literally in the evolution of low-slung, flat-footed (or *plantigrade*) animals such as amphibians and lizards, to those that walk on tip-toes (*digitigrade*), such as birds and hoofed mammals. This process was one of becoming increasingly emancipated from close adherence to the Earth's surface. Birds and humans have so far been most successful at this by freeing their front limbs from the Earth entirely. It is more than mere linguistic coincidence that "plantigrade" (which means feet "planted" fully on the ground) also has the connotation of no higher than "plant grade," i.e., earthbound. Despite our having free hands, the human foot is plantigrade because both heel and toes contact the ground when we are standing and walking.

FIGURE 1.3   A binary biped. Artwork by Paul Koroshetz.

THE BEAT OF THE FEET

Whenever the moon and stars are set,
Whenever the wind is high,
All night long in the dark and wet, a man goes riding by.
Late in the night when the fires are out,
Why does he gallop and gallop about?
Whenever the trees are crying out loud,
And the ships are tossed at sea.
By, on the highway, low and loud, by at a gallop goes he.
By at a gallop he goes, and then, by he comes back at the gallop again.

Robert Louis Stevenson, "Windy Nights"

Walking or running is a very rhythmic activity. The tempo of this beat of feet decreases or increases as the track maker slows down or speeds up. In most cases the heartbeat also slows down and speeds up with these changes in pace, so that the internal and external music of a body synchronizes naturally.

The locomotion of quadrupeds is more complicated than that of bipeds because more feet are involved. If one has ever tried to watch the sequence

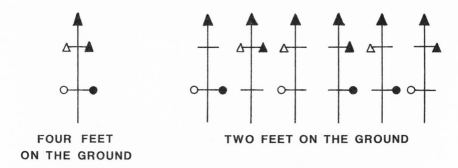

**FOUR FEET
ON THE GROUND**

TWO FEET ON THE GROUND

FIGURE 1.4   The beat of the feet: Trackway notations such as these, devised by Eadweard Muybridge, resemble a musical score.

of footfalls of a dog or cat, one soon notices that without a video camera it is hard to keep track of the cycle of footsteps, even at slow walking speeds. If we study the locomotion of the horse, we learn that it can progress at a walk, pace, trot, canter, or gallop—all different patterns of footfalls. Although the horses are blissfully ignorant of the definitions given their types of locomotion by humans, people with an equestrian bent have become engrossed in the intricacies of equine deportment, and strict adherence to protocol is desired in such things when it comes to deciding the distribution of Olympic medals (which, incidentally, go to the rider, not the horse). Each gait has its own distinct rhythm, cadence, and characteristic sequence of footfalls. The trackway records footprints as notes in space, comparable to a musical score. Though the sound wafts away from the trackway as it is registered, the horseman may hear the gait by listening to the beat. We too may attune our ears to a walk or trot, or thrill to the sound of a gallop. The beat and rhythm of horses in motion was incorporated into poetry before train tracks brought us the rhythm of the rails. Track making can be heard as well as seen, and the sound and vibration of footsteps has been recreated in many a drama.

In the late nineteenth century, Eadward Muybridge pioneered the use of photography to make detailed studies of animal locomotion. In doing so he developed a graphic method for recording the sequence of footfalls or beats made by animals as they progress with different gaits.[17] Learning his system of notation for recording a walk, trot, or gallop is like learning to write a musical score. These symbolic abstractions turn the *real* trackway pattern into musical scores designed to explain the timing of footfalls. Following such a score, one may begin to make the appropriate foot motions. It's a silent dance at first, but becomes audible once the foot tapping starts. With the right sheet music one may rerun the Kentucky Derby or Paul Revere's historic ride.

THE TRAIL THROUGH TIME

Is there any conceivable reason that evolution, which has labored so mightily for 15 billion years and produced so much wonderment, would just up and abruptly cease? Are there not higher spirals lying ahead? If we have discerned even the vaguest features of time's arrow, can we not stand on tiptoe and foresee dimly the arrow's arc into tomorrow?

Ken Wilber,
Sex, Ecology, Spirituality: The Spirit of Evolution

Paleontologists deal with long periods of time. We talk of millions of years as if they were mere days or weeks. In our attempts to explain such incomprehensible spans of time we collapse 4.5 billion years of Earth history into a two- or three-month curriculum. Sometimes we condense time even further and express Earth history in terms of the hours of a single day. In this 24-hour day of Earth's history, the first amphibian walked on land just before 10 o'clock in the evening. The dinosaurs roamed the Earth from 10:50 p.m. until around 11:40 p.m., and Lucy and her relatives left their tracks on the African savannah a mere minute before midnight. Neil Armstrong stepped on the moon in the same split second that the clock chimed midnight. It's all a mental exercise in time travel.

Like detectives, geologists must be good chronologists. We can go back exactly 155 million years to the age of brontosaurs and date a track-making event to the nearest million years or so. In our 24-hour day a minute represents 3 million years, so going back 155 million years is the equivalent of pinning down activity at a crime scene to a 20-second interval between 11:10 and 11:11 P.M. In terms of our 24-hour-clock time scale, geologists can sometimes pin down events with what amounts to split-second accuracy.

Although physicists and metaphysicians may argue that linear time does not really exist, geologists cannot ignore accurate chronology. They see prehistory as a sequence of distinctive events, measured in a number of different ways. In our short life span we measure events in days, months, and years. For geologists, the most satisfactory time-measurement methods are based on natural cycles given us by the celestial mechanics of the Solar System. They group these cycles by length: Seventh-, sixth-, fifth-, and fourth-order cycles are those of about 26,000, 42,000, 109,000, and 413,000 years, respectively (cycles will be discussed in more detail in Chapter 4). One such longer cycle is the precession of the equinoxes (about 26,000 years, a seventh-order cycle), based on the change in angle of the axis around which the Earth rotates. Twenty-five hundred precessional cycles have taken place since the extinction of the dinosaurs 65 million years ago. Such celestial periodicity affects climate, and may have an effect on species extinction and evolutionary change. Such changes in turn are the basis of fossil zones, the geological equivalent of historical dynasties. Tracks are sometimes all that remains of certain dynasties.

Geologists do not date every track or fossil they find. In practice they rarely date the fossils at all. Other geologists known as geochronologists date the rock layers, thereby establishing the age of the fossils they contain. It is sort of like numbering the pages of a book; we often find tracks on pages that are already numbered. Fission-track dating and other radiometric methods use the statistical rate of decay of unstable elements such as uranium. Carbon 14 (C_{14}) is useful for recent archaeological dating of organic materials such as wood, leather, and bone (but usually not rock). Whether we are dealing with celestial cycles or the ticking of a radioactive clock, we are learning to transcend time and establish chronological timeposts on the eternal trail.

The Russian philosopher-mathematician Peter Ouspensky (1878–1947), in his work *Tertium Organum,* expressed his view that time is merely another, fourth dimension of space "imperfectly sensed."[18] This is not so hard to comprehend. Viewed from an evolutionary perspective, organisms have increased their ability to explore space. The spatial realm of a worm or centipede is very limited compared to that of the albatross or astronaut. Through an improved ability to travel and perceive a larger spatial realm we evolve a greater overview, becoming more conscious of expanded dimensions. Like all sciences, paleontology and tracking have undergone a similar evolution into expanded dimensions. Having begun with a focus on individual organisms, these fields broadened into the macrocosm to consider populations, communities, ecosystems, and the evolutionary relationships of all species in evolutionary time. Paleontology and the life sciences have also delved into the microcosm, going deep inside the organism to explore organs, tissues, and genetics in the dimension of developmental time. Trackers have expanded their vistas from studying individual tracks to studying track assemblages, or populations, that provide insights into the broader picture of ancient animal communities and ecosystems. Most recently, the recognition of dinosaur "freeways,"[19] vast trampled terrains that can only be viewed in their entirety from space, establishes a new global perspective on the wide-ranging activities of Earth's former inhabitants.

The progressive transcendence of spatial dimensions by evolution and knowledge continues relentlessly and is now breaking down our traditional perception of time. Since the nineteenth century we have redefined the chronology of the history of the Earth and universe and have confidently constructed an elaborate evolutionary timescale using both rocks and molecular paleontology. Our newly educated temporal consciousness travels back and forth from the present to the Mesozoic and Paleozoic with the same confidence displayed by our ancestors when they first explored new dimensions of space on the Cambrian seabed or the virgin shores of Devonian continents. Like surfers who have learned to ride the waves of time, we now journey back through vast celestial cycles with the nonchalance of Sunday drivers on a ride through rolling countryside. Along the way we investigate and establish geological dates and units of time that we happily rely on as temporal signposts—"timeposts"—that are just as trustworthy as the spatial map coordinates of longitude and latitude, signposts that science established just a few generations ago. As chronological markers on our paleontological timescape, such points in time are just as reliable and real as dusk and dawn or the swing of the clock pendulum. There is an intimate relationship between signpost and timepost, for we measure distance, quite literally, in degrees, minutes, and seconds and in the "time" it takes to get there. A journey can be three hours "long." Conceptually the time-space continuum is already ingrained in our language.

Surely Ouspensky and like-minded thinkers have a compelling perspective when they suggest that the temporal dimension, of which fossil footprints are an obvious material manifestation, is simply the unfolding perception of another spatial dimension "imperfectly sensed." In this Ouspenskian sense the temporal dimension is merely a manifestation of the evolution of our intellects and consciousness. We viewed the oceans and the heavens with wonder and apprehension before setting sail and taking flight into these realms. Does this then mean that because we stand at time's portal and look into eternity, we are soon destined to travel in this realm of time? Have we not already taken the first steps?

PIONEERS ALONG THE TRAIL

Oh how I love to travel back
And tread again that ancient track.

Henry Vaughn

Let us now turn to the trail taken by trackers in establishing the science of vertebrate ichnology as a respectable branch of modern paleontology. Rev. Edward Hitchcock (1793–1864) is rightly regarded as the father of vertebrate ichnology. Between 1836 and 1864, he devoted the greater part of his career to the description of Early Jurassic dinosaur tracks from the

Connecticut Valley region, and amassed a classic fossil collection still on display at Amherst College, Amherst, Massachusetts. At the time he believed that the majority of tracks were those of giant birds, a conclusion that turned out to be really not too far from the mark, since we now know that the dinosaur in question was closely related to ancestral birds.

Hitchcock's work was well respected by leading contemporary geologists and paleontologists of the day, notably Charles Darwin, Charles Lyell, and Rev. William Buckland.[20] We must remember that at that time, Early Jurassic dinosaurs were not known to science, and giant fossil birds, and extraordinary marsupials from Down Under, were just becoming known. So the interpretations followed the scientific perceptions and paradigms of the day. Buckland, a celebrated professor of geology at Oxford University in England, led the way, describing British tracks found in Great Britain and presiding as master of ceremonies over experiments in which tortoises (primitive reptiles) made tracks in dough substrates on his kitchen table.[21] These were heady days for trackers on both sides of the Atlantic and they firmly believed that the new science of ichnology was destined for continued acclaim and approbation.

But early success led to an early slump. Later discoveries of the skeletal remains of dinosaurs and other vertebrates showed that the interpretations of both Hitchcock and Buckland had been wrong, at least in regard to who made the tracks. The paleontological mainstream lost faith in footprints and the subject soon languished in obscurity. Little work of note was published for the remainder of the nineteenth century, and even the revisions that reattributed Hitchcock's bird tracks to dinosaurs were somewhat turgid and perfunctory.

The fossil footprint renaissance started just recently. Discoveries and casual reports in the 1930s in the western United States and elsewhere suggested some untapped potential and the glimmerings of interest among professional paleontologists, who could not deny the value of rich track yields from some bone-barren rock formations. It was mainly in Europe that the faith was kept alive, by an assortment of dedicated and mostly forgotten enthusiasts. Many who wrote obscure reports on fossil footprints were either amateurs or scientists of other persuasions whose well-trained clerical instincts assured posterity that new finds were recorded in local natural history journals.

Not until the 1960s and 1970s did a few specialists in the field of fossil footprints emerge. I once dubbed these gentlemen the magnificent seven. With the exception of Don Baird, a former Princeton paleontology professor, and Rudolfo Casamiquela, a pioneer South American geologist, all are Europeans: George Demathieu (French), Paul Ellenberger (French), Hartmut Haubold (German), Giuseppe Leonardi (Italian), and Bill Sarjeant (English). Ellenberger and Leonardi also happen to be priests. As important as their individual contributions is *how* this group has worked together. Collaborating in several languages to document sites all over Europe, the New World, and southern Africa, they learned to begin to speak the same scientific language whenever possible. They held the faith while indifference about tracks still prevailed. None has become a paleontological superstar, yet all are known and respected in the paleontological profession. They helped lay the foundation for the current renaissance of the 1980s and 1990s.

The late sixties saw a paradigm shift in the field of biology toward ecological consciousness and holistic thinking.[22] In the process the dinosaur renaissance was born, and these once-defunct evolutionary failures suddenly became athletic, intelligent, and scientifically marketable. Following right behind the track makers in the scientific and public consciousness came their tracks. Awareness of the value of footprints for providing information about dynamic aspects of dinosaur life awakened, and the pages of prestigious scientific journals aired debates about dinosaur speed, locomotion, and social behavior. Today, dinosaur trackers are still enjoying the approbation of recent years. After almost 150 years of neglect the science has enjoyed its first truly golden age since its inception in the 1830s.

LEADING WITH THE LEGS

Nature talks in signs, and to understand its language, one has to pay attention to form.

Jeremy Narby, *The Cosmic Serpent*

It has been noted that major evolutionary change is most obviously manifested in the limbs. But limbs and the feet that make footprints are part of the whole organism, by which we mean not just the whole anatomy, but the anatomy plus the whole physiology, nervous system, metabolism, coloration, habitat, spirit, and behavior—the organism from soul to skin. If we truly understand the holism message, we can learn to read the whole from any of the parts. To support this confident assertion, I shall start with tracks.

We are slowly beginning to realize that living systems function as coherent integrated systems. Mae-Wan Ho has spearheaded a revolution in biological thinking about organisms. She and her former colleague Brian Goodwin make a plea for a science dedicated to the "knowledge of the organic whole."[23] Their conviction is grounded not in mystic intuition (or at least not this alone), but rather in hard scientific evidence. In her seminal work Ho reiterates the known maxim that living organisms defy entropy, the tendency of systems to run down, by transcending the Second Law of Thermodynamics, which states that all inanimate matter decays to equilibrium or "heat death." By contrast, organisms self-organize to higher levels of complexity and order. This "negative entropy" has been equated with information,[24] or intelligence, and is explained by the organism's ability to store energy at the same time as energy flows through it. One manifestation of stored energy is the emission of biophotons, the phenomenon of bioluminescence, or "inner light," which has recently been studied intensively. We are all literally vessels that simultaneously absorb, store, and emit light.

Like an eddy in a river, an organism's life cycle is a smaller circular current connected to energy flow in a complex system, rather like an excursion to and from the main stream. Each eddy is a thermodynamically free closed, self-sufficient energetic domain. This highly emancipated state is in essence life. On the biophysics level we are polychromatic organisms, radiating biophotons. Such findings have implications for research into energy fields or auras.[25] Organisms are what Ho calls polyphasic liquid crystals—phases of matter in between the solid and liquid state, with molecules in perfect harmonious alignment.

The nice thing about the holistic way of thinking is that we can move from biophoton to whole organism and back while still looking at one whole entity. We can look from track to organism to biosphere, without ever crossing a boundary in the biospheric spectrum of wholeness. As a complement to Ho's approach, it is instructive to examine the work of Wolfgang Schad, mentioned earlier in this chapter, who developed a similar perspective by studying mammals as whole organisms. His approach derives from the pioneer scientific studies of the nineteenth-century German poet-philosopher-naturalist Johann Wolfgang von Goethe, which are currently being reexamined by some of today's leading philosophers of science, such as the English physicist Henry Bortoft.[26] Schad's and Goethe's approaches have in common that they are both phenomenological—in plain English, fundamental or very down-to-earth. The phenomenological approach is not a scientific "method"; rather, it emphasizes direct perception of the object at hand. It is a "way of perceiving" that seeks to engage the observer in the observational process, from which knowledge arises, rather than detaching the "objective" observer from the observed, in the realm of hypothetical models and preconceptions. This latter approach can dull the very observational process by which scientists set so much store, by laying down purportedly objective rules that constrain direct perception of the object at hand. In short, *how* we think and observe is fundamentally important.

Schad demonstrates that we may develop greater insight into how animals are constructed if we look at them holistically, looking for the most obvious patterns of organization. His holism starts at the fundamental level of the senses. He shows, first, that the senses of all mammals, including humans, are distributed along an anterior–posterior (or head-to-tail) axis, with different types of animal emphasizing the anterior or posterior poles to varying degrees. In humans, for example, the anterior senses of sight and hearing extend to the limits of the visible universe in the first case and for many miles in the case of hearing. The organ of smell, the nose, is in the center of the face, and like the respiratory and circulatory systems at the center of the body, has an intermediate or central quality, which is rhythmic, as explained below. Taste and touch are associated with the limbs and alimentary system and are located posteriorly in both the head and the body and operate only in close or direct contact with the objects they sense.

Second, Schad explains that senses can be understood in terms of whether they receive long-range signals (outward-directed senses such as sight and hearing) or short-range signals (inward-directed, such as taste and touch). They also can be understood in their relation to motion: Are the sense organs static, or do they move? There is a built in polarity to the spectrum of long-range (outward) and short-range (inward) senses, and a countercurrent, or reverse polarity, pertaining to motion. This is because the anterior senses are associated with static organs, which hold still to receive long-range stimuli, whereas the posterior senses are associated with mobile limbs and digestive or metabolic organs, which move constantly in direct contact with the environment. This holds equally for the head, with rigid anterior cranium and posterior jaw, lips, and tongue, which are the mobile, metabolically connected limbs of the head. Schad's holistic synthesis can be laid out as follows:

OUTWARD		INWARD
Anterior	SENSE—(speech)—NERVE	system
Central	RESPIRATORY—CIRCULATORY	system
Posterior	LIMB—(reproductive)—METABOLIC	system

On the left side of the scheme we have the outward-directed senses, those that receive information and material. On the right are the inward-directed systems, which digest or process information and material. In Schad's scheme the speech system mediates centrally between sense and nerve systems, and the reproductive system is situated centrally between the limb and metabolic systems.[27] Other vital currents and polarities include the shift from more static to more mobile as one goes from top left to bottom right. A fuller appreciation of Schad's perspective can be cultivated by understanding the principle of polarity. Emphasis at one bio-

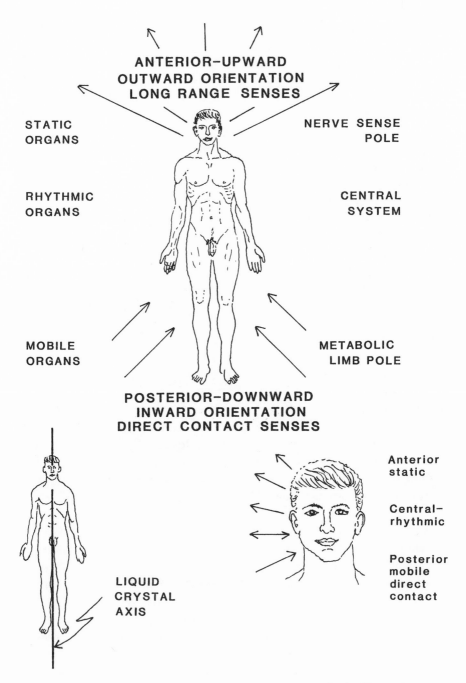

FIGURE 1.5 Inward- and outward-oriented senses in the human body: Observation reveals similar patterns in most mammals and interesting variations in other animals.

logical pole means a corresponding deemphasis at the opposite pole: a principle of compensation first noted by Goethe in 1795! You can develop some of the senses well in some areas most of the time, but you can't develop all senses, in all areas, all of the time. Let us now see how this applies to the biology of familiar mammals.

Schad looked at the senses of three major groups of mammals—the rodents, carnivores, and hoofed mammals (ungulates). He found that the anterior senses of the rodent are highly developed: its eyes and whiskers are large and it is always in constant frenetic motion reacting nervously to every stimulus in the environment. Yet despite this constant motion of its highly outwardly directed senses, it never travels far from the safety of its home and is very much at the mercy of fluctuations in the environment. By contrast to the small rodent, which has an anterior physiological pole, the ungulate is large and placid and has its dominant senses centered around what Schad calls the posterior physiological pole, or center of gravity—that of the digestive system and limbs. The ungulate is inward and literally ruminating in its sense orientation, and very emancipated from the environment. With no home base it wanders and migrates freely, sometimes over huge distances. The central group, the carnivores, are intermediate in size, with a range of intermediate characteristics that can be described as rhythmic, mirroring the beating of the heart and the inhalation and exhalation of breath. They do not exhibit a homogenized blend of polar behaviors. Instead, they alternate rhythmically between alert nerve sense and placid metabolic behaviors and characteristics. They stay at home in a den like rodents one day, and wander like ungulates the next. Similarly, they doze placidly for hours, then spring suddenly into nervous activity. Schad suggests that there are three main areas of emphasis: the nerve-sense (as exemplified by the rodents), the central rhythmic (carnivores), and the limb-metabolic (ungulates). These reiterate within groups and subgroups again and again in a type of fractal pattern that appears universal to life systems.

When we examine other fundamental aspects of the biology of these familiar mammals, we soon see that the system is thoroughly holistic. For example rodents have a short gestation period and very immature (*altricial*) young, which are expelled "outward" very early in life. They are literally naked in the face of a hostile environment, and may even be eaten by their own parents. They therefore succumb easily to death. The ungulate, on the other hand, has a long gestation period and is retained inwardly within the womb until it is ready to literally stand precocially (with independence) on its own feet. It can be born anywhere and within minutes can run with the herd. It does not easily relinquish its life. The carnivore is born in the intermediate, semi-altricial, condition. In behavior it alternates between reckless endangerment and aggressive defense of its life.

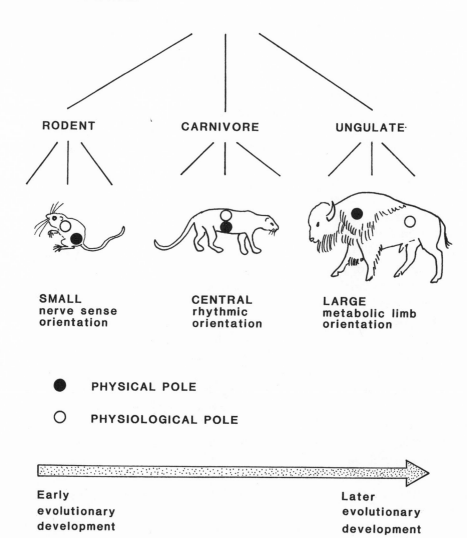

FIGURE 1.6 The spectrum of morphology in rodents, carnivores, and ungulates demonstrates the complementary polarity of physiological and physical centers of gravity. Based on the work of Wolfgang Schad.

Paradoxically, the outward nature of the rodent is expressed in its solitary habits and the inward nature of the ungulate in its gregarious behavior. Mice may live within feet of each other and may be too neurotically preoccupied with the environment to ever interact. By contrast, ungulates, such as buffalo, run in herds of millions so that other buffalo constitute their environment as much as the prairie itself. The carnivore is a lone wolf one day and a pack animal the next. Even coloration can often be understood more clearly when seen through Schad's lens. Rodents' coats generally have a strong contrast between dark backs and light undersides, whereas many carnivores show rhythmic alterations of spots and stripes, and the larger ungulates are often long-haired and dark all over. We can begin to see how the whole spectrum of color in animals and animal groups might be part of a coherent system. Ho's work on the polychromatic (colored) characteristics of organisms viewed by biophoton researchers only serves to emphasize the validity of this point of view.

The professional biologist may at first protest that there are many exceptions, such as large rodents and small ungulates, but in a holistic system these may not really be exceptions. Schad shows very elegantly that the threefold divisions repeat again and again, giving us the wonderful diversity we observe in the mammal world. So among rodents we have nerve-sense–dominated mice, and larger metabolic forms like the dark-colored porcupine and capibara. Similarly we find that there are small, nerve sense–dominated gazelles and large, metabolic bison. Among carnivores there are rodentlike stoats and weasels and ungulatelike bears. This is to say nothing of the contrast between small nerve sense–dominated dolphins and porpoises and gigantic metabolic whales. This view helps to explain the convergence of types like the one suggested by the mouse-stoat comparison. There are myriad other examples reflecting the repeated flow of evolution down one of the three main pathways. The large bison is like the large grizzly bear in having a dark shaggy coat, with large front limbs, emphasized shoulders, and bowed head and neck. Both animals represent the metabolic extreme of their respective groups.

Keeping in mind all these currents of size, sense, color, behavior, and physiology distributed throughout species, we should turn now to actual morphology, which is all the poor paleontologist has to go on. In Schad's scheme we find that for every current or polarity there is an equal and opposite "compensating" current and polarity, if only we know what to look for. Ho calls these "coupled cycles" and describes them as the "ultimate wisdom of nature." So the rodent, with emphasis on its anterior physiological center of gravity, has well-developed, rapidly growing "anterior" incisor teeth; the carnivore has well-developed "central" canines; and the ungulate has broad "posterior" molars. There is also what we might call a countercurrent, or reversed polarity, in the overall shape of the body, for the rodent has a small head, short neck, long tail, and long hind legs, and frequently sits up on its haunches. So its well-developed anterior physiological (nerve sense) center of gravity corresponds to a well-developed posterior physical center of gravity. In ungulates, by contrast, the well-developed posterior physiological (metabolic) center of gravity is counterbalanced by a well-developed anterior physical center of gravity expressed in a short tail, large head, usually with horns and long neck, and well-developed fore limbs, and shoulders. The carnivore as we know from countless literary sources seems to have the perfect balance of form, with its physiological and physical centers of gravity in the same central location.

Now we can begin to see how all this relates to limbs and tracks. Nerve-sense animals have well-developed hind feet. In the case of rodents they have relatively large feet that are widely splayed, with five toes. They are small and light and travel on the surface of things, notably staying away from exposure in open areas. They spread a little weight over a lot of space and so leave few well-impressed tracks. Ungulates have relatively small feet, with only two toes in the case of most horned animals. They therefore concentrate a lot of weight in a small area, and so leave deep tracks. But this is only the beginning of what we can do with tracks.

In the ungulates there is a correspondence between hooves and horns. The so-called odd-toed ungulates, such as horses and rhino are "narrow" animals in whose feet the central axis is emphasized because there is a toe in the middle; they have no horns, or one horn that is centrally located, as in the case of the rhino. By contrast, the even-toed, cloven-hoofed ungulates— goats, deer, bison—have two laterally located horns. As above so below. But the holistic connections do not end there. Small nerve-sense ungulates like gazelles have narrow, elongate, straight horns, hooves, and tracks, whereas in buffalo or musk ox the strongly curved geometry of the horns reflects the shape of the tracks. In the rare case of a sheep that was suspended off the ground, the hooves grew in the same form as the horns. Thus the tracks tell the tracker what other parts of the animal looks like, just as the claws on rodents and carnivores and the hooves of ungulates reflect the shapes of their teeth, and hence shed light on diet.

As we follow the eternal trail, I shall show how these holistic principles apply very nicely to dinosaurs and their ancestors and descendants, as well as to mammals. So we shall track the whole animal, not just its feet. If we take a Schadian perspective, we go beyond just seeing the organism, or even the species, and recognize holistic relationships at all higher levels of animal classification, be it rodents, carnivores, ungulates, mammals, or specific families or subfamiles within such groups. Indeed, the biosphere as a whole is an integrated system that should be viewed holistically. Schad has clearly demonstrated how holistic direct observation of the organism can be its own reward. He has also noted that many evolutionary novelties are first expressed in the limbs of animals. Obvious examples include the transformation from fish to amphibian, or from reptile to bird.[28] Similarly, our hominid ancestors stood and walked upright long before our brains ballooned out. So evolution often leads with the legs. This suggests that the track record may be the first place where new evolutionary trends are recorded. Furthermore, these significant evolutionary changes are thought to take place most frequently in shoreline settings, which is ex-

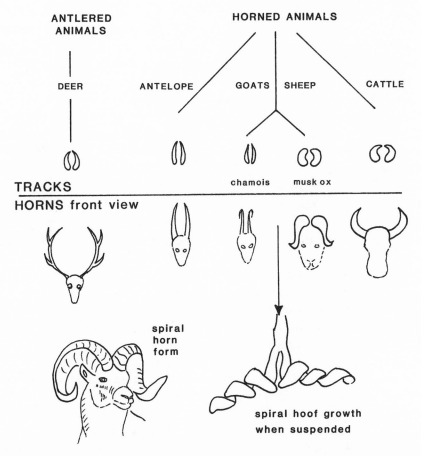

FIGURE 1.7   Horns match hooves in number, narrowness, and shape, and may even grow in spirals in some cases.

actly where tracks are most often preserved. So tracks may also record where many of the latest fashion trends in legs and feet originate.

MALE FEET AND FEMALE FOOTPRINTS

If you would hit the mark, you must aim a little above it; every arrow that
flies feels the attraction of earth.

Henry Wadsworth Longfellow, "Elegiac Verse"

To make an impression someone has to be listening. The study of tracks is about the interaction between two entities, the organism and Mother Earth, our receptive partner from which all life arises and to which all life returns. The foot or appendage that prods or pokes the Earth represents the male force of masculine energy (the yang), busily rushing about above the surface. The footprint is the embodiment of feminine receptive Earth (the yin), which registers the impressions of busy organisms scurrying on their ephemeral journey from birth to inevitable reunion with Mother Earth. For every foot there is a footprint, for every bolt a nut, for every ball a socket, for every roaming spirit a home to which we may return.

Every footfall is a sacred communion with the feminine principle of the universe. It is a source of wonder that so-called primitive cultures understood this so well, whereas we seem largely oblivious to what we are doing to Mother's skin. Gender is one of the most fundamental polarities expressed in nature, and is ingrained in most languages. So trackers should know that males and females of most species are slightly, if not very, different—a phenomenon known in biological jargon as sexual dimorphism. Schad's work does not explicitly explore the masculine/feminine polarity in detail, so I shall.

It has been known for some time that lineages of animals often evolve from small, unspecialized forms to larger species. This "they get bigger with time" idea has been expressed in Cope's law or rule, formulated by the famous nineteenth-century American dinosaur hunter Edward Drinker Cope, which states basically that body size increases in evolutionary lineages.[29] What is not well understood is how this process works. Using Schad's groupings of rodents, carnivores, and ungulates, we may perhaps suggest an answer. A recent book by the paleontologist Ken McNamara, *The Shapes of Time,* shows how a phenomenon known as *heterochrony*—literally, "different timing"—helps explain the pathways that species follow in evolutionary time.[30] Heterochrony can most easily be understood as the switching on or off, or delay or acceleration, of growth processes during development. If an animal stops growing or slows its growth early in life, it remains small, like a rodent. If it prolongs or accelerates its growth, it gets bigger, like an ungulate. Such processes tell us that time is an important dimension in generating the multitude of shapes seen in nature.

McNamara shows us that while the hominid skull has grown progressively larger, the jaw has grown smaller, a fine example of a coupled cycle of countercurrents that simultaneously emphasizes and deemphasizes opposite poles of the skull. The principle of heterochrony helps us understand why women are smaller than males, with prettier, less hairy, more childlike faces and voices that don't break, or why females have narrow shoulders and broad hips while males are broad-shouldered with narrow hips. Growth in one sector ended earlier but in another kept going. Almost any systematic shift in morphology—the form, size, and structure of body parts—may be attributed to the influence of timing. When we look at the time factor in the evolution of rodents, carnivores, and ungulates, we see that within each group, the smaller, primitive, or ancestral forms tended to flourish early; and the largest representatives appeared later. The same patterns may be observed in the evolution of dinosaurs and other groups. Thus we get great diversity arising from variations on a relatively simple theme. Such subtle variations on themes produce *convergence*—a phenomenon well known to biologists, which we shall explore in subsequent chapters.

A lineage of mammals or dinosaurs seems to begin as a small, unspecialized, outwardly oriented group that is highly sensitive to its new environment. Its members evolve to produce some lineages of larger, more inwardly oriented species that become more and more emancipated from the environment. Paradoxically it is these large emancipated species that appear to be most specialized, and they are conspicuous by their absence when the cycle is completed and they become extinct. It is among small unspecialized forms that we then appear to find the ancestors of the new lineages that replace them. The cycle could be seen as a rapid outward expansion, followed eventually by a slow return to the Earth. In terms of human experience this would be masculine yang energy giving way to feminine yin forces. The countercurrent, leading to smallness and sensitive dependence on the environment, can be seen as an expression of the feminine principle, yin, which would in turn give way to a subsequent emancipation from Mother Earth a sign of the masculine principle reasserting itself again. We can learn to look for and appreciate such coupled cycles of mutually reinforcing currents and countercurrents at all levels in evolutionary space and time.

It is almost as if life starts each new cycle of evolution with a diminutive, cautious, but excited tasting and testing of the new environment that leaves little evidence of tracks. It is evolutionary springtime. But as the lineage evolves it builds up to a swelling, gigantic, steam-rolling confidence that attempts to test its independence from the rest of the biosphere, leaving the scars of many heavy footfalls. The Earth is drained of her nutrients and resources as the organism grows to maximum satiation. Summer is over and it is time for fall and the lean months of winter. But as Mother Earth reclaims her own she again becomes pregnant in preparation for rebirth in the new spring.

All such polarities, cycles, and seasons can be seen in our evolving track record: inauspicious subtle beginnings give rise to unsubtle endings and subtle new beginnings. Like waves that rise and fall, as species become too big and too successful they pay the price of posing too great a challenge to the biosphere, and it seems the bigger they are the harder they fall. As we shall see in the later chapters, these masculine/feminine, small/large, expansive/contractive cycles continue to reiterate endlessly on down through time—even to this current millennial juncture, as masculine gives way, once more, to feminine.

chapter two

PALEOZOIC PRELUDE

ACT II: CAST OF CHARACTERS

In Act I there were few characters, and since they are members of our own species they needed no formal introduction. The cast of characters for Act II is more varied and includes a prize-winner dubbed "the world's greatest observer," a couple of 250-million-year-old sausage dogs, and Oxford's most famous nineteenth-century paleontologist, who praised fossil footprints in terms of "the power, wisdom and goodness of God." God, or some semantic equivalent, manages to get into most chapters and evolves along with the other characters, sometimes in the background, but sometimes in the thick of the latest controversies. We shall have our first encounter with creationist interpretations of fossil footprints, and find that a new breed of scientifically improved creationism is emerging.

As we look back in deep time to the Paleozoic Era (550 million to 250 million years ago) of ancient life, a few geological timeposts and biological signposts might prove helpful for those not steeped in the vocabulary of fossils and rocks. This era can be divided into two phases, the Early Paleozoic and the Late Paleozoic, each one in turn containing three time periods. During the early days the world was dominated by animals without backbones, and the most abundant of these invertebrates were the arthropods—joint-limbed, many-legged organisms with a hard exoskeleton—and they still are today; the phylum Arthropoda includes insects, crustaceans, spiders, and centipedes. Early arthropod protagonists were pirouetting trilobites and roly-polies with no apparent sense of direction. But perhaps we cast unwarranted aspersions, for we may yet discover a sense of purpose and design in the earliest of animal twitchings and ramblings.

In the Early Paleozoic (550 million to 400 million years ago) the only vertebrates that existed were fish, which clearly were not designed to leave footprints, despite claims to the contrary. Fish do, however, give us important clues to the early evolution of locomotion in our vertebrate ancestors. There is a pattern to the evolution of movement and locomotion, and this helps provide an intrinsic explanation of animal design. This theme of holism will be reiterated throughout the book, stressing that organisms are holistic self-organizing systems that are their own explanation. They don't need us to explain their existence, only to appreciate them!

The Late Paleozoic spans a slice of time from approximately 400 million to 250 million years ago (about 10:00 until 10:45 p.m. on our 24-hour clock). The Late Paleozoic can be divided into three periods: the Devonian (named after rocks found in Devon, England) the Carboniferous (refers to extensive coal deposits), and the Permian (named after a town in Russia). These three periods have been dubbed the ages of fish, amphibians, and reptiles, respectively. During the Late Paleozoic, vertebrates with four legs (*tetrapods*, literally "four legs") were establishing themselves on land. Many were not particularly exotic; we can think of them as salamanderlike amphibians and, later, alligatorlike reptiles that hung out mainly in wet swampy areas. They did sometimes run into some remarkable invertebrates, including millipedes that were more than 18 inches (45 cm) in diameter and over six feet (1.8 m) long.

The tracker's job of matching tracks with track makers will start with invertebrates and early vertebrates. Of evolutionary and historical significance to our own mammal line are a group that we can refer to as protomammals. These distant ancestors came in a variety of shapes and sizes, ranging from famous fin-backed varieties to sausage-dog designs. Their tracks suggest behavior ranging from social pair bonding and home building to antisocial predation, the ability to run, and a strong preference for desert habitats.

THE DEEP STRUCTURE OF SEDIMENTS AND STRATA

I am the daughter of Earth and Water, And the nursling of the sky;
I pass through the pores of the ocean and shores; I change, but I cannot die.

Percy Bysshe Shelley, "The Cloud"

Long before sand and mud on the seashore was disturbed by the feet of dinosaurs or hominids, it was subject to the influence of waves, tides, rain showers, and other natural phenomena. The oldest rocks in Earth's crust are of the *igneous* variety—"born of fire." With a little geological training we can trace their paths across the face of the Earth, for continental drift and polar wandering pathways are well-known to geologists.

Deep in Precambrian time, between 4.5 billion and 4.0 billion years ago, the molten fireball Earth cooled. Unfortunately, the process left no tracks. Any rivers of lava from this epoch are long gone. But as the earth settled down, exhaling gases and finding its early equilibrium, an atmosphere began to accumulate. Gases and water vapor condensed into liquid moisture that cooled the hot rocky surface. After her early exertions, Earth broke sweat. At first, droplets could not settle on the burning surface without being driven off as steam, but before long, geologically speaking, the first drop of dew condensed on a rock face reflecting the light of the moon and stars. As it ran like a tear down her cheek, Earth felt the imperceptible trickle of her first miniature river.

The trail of Earth's first dewdrop is not recorded, nor can we say precisely where and when the first two dewdrops converged to become the first confluence of tributaries. We know, however, that these trails of dewdrops soon became a flood of surface water. With the flow of rivers and the wash and swash of tides and waves, the angular teenage edges of Earth's rocky crust began to mature and round out. Soon, rivers, lakes, and proto-oceans were recycling crystals into grains of sand and flakes of mud, redistributing them as a new type of geological phenomenon—*sedimentary* rock.

Just as igneous rocks formed from a partnership with fire, sedimentary rocks formed from a partnership with water and sometimes with wind and air. These rocks' deep structure reflects a fundamental integration with the dynamics of tide and wave, with the cycles of climate and current flow. Ripple marks record the speed of currents over beach and riverbed and the pounding of waves on the shore. The trained sedimentologist can easily pick out the traces of the gentle ebb and flow of tides, and distinguish their signature from the sedimentary upheaval created by a violent storm.

Perhaps the most evocative sedimentary structures are raindrop impressions. These may allow us to tell the difference between a gentle shower and a violent hailstorm. With a little practice we can learn to read the rhythm of the rain, as the Bushmen (as Europeans called the San people) of southern Africa have done. In *The Harmless People*, Elizabeth Marshall Thomas describes how different forms of rain have different meanings for the Bushmen: "Great destructive thunderheads that bring hailstorms or electric storms [are male]. The soft gray clouds that carry land rains which blow like mists into the plants of the veldt and make them grow are female. The rain of male and female clouds is itself male or female depending on the clouds' gender or whether the rain is creative or destructive, but if you can't tell from the rain as it falls, you can tell from its footprints, for male rain leaves sharp pierced marks in the dust and female rain leaves wide soft splashes."[1]

Here we are reminded of the relationship between yang, aggressive male energy of "banners waving in the sun,"[2] and yin, submissive female energy, the word originally meaning cloudy or overcast, "that yields to and completes the yang initiative."[3] It seems that Bushmen shared the insights of Taoists in recognizing occasions of stormy yang energy falling to a receptive yin Earth, and occasions when the rain clouds were pregnant with soft yin energy. Such metaphors are wonderfully apt for geology, for as the wild yang energy of squalls and tidal waves beats on the face of Mother Earth, so her resilient crust absorbs the flailing energy and conserves a pattern in what geologists call the "architecture" of sedimentary layers. We should perhaps heed the popular refrain and learn to listen to the rhythm of the falling rain.

The accumulation of sedimentary rocks gives rise to layered rock sequences, or strata, the study of which is the science of stratigraphy. Each layer is a discrete stratum, or "bed," of rock, and each represents a unit of time, a step in geological and evolutionary history. Some layers also contain the first fossil signatures of 3.5-billion-year-old life. Exactly how life evolved is an elusive mystery, as is the definition of life itself (dictionaries of biology are of little help!). Life represents the emergence of a unique level of complexity and novelty in the cosmos, the transformation of dust into "vital dust," as Christian de Duve terms it, the emergence of biosphere from the physiosphere. William Bryant Logan put it poetically: "[O]ne morning in the Archean Era, an assembly of chemical compounds, possibly clay, obeying some divine suggestion, threw an envelope around itself and began to live. It now had an inside and an outside." Clay is a most interesting substance, not least because both scientists and philosophers regard it as the birthplace of life. "Adam" is the Hebrew word for red clay, and clay, Logan notes, "may quite literally have been the matrix [the old word for "womb"] that spawned all the creatures now inhabiting the earth." From the tracker's point of view, clay is the substrate in which all the best tracks are registered, and it is the substrate in which the human mind first registered its written records in cuneiform script (discussed in Chapter 9). Clay is unique in its ability to absorb water as an integral part of its chemical makeup. In the view of the biologist Hayman Hartmann, cited by Logan, clay is alive. "There are only two things in the universe that require liquid water for their existence: organic life and clay."[4]

In this clay medium the first life to stir was vegetative, algal, bacterial. Although too static to leave trails, algal structures known as *stromatolites* "grew" into layered mounds. In homage to the life-giving sun these algal communities rose in slow motion toward the light, their hair-like algal strands literally standing up by day and lying down at night. Clustered together like so many high-rise buildings, they have been likened to miniature cities, built like many of their modern counterparts on prime real estate with easy access to the bounty of the sea.

CRUISING THE CAMBRIAN: FOSSIL ART IN THE MAKING

Lady Ichnology is coming of age.
Dolf Seilacher

FIGURE 2.1   A yin/yang view of rain, weather, and Earth inspired by the worldview of the Bushmen. Artwork by Paul Koroshetz.

There is no Nobel Prize awarded in paleontology. The next best thing, however, is the Crawford Prize, awarded in 1992 to Dolf Seilacher, a former wireless operator in the German wartime navy. The award recognized his scientific contributions in paleontology. His work has focused primarily on the paleontological bounty of ancient seas that date back some 500 million to 600 million years, to the Precambrian-to-Cambrian transition, when life first arose. He pioneered the study of marine trace fossils (mainly burrows and other invertebrate trails), and investigated the anatomical construction of the first multicellular creatures. In doing so he showed that they were vegetative in form, and not like modern jellyfish or worms, as once thought. He has helped to demonstrate what things are not. He embraced the bizarre fringes of paleontology and bridged the gap between the world of trace fossils, or what some have called "ghost fossils," and solid material-body fossils. He also used his prize money to help bridge the world between art and science, by promoting a traveling exhibit entitled Fossil Art.[5]

Della Willis has called Seilacher "the world's greatest observer."[6] Seilacher's secret is that he spends time with his fossils and carefully draws, or "traces," every contour of his specimens. Drawing is a kind of tracking (from the French *tracer*, "to draw") that allows one to really get to know one's fossils intimately, to observe every subtlety and nuance and get beneath the animal's skin.

The transition from static stromatolite cities to traces of the activity of mobile animals in the latest Precambrian and Cambrian appears to represent a quantum leap in biotic organization. From a world of vegetative tranquillity, multicellular life emerged, constructed according to entirely new blueprints. Enter nervous systems, eyes, limbs, and mobility. But this apparent evolutionary progression from static to mobile is deceptive. In his poetic ruminations on the work of the renowned biologist Lynn Margulis, William Irwin Thompson points out that the humble bacterium, the spirochete, represents the first organism to display a sense of direction.[7] These snakelike microorganisms twitch randomly when floating in a fluid medium, with no sign of anterior or posterior orientation. But when attracted to the surface of other microorganisms they develop directional motion, and one end becomes static (the anterior "head") and the other mobile, with a flagellating "tail." Here we see parallels with a Schadian perspective on polarity in organisms. There are significant implications for consciousness. The once directionless, unoriented spirochete has developed an awareness of the other. It not only attaches to another organism, but works in unison with a multitude of its peers to beat in rhythmic symbiotic harmony with its host, to generate feeding currents and motion. It quite literally has a sense of community and collective purpose.

The first large, enigmatic multicellular organisms of these times are known as the Ediacaran biota, named after the site in Australia where they were first discovered, the Ediacara Hills. Many species consist of large, flat, disc-shaped creatures, sometimes compared with jellyfish. Most lack any obvious anterior or posterior orientation, yet they have form, having been labeled "radial," "unipolar," or "bipolar" (the latter conjuring images of Precambrian personality disorders that perhaps led to extinction!).[8] Seilacher sees these creatures as air mattresses, or balloonlike creatures, and calls them "pneu," implying that they grew by a process resembling inflation.[9] The form of some of Seilacher's beloved Precambrian organisms is reminiscent of magnetic fields, and reminds us of Mae-Wan Ho's holistic view of organisms. What maintains and sustains this field structure is a difficult problem often avoided by biologists. With little consensus on how living creatures grow, understanding extinct species is clearly a challenge. Nevertheless, new insights are beginning to emerge.

Since Goethe pioneered the science of plant growth, or "morphogenesis," in the late eighteenth century, the Germans have continued to publish interesting work in this field.[10] Scientists are divided as to whether organisms grow as the result of the "drive" of intrinsic, internal instructions provided by the genetic code (microcosm), or whether external, extrinsic factors in the environment (macrocosm) generate a "teleological pull" (that is, an attraction toward a predetermined form or design). The former cause-and-effect mode of thinking—appropriate to traditional physics and material science—presumes that all effects are the result of prior causes or conditions, i.e., the present is caused by the past. The latter, teleological, mode of thought, which suggests that life responds to goals and objectives (the impulses of the will or soul), is derived from psychological thinking, i.e., the present is determined by the future. In the realm of biology, neither explanation is entirely satisfactory. Species are whole organisms that function to integrate the purely physical, bodily realm, and the higher psychological realms of mind. As the biosphere mediates between the physiosphere and the noosphere, so biology occupies an intermediate or central position between physical sciences and the psychological realm.

To understand the biological realm more fully we must pay attention to the whole context of the dynamic life process, rather than merely looking at organisms as objects or bits of "text" that populate the world around us. This phenomenological perspective recognizes the "etheric" world not as a new hypothesis but as a different way of seeing.[11] It is more fundamentally perceptive because it pays attention to the observational process, rather than claiming detachment and objectivity. This may seem odd at first because we are conditioned to think that scientific objectivity is indisputable. But herein lies a delightful paradox. To claim objectivity is to imply that one is not really fully using ones senses, but deliberately detaching from them like some sort of machine or automaton. By contrast, when deliberately developing conscious perception, by which knowledge is acquired in the first place, we free ourselves from the ludicrous constraint of deliberately restricting our observations. A holistic view allows that valuable scientific insight can be obtained by both traditional objective and phenomenological approaches. The latter may be considered the more novel paradigm in the context of how we presently do science.

From this rewarding phenomenological perspective Schad has succinctly "hit upon the central characteristic of the life processes: *as such they are determined not so much by previous or future conditions, but rather at each moment by their own present....* One must look beyond causal or teleological relationships; one must seek above all the *simultaneous relationship between phenomena.* That two phenomena condition and promote one another [and] are mutually explanatory, is the biological process fundamental to all organisms" (emphasis added).[12] Here again we see what Ho called nature's ultimate wisdom of manifold coupled cycles.

So in the study of tracks, the narrowness or breadth of a foot, hand, or body is expressed in coherent biological relationships throughout the body of individuals or species. Tallness does not "cause" slenderness any more than slenderness causes tallness, and shortness does not cause stockiness any more than stockiness causes shortness. Modern medicine is beginning to recognize the simultaneity of responses to stimuli, for example in the brain and gut, or simply, as is so often said, in the mind-body connection. "Dis-ease" in the mind is not caused by a dis-ease in the body any more than dis-ease in the body is caused by dis-ease in the mind. They occur simultaneously "because" this is the nature of biological function. The very word "because" takes the cause out of the past into the present realm of being. If a rock, A, falls from the sky to kill B, A may cause B's death. But in the case of two components of the biosphere, A is because of B and B is because of A. She cannot be my mother unless I am her son and I cannot be her son unless she is my mother. Biology!

For somewhat different perspectives on morphogenesis we can turn to the controversial but intriguing ideas of the biologist Rupert Sheldrake, who noted: "As plants and animals develop . . . their form and organization become increasingly complex. How this happens is a mystery."[13] But as Seilacher has demonstrated, understanding how a creature is constructed helps us better understand its behavior. As Sheldrake points out, growth is a type of behavior repeated "habitually," generation after generation. He describes this as "morphic resonance." The central idea here is that creatures behave in a certain way not just because of their shape, but because growth itself is a specific behavior that produces shape. So which came first—behavior or shape? The answer is that both evolved together in the growth process. It is a coupled cycle.

Trace fossils help us understand this. A simple pencil-shaped worm could move in any number of directions; in fact, though, similarly shaped worms develop very different and distinct behaviors that result in traces that look like writings in quite distinct languages. So different behaviors and shapes of tracks or burrows arise from a single animal form. This leads Seilacher to speak of different "programs" that govern different behaviors among animals that are essentially similar in their basic design. We shall see examples of this again among vertebrates, which though their feet may be similar have certain characteristic behaviors that impart different and specific signatures to their trackways. Accomplished trackers, such as Bushmen, read the animal's "nature," or spirit, in such clear or subtle signs. Wildebeests ("wild beasts") are so named for their wild, playful spirit.

Seilacher reports fascinating examples of changes in behavior during the evolution of certain types of trace fossil. For example some types represent animals that meandered around on or near the surface in the Cambrian, but by the late Paleozoic these same forms had progressively contracted their foraging activity into a central area, where they burrowed deeper and deeper in tighter and tighter spirals.[14] This shift in programming can very easily be interpreted as an example of heterochrony ("different timing"), the slow or fast evolutionary growth pathways that species follow, discussed in Chapter 1) in which the tight spiraling behavior began earlier and earlier in life. From a Schadian perspective this evolution represents a shift from the "outward" nerve-sense exploration of the surface of the environment to a deeper, more "inward" and centralized infolding of the system. It seems more than mere coincidence that in heading toward this inward metabolic pole, the burrow system takes on the spiral shape of the organ most closely associated with metabolism, the intestine.

By the Late Precambrian, evidence of vital motion in large organisms begins. Burrows and trails show us that creatures were on the move some 550 million years ago. The first stirrings of multicellular life excite trackers like Seilacher, for this is where the organic part of the eternal trail really takes off. Animal mobility meant that locomotor evolution had begun in earnest, giving us what we now call the track record.

Not long after the first signs of latest Precambrian fidgeting comes the so-called "Cambrian explosion." The most famous denizens of this period are the trilobites, early arthropods sometimes more than a foot in length that resembled giant wood lice. They plowed along the seafloor leaving unmistakable trails known to trackers as *Cruziana* (pronounced cruise-e-AH-na). What could be more appropriate than to label the first vigorously mobile creatures as "cruisers?" As they crisscrossed the Cambrian seafloor (*cruz* means "cross") they tried out their newfound mobility with apparent glee, and celebrated a true crossroads in the history of locomotion. Some were so ecstatic that they danced in circles, producing what Seilacher, in his Fossil Art exhibit, describes as "trilobite pirouettes."

One of Seilacher's talents is to think like a trilobite, or worm, and imagine himself plowing through the muddy darkness. Perhaps his communications training helps him " feel" his way through the "wireless" murky underworld of Paleozoic deep time, as he once did in the wartime waters of the North Atlantic. I have teased him in print by noting that he is one of the few trackers who can really "get under the skin" of a worm, "program" himself to think like one of them.[15] Whether the worms veer left, or right, up or down, or completely "round the bend," Seilacher's talent lies in adopting invertebrate consciousness and thinking like the track maker.

The *Cruziana* trails in his exhibit seem to have been made by a trilobite that moved in clockwise circles. "The caprice of an artistic loner?" he asks. "Probably not."[16] My guess is that with their newfound mobility, creatures developed an instinctive sense of home base. As with the rambling forms that gravitated more and more to home base with time, there is a fundamental polarity in the system. The journey is the mobile phase, home is

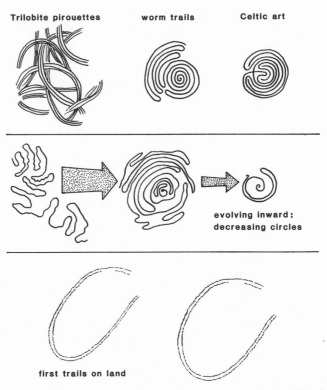

Trilobite pirouettes   worm trails   Celtic art

evolving inward:
decreasing circles

first trails on land

FIGURE 2.2   Trilobite pirouettes, now savored as "fossil art," like worm trails produce fascinating designs reminiscent of Celtic art. Some trails evolve in time from wide-ranging patterns to ever-decreasing circles. The first myriapods on land walked in circles, which is a good way to get back home.

the static point. Yang/yin. More flexible creatures, such as worms, are capable of doubling back and systematically harvesting areas of the seabed so as to leave no grain of sand unturned. In doing so they produced a very geometric, precise form of fossil art that closely resembles an intestinal map or an agricultural landscape. Biological form reiterates itself on many scales. Such geometric designs, in the form of aesthetic spirals, resemble Neolithic and Bronze Age rock carvings and later designs found on coins and other artifacts.[17] Imagine the impact that such natural art would have had on the first rock carvers and rock painters. Such intricate spiral designs, "etched" or entombed in the rock, surely aroused a supernatural sense of awe and wonder, giving the observer the sense that someone or something had been there before.

WALKING IN CIRCLES:
THE FIRST FOOTPRINTS ON LAND

The nature of God is a circle of which the center is everywhere and the
circumference is nowhere.

Anonymous

Long before the first fish swam in the sea, the oceans were home to thousands of species of arthropods; the number of individuals ran to many billions. They have been abundant ever since their first appearance, about
550 million years ago in the Cambrian period. There is also compelling
evidence that arthropods were the first animals to walk on land. In England's picturesque Lake District, where daredevil speedboat pilots once
skimmed the lakes to set world speed records, geologists have found trackway evidence of another, much older, world land speed record, set some
450 million years ago, in the Ordovician period, in the Early Paleozoic.

In the rural backwaters of the River Lickle and Sour Milk Gill, we find
narrow trails, little more than .2 inch (.5 cm) wide, made by arthropods
known as myriapods (literally, "countless feet").[18] These include centipedes, millipedes, and forms like the common English wood louse,
known to Americans as "roly-polies" and to biologists as the genus *Onis-
cus*. What their small tracks lacked in size they made up for in abundance.
With an average of 15 to 25 per inch (6–10/cm) in each of two parallel
rows, each individual could make some two thousand footprints in a one-
yard excursion on land. The trails have been named *Diplichnites*, in reference to the "double trace" row of footprints.

Many of these trails prescribe circular paths, but we really don't know
why. Given that there is a purpose to animal behavior, we can speculate
that perhaps, like trilobites, these animals were just establishing a sense of
direction by finding their way back to their starting point. Perhaps there
is no obvious explanation. In Stanley Burnshaw's story of the millipede
who asked God how all his legs worked, God's reply was "Don't think—
just walk."[19]

It is fun to suggest that these ancestors of wood lice were stunned by the
sheer enormity of their pioneering explorations on land. Dazed by sudden
exposure to atmospheric pressure, they were still finding their land legs.
(But no sign of the woods!) It is an odd coincidence that in homeopathic
medicine, the wood louse provides a remedy that cures violent pulsation
of the arteries and cramping of other organs of the vascular system. I
thought this connection was too tenuous to be worth mentioning, but

when I asked a paleontology class why the first wood lice on land walked in circles, they answered that the lice must have been dazed by the difficult adjustment to a new pressure regime. As if our poor wood lice did not have enough trouble with their exploratory walks on land, the gods of metamorphism and rock deformation have squeezed their nearly circular signatures into distorted elliptical shapes. In a kindly gesture, the authors of the track study used a computer to restore the deformed trackways back to their original subcircular shapes.

FISH OUT OF WATER: GIANT STEPS FOR VERTEBRATES

When you were a tadpole, and I was a fish, in the Palaeozoic time,
And side by side in the ebbing tide, we sprawled through the ooze and slime.

Langdon Smith, "A Toast to a Lady"

Paleontologists all know that amphibians evolved from so-called "lobe-finned" fish, like the famous coelocanth, by developing limbs from their stout fins. The traditional Darwinian view is that they needed limbs on land, and so evolved them by natural selection when they found themselves in shallow rivers and lakes that were drying out. But what about intermediate forms between fish and amphibians? Was there ever a transitional evolutionary stage—a fish capable of progression on land? The modern mud skipper and the "walking" catfish get around quite easily, and in terms of speed and mobility put tortoises and other slow-moving tetrapods to shame. But these species are not the ancestors of amphibians. According to at least one tracker, however, not only did intermediate forms exist, but their tracks are also known.

An early report from Devonian strata on the Island of Hoy in the Orkney Islands archipelago off the coast of Scotland mentions a long groove, or "belly-drag," trace with the impression of fins or finlike footprints on either side.[20] The trace has been interpreted as evidence of the activity of a fishy ancestor of the amphibians and relative of the famous primitive lobe-finned coelacanth. Prior to the 1930s, this "living fossil" was thought to be extinct, but then was found alive and well off the east coast of Africa, and most recently in Indonesian waters. It is appealing to entertain the thought of finding the trackway evidence of the first fish out of water, but wishes don't always come true. The original Scottish specimens were never adequately studied or preserved in a museum, and so fail to constitute convincing evidence of early vertebrates on land. But there is some consolation for British trackers in recent finds from another Scottish site, which suggest an amphibian trackway.[21]

In 1935, the same year that the old fish of Hoy interpretation was published, the American paleontologist Bradford Willard reported another fishy trail from Devonian strata in Pennsylvania.[22] Willard hypothesized that this track maker was half fish, half amphibian, and dubbed it "Ichthyopoda" (literally, "fish with feet"), and named the tracks *Paramphibius* (meaning "toward an amphibian"). In a classic rebuttal, the paleobiologist Kenneth Caster, of the University of Ohio, debunked the fish-with-feet fairy tale by showing that the trails had been made by large horseshoe crabs.[23] These arthropods, also known as limulids, have been found literally dead in their tracks in these rocks. We must concede that arthropods, not tetrapods, probably led the way onto land. Nevertheless, some Devonian amphibian trackways really do exist, notably at two sites in Australia and one in Ireland.[24] All were described as very important because they appeared to be tens of millions of years older than any previously known amphibian skeletal remains.

•   •   •

The oldest known true tetrapods are a handful of Devonian amphibians including *Ichthyostega*, *Acanthostega*, and *Tulerpton*. All of these animals possessed limbs and feet rather than the fleshy or lobelike fin appendages of their fish ancestors. Thus it is to be expected that we should begin to find trackways of the first walking tetrapods in the Devonian. But tracking Devonian amphibians is not an easy business if for no other reason than that they had very odd feet, sometimes with six, seven, or eight digits. This large number reflects an inheritance from the many bones in the fins of fish. Though we might hope to find tracks with six to eight toe impressions, so far no such examples are known. In fact, most tracks represent rather nondescript round holes. This shows that as the pointed fish fin transformed into the longer amphibian appendage, the toe bones became shorter and fewer, and the foot wider. A similar pattern is seen repeatedly in the evolution of feet.

However, just when we thought we knew how and when vertebrates first got their feet on the ground, the paleobiologist Jenny Clack, of Cambridge University, has cautioned us to think again. "Devonian tetrapods are unlikely to have made any of the known tracks, unless they were produced underwater," she asserts.[25] Moreover, the rock dating is dubious, so the footprints have not been proven to be much older than the bones. Indeed, it was probably some 25 million years after paddlelike limbs evolved in water that they were put to use as walking devices. Our view of

Devonian amphibians has therefore changed radically. They were perhaps somewhat like seals, leaving hard-to-interpret traces on the bottom. It was not fish out of water walking on fins that led to legs, but underwater fish developing their fins that promoted a new type of limb.

The implications of these observations for the study of tracks is considerable. An animal like *Ichthyostega* is really a fish with limbs. Put another way, if we found only the skull and non-limb portions of the skeleton we would regard it as a fish; this has actually happened. This means that evolution often is most progressive in the limbs. If limbs lead the way (as posited in Chapter 1) then tracks may give us advanced warning of things to come. This would seem to be the case if animals were walking underwater long before they ventured on land.

Did amphibians develop limbs with less bones than in fish fins, as a simple adaptation to the environment? According to the heterochrony expert Ken McNamara, many animal lineages begin with species that have many bones or skeletal segments and lead to species with less.[26] Such cycles or trends of reduction in number of bones are seen repeatedly in vertebrate foot skeletons. We see a trend toward the reduction in the number of bones in the amphibian skull as well as in the feet, but it happened in the feet first. This makes it look as though the evolution of limbs is as much an intrinsic biological phenomenon as a response to the environment.

By the time amphibians were making a habit of walking on land, in the Carboniferous period, it seems they had settled on a reduced plan of five digits. This blueprint, inherited or modified by all tetrapod descendants, is quite possibly the basis for our present decimal system. Had Devonian amphibians retained their extra digits and passed the blueprints on to us, we might be counting and measuring in multiples of seven or eight, and running 70- and 8,000-meter races in the Olympics.

MONSTER MILLIPEDES

MILLIPEDE Any of a class (Diplopoda) of myriapod arthropods having usually a cylindrical segmented body covered with hard integument, two pairs of legs on most apparent segments, and unlike centipedes, no poison fangs.
Webster's Tenth Dictionary

Just as the first amphibian tracks are of Devonian period, the first reptilian tracks date from the Carboniferous period. The Carboniferous landscape is one of lush tropical forests or "coal swamps" (the foliage was the raw material of our coal, from which the Carboniferous get its name). The forests swarmed with giant dragonflies, huge centipedes, grotesque spiders, and amphibians. The forests pumped new levels of photosynthetic oxygen into the atmosphere, leading some science journalists to rechristen this the Oxygeniferous period, instead of the "age of amphibians" or the "age of coal swamps." Levels of oxygen have been estimated at almost twice the present atmospheric level, 35-40 percent compared to the current 21 percent, and this may have contributed to the extraordinary gigantism of myriapods 18 inches (45 cm) in diameter and six feet (1.8 m) long, spiders more than a foot (30 cm) in length, and dragonflies with two-foot (60 cm) wing spans. Before we get heady with excess oxygen we should interject a note of caution. James Lovelock, in his classic work *Gaia*, suggested that the atmosphere cannot contain more than 25 to 26 percent oxygen without causing spontaneous combustion of plants and other organic material.

Arthropods have always been pioneers, and their behavior during the Carboniferous was no exception. Large amounts of decaying vegetation produced the forest-floor litter on which they fed, and what was left produced the abundant coal. In these heady days reptiles may have benefited from the new levels of oxygen coursing through their nostrils and arteries, aerating their expanding lungs and brains and prodding them to higher consciousness. The whole Carboniferous world pulsed with new life, as plants and animals arose from the water to the atmosphere and learned to adapt to the problems of desiccation, temperature fluctuation, and gravity by protecting their internal fluids and developing the mechanical strength to stand tall. Plants sturdied themselves with wood, waxy desiccation-resistant cuticles, and circulated fluids (xylem or sap and phloem). Invertebrates did much the same thing. To avoid drying out they improved their circulation systems and thick skins, also known as cuticles. (Incidentally, plant and arthropod cuticles are made of similar materials.) Some scientists have likened life on land to a complex oceanic wave that flooded ashore. This so-called "hypersea" was an upwelling of minerals and interconnected, nutrient-rich, fluid-filled capsules that spread like a green tide across the once arid and inhospitable landscapes.[27] Certainly forests became the lungs of the planet, and the flow of fresh and saltwater was analogous to the organisms' interconnected arterial and venous circulation.

In the track record we find extraordinary examples of the trackways or trails of giant millipedes. The first, discovered in the 1970s, comes from an old sandstone quarry on the rain-drenched island of Arran in the waters off the southwest coast of Scotland.[28] The largest trackway is 20 feet (6 m) long and an average of 14 inches (36 cm) wide, and another, also from Scotland, reveals a trackway 18 inches (46 cm) wide. Another giant millipede trail reported from New Brunswick, Canada, is 12 inches (30 cm) wide and was left by an animal meandering its way through a stand of

FIGURE 2.3   Reconstruction of the giant myriapod *Arthropleura* leaving a trail as it walks through a grove of horsetails. Courtesy of Annamarie Burzynski.

horsetails.[29] Unlike the roly-poly trails, these indicate truly monstrous arthropods. The track maker was probably *Arthropleura,* a giant myriapod that grew to be several yards in length. This is a rare example of a particular genus being matched with its trail. Such a wide-gauge trackway is more like a tram railway than a millipede trail, and has fooled more than one paleontologist into thinking he was dealing with tracks of large amphibians or reptiles. The Carboniferous was evidently the only time in Earth history when monster millipedes reached such a zenith, leaving trails over wide areas.

THE FIRST REPTILE TRACKS

"Are you a botanist, Dr. Johnson?"
"No, Sir, . . . should I wish to become a botanist, I must first turn myself into a reptile."
    Samuel Johnson, reported in Boswell's *Life of Johnson*

Carl Sagan, in his book *The Dragons of Eden*, proposed that in the Carboniferous vertebrates became more complex neurologically than genetically, so their ability to carry information in the nervous system outstripped the information- carrying capacity of their genes.[30] Such developments, he hypothesized, were partly a natural response to the new demands of living on land and standing up against the force of gravity, rather than swimming in the near weightless medium of water. Such improved posture and locomotion required improvements in the skeleton, muscles, and circulatory system that fed back into a more sophisticated brain and nervous system.

The first Carboniferous reptile footprints ever found were discovered in Nova Scotia, in 1841, the same year that the great British zoologist Richard Owen named the Dinosauria. At this time Carboniferous tetrapods were entirely unknown. This classic example of footprints being found before the discovery of a potential track maker caused the Canadian geologist John William Dawson to write that "like Robinson Crusoe on his desert island we saw the footprints before we knew the animal that produced them."[31] Also in 1841, the geologist William Logan, who later became the first director of the Geological Survey of Canada, excavated a block of tracks and took it to London to show to leading British geologists. Owing to the prevailing opinion that only fishes existed during the Carboniferous, Logan's evidence was apparently not believed, even though geologists could see the evidence with their own eyes.[32]

In 1844 the first Carboniferous skeletal tetrapod remains were discovered in Europe, and in the 1870s, Dawson found and named the first such skeletal remains from Nova Scotia. The fossils included the remains of a small lizard-sized creature found inside a fossil tree stump. It appears that periodic floods in the coal swamp buried the lower parts of standing tree trunks in sediment. As new forests grew, the old tree stumps rotted out, forming habitats or, in some cases, death traps for small amphibians reptiles and arthropods. At least eight amphibian and three reptile species have been recovered from about 30 tree stumps. There is, as Johnson implied, a close evolutionary relationship between forests (Earth's botanical lungs) and early amphibians and reptiles, bearers of the first animal lungs.

In 1872, Dawson christened the creature *Hylonomus* and classified it as a reptile. Ten years later he named the tracks *Hylopus*, meaning "*Hylonomus* tracks," again implying that we know the track maker. This interpretation may not be correct, but *Hylonomus* certainly had the means and opportunity, and the relationship is now enshrined in good paleontological Latin, whether we like it or not. Since Dawson's day *Hylonomus* has consistently been regarded by many as the oldest reptile, and has only recently been challenged by another reptile, christened "Lizzie," from the Carboniferous of Scotland.[33] Dawson was the first to explicitly state that tracks indicate that "reptilian animals existed in considerable numbers throughout the coalfield of Nova Scotia from the beginning to the end of the Carboniferous Period."[34]

GIANT SWAMP DWELLERS

An exceptional paleontological site discovered in 1994 at Brule, Nova Scotia . . . contains a world class paleobotanical and vertebrate trackway record.

J. H. Calder

The Carboniferous also produced its share of large amphibians and reptiles. The largest known tracks, from a locality named Horton Bluff in Nova Scotia, have been given the name *Baropezia* (meaning "broad foot"). They form part of a very long, 200-foot (60 m) trackway with individual footprints up to 5 inches (12 cm) long, indicating an animal on the order of 16 feet (5 m) in length, with feet large enough to completely obliterate small creatures like *Hylonomus*.[35] (Perhaps they had good reason to hide in rotten tree stumps!)

The tracks are hard to study as they occur in rocks in the intertidal zone and so are mainly covered with mud except after severe storms. Despite this they are exposed for about seven hours as the tide ebbs. A team from the Nova Scotia Museum in Halifax was able to use this seven-hour window of opportunity to clear off the trackway and make a fiberglass replica of 27 tracks.

Elsewhere in North America footprints of a giant amphibian in Kansas about 6 inches (15 cm) long were reported in 1922; the stride is about 18 inches (45 cm).[36] The lack of tail drag marks suggests a relatively upright posture, well adapted for traveling on land. According to this report, the animal might have weighed as much as 400 to 500 pounds, though this is perhaps an overestimate. The presence of similar large animals in the coast of Northumberland in England was suggested in 1995 when trackers reported footprints of a giant amphibian with a trackway 26 to 28 inches (65–70 cm) wide, indicating an animal larger than anything known from the bone record at that time.

Back in Nova Scotia, a recent discovery reveals dozens of trackways up to 40 feet (12 m) long in a forest setting among conifer tree stumps and fallen trees.[37] The footprints include parallel trackways of amphibians, possibly suggesting gregarious behavior. The site, like many on the shores of Nova Scotia, is subject to brutal erosion by sea ice in the winter, but it is this same destructive erosion that brings spectacular footprints to light in the first place. Clearly there is much to study in this relatively remote region.[38] Among topics ripe for investigation are the various avenues by which early amphibians and reptiles improved their locomotion. Tracks suggest that some groups were already relatively erect and upright in the Carboniferous. Such evidence is rarely incorporated into museum reconstructions, which prefer to equate Carboniferous swamps with primeval creatures that could only slither and sprawl. The track record shows that we underestimate the ability of ancient tetrapods to stand upright.

DIMETROPUS: TRACKS OF OUR EARLIEST MAMMALIAN CELEBRITY

In the early morning . . . *Dimetrodon* still had a low body temperature after the cold night. . . . Without a sail [fin] . . . it would take . . . 12 hours of basking in the sun to increase its temperature. . . . With its sail, the same animal took three hours. . . . As a top predator, these few extra hours may have permitted *Dimetrodon* to grab its still torpid prey.

Michael Benton, *Vertebrate Paleontology*

Most lay readers probably know little or nothing about Paleozoic tetrapods. One creature however, the fin-backed reptile *Dimetrodon*, with its unmistakable fin, or "sail," is irresistible to authors and artists compiling books on prehistoric animals. As a result, *Dimetrodon* is a celebrity and makes many frequent cameo appearances alongside various other famous dinosaurs. The showy sail (also seen in the genus *Edaphosaurus*) not only attracts attention, but also demands some sort of explanation. Many interpretations have been proposed, including the far-fetched possibility that these animals really did launch themselves into Permian waterways and sail off into the sunset. But most paleontologists would scoff at this notion; they generally agree that the sail was a radiator or device for regulating body temperature.

Dimetrodon was a large creature more than 6 feet (2 m) long that has in the past been referred to as a "mammal-like reptile" or "protomammal." Mammal-like attributes included teeth crudely resembling incisors, canines, and molars. The ability to regulate body temperature, or thermoregulate, is also a mammal-like characteristic, ultimately leading to a so-called warm-blooded metabolism.

But what of the tracks of such a famous prehistoric celebrity? In 1940 the famous vertebrate paleontologist Alfred Sherwood Romer gave the

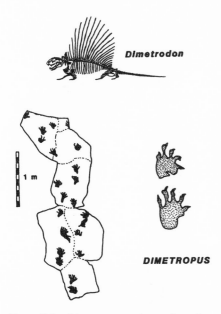

Dimetrodon

DIMETROPUS

1 m

Figure 2.4 *Dimetropus* tracks attributed to *Dimetrodon* are widely known in North America and Europe. As a protomammal, *Dimetrodon* begins to stand taller than its predecessors through improved stance and remarkably elevated neural spine.

name *Dimetropus* (meaning "*Dimetrodon* footprint") to Permian tracks from Texas.[39] These tracks are generally large, up to 9 inches (22 cm) in length, and fit the foot morphology of *Dimetrodon*. This is yet another example of a track's being confidently attributed to a particular genus of track maker by a leading paleontologist. Moreover, *Dimetrodon* had both the means and the opportunity to make *Dimetropus* tracks. Recent discoveries in New Mexico and France have produced extensive *Dimetropus* trackways that provide new information on locomotion and ecology, and show that *Dimetrodon* and its relatives probably walked more erect than most museum reconstructions suggest.

One of the most spectacular and controversial track finds in recent years has been the discovery of abundant Early Permian footprints in the Robledo Mountains of New Mexico. The story has been told by Jerry MacDonald in his eccentric autobiography, *Earth's First Steps*, in which he recalls years of exploration and strenuous excavation in search of the "mother lode" of track-rich strata that he was convinced lay out in "them thar hills."[40] MacDonald's search began in a climate of skepticism and disbelief among professional geologists and members of the local community, who apparently held that tracks were diminutive and rare. Thus MacDonald's discoveries of large and extensive trackways sent him into an ecstasy of ichnological superlatives. More significant, his passion took him all the way to Washington, D.C., where he successfully lobbied for federal money to study and preserve the tracks.

To add poetry to the story, MacDonald labeled the wafer-thin layers of clay that separate sandstone layers as "God's Saran Wrap." These clay "drapes," as geologists call them, are the result of the settling of fine mud on surfaces, after floods and rains, and provide the ideal medium for preserving tracks, rather like printer's ink rolled on glass to take fingerprints. MacDonald also worked into his tale local Native American legends about "spirit animals that walk through mountains"—which might well have arisen from their interpretation of fossil trackways. MacDonald is clearly an original, without whom the Robledo mother lode might never have been found. Twenty-four authors have since contributed to a more measured scientific assessment of the tracks, published by the New Mexico Museum of Natural History and Science in Albuquerque, where the massive collections are housed.[41] These volumes leave the reader with no doubt that the Robledo tracks are spectacularly abundant, and that they provide a remarkable window into the Permian world of the Southwest.

PERMIAN MURDERS AND DIRTY DEVILS

The creature's at his dirty work again.
Alexander Pope

Not only modern twentieth-century celebrities are implicated in murders, allegedly leaving their footprints at crime scenes. Murderous scenarios evidently played out in the Permian implicate the prehistoric celebrity *Dimetrodon*. Let the record show that the prosecution's Exhibit A is labeled NMMNH P–14653.[42] This critical piece of evidence in the New Mexico Museum of Natural History collections shows the trackway of the *Dimetropus* type, from the Robledos, intercepting the trackway of a small vertebrate. At the interception point the heretofore clear trackway of the small animal apparently disappears.

Was the large track maker guilty of murder? The defense might claim that this was merely a case of the lesser crime of kidnapping, for direct evidence that the prey was dispatched to the next life is lacking. The defense might also claim that *Dimetrodon* was not the track maker, by calling the testimony of expert witnesses into question. But the prosecution would counter that the expert's credentials were impeccable, and that there is reasonable certainty that the shoe fits *Dimetrodon*. On the basis of its teeth, we also have clear evidence that *Dimetrodon* was a carnivore and evidently one of the top predators of the day. It would hardly have snapped up a small morsel just to let it go again. As a carnivore it had both the means and the opportunity.

Another Permian trackway that was evidently terminated in midstep was reported from Utah in the early 1990s. Though the track specimen has been submerged below the muddy waters of the Dirty Devil River branch of Lake Powell, photographs and a rubber mold were preserved for posterity by Utah's then state paleontologist, Jim Madsen.[43] The evidence suggests a protomammal predator about the size of a dog of medium build snapping up a much smaller reptile. This apparent evidence of a death-dealing dirty devil is about as close as we come in the Permian to the blood-and-guts gore of the raptor-ravaged Mesozoic. (In fact despite all the Hollywood Jurassic Park hype, there is no convincing footprint evidence of dinosaurs caught in the act of committing murders.) All reported terminator trackways, as aficionados call them, are found in the Permian, and all available evidence suggests that the terminators were protomammals. Our earliest ancestors apparently did not keep their paws clean. But before this verdict goes to press the reader should know that the case has been appealed. Claims that the evidence has been misinterpreted have just been published and we must consider the case again open.[44]

THE MAMMAL UNDERGROUND:
OF BURROWS AND BUSH PIGS

The mammal like reptiles of South Africa may safely be regarded as the most important fossil animals ever discovered, and their importance lies chiefly in the fact that there is little or no doubt that among them we have the ancestors of the mammals, and the remote ancestors of man.

Robert Broom,
The Mammal Like Reptiles of
South Africa and the Origin of Mammals

The Karoo National Park region of South Africa is one of the world's best hunting grounds for Permian protomammal fossils. They are of special interest to humans because they are the earliest ancestors on "our" branch of the family tree: the line with obvious mammal characteristics. Southern Africa owes much of its early paleontological fame to a Scottish doctor by the name of Robert Broom, who became a world authority on protomammals. He studied medicine at Glasgow University toward the end of the nineteenth century, but found that he was fascinated with the evolutionary origin of mammals.[45] His passion first took him to Australia in search of the platypus and other marsupials, but when he saw the first protomammal fossils from southern Africa he quickly decided where his paleontological destiny lay. His lifestyle was described as uniquely "Broomian." He adopted the strategy of practicing medicine in rural areas, where he could easily combine his ministrations with fossil exploration. This habit caused him to go in the field in formal attire, with stiff collar and black tie, regardless of the torrid climatic conditions. Our earliest ancestors were tracked down by the best-dressed paleontologist on the African continent.

Biologically a mammal is a mammal because it suckles or nurtures its young, forming a strong bond between mother and offspring that is symbolic of a social warmth and affection not usually associated, in our "mammalocentric" minds, with "cold-blooded" fish, amphibians, and reptiles. We should not, however, malign three entire classes of vertebrates as delinquent parents. Many naturalists have observed examples of parental care among fish, amphibians, and reptiles. Konrad Lorenz tells a classic story of a jewel fish's rounding up its young and returning them to the underwater "nest" with such "thought" and concern that students witnessing this aquarium scene broke into spontaneous applause.[46] The Surinam toad carries its young on its back, and the Australian crocodile defends its nest most vigorously. Hence we find hints of some degree of social activity and even sophistication among our Permian ancestors.

One of the most spectacular features of the geology of the Karoo area of southern Africa is that large areas of the Permian landscape have been exposed by present-day erosion. Some of these "paleosurfaces" (literally, "ancient surfaces") in the Karoo reveal entire river systems complete with meandering channels and a floodplain with abundant tracks.[47] Aerial photographs of these Permian landscapes are hard to distinguish from modern river landscapes. We can look back in time and see the Earth's surface as it was more than 250 million years ago. Here we see footprint evidence for the activity of a whole range of vertebrate and invertebrate track makers, culminating in the intense trampling of muddy areas around shrinking water holes. South African geologists have identified the tracks of several types of protomammals. These include the large herbivorous dinocephalians, meaning "terrible heads," with tracks a foot long; the stubby-toed reptile *Bradysaurus;* and the genus *Diictodon,* a member of the dicynodont tribe (meaning "two wolflike teeth") that has been described as a Permian "bush pig" resembling a "lizard-skinned sausage-dog with a large tortoise-like head."[48]

At some sites we find complete *Diictodon* "bush pig" skeletons buried in their burrows. These individuals were caught napping during floods. The burrows are helical or corkscrew-shaped and describe two perfect clockwise spirals before descending to a broad, tongue-shaped terminal chamber. The South African geologist Richard Smith compared these Permian burrows with almost identical types from Nebraska that have been attributed to a 25-million-year-old beaver known as *Paleocastor* (literally, "ancient beaver"). Such burrows have been given the name *Daimonelix,* meaning "demon spiral" and loosely translated as the "devil's corkscrew."[49]

An evocative image was created by the discovery of the fossils of a pair of Permian "bush pigs" or "sausage dogs" (or what Glasgow comedian Billy Connolly calls "draft excluders") that had been buried alive in their

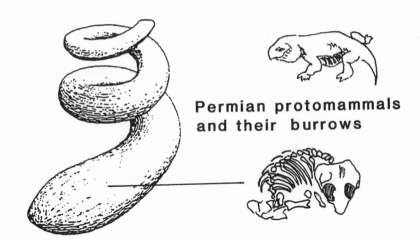

**Permian protomammals
and their burrows**

FIGURE 2.5 "Devil's corkscrew" burrows made by the mammal-like reptile *Diictodon*, from Permian South Africa, show an inward exploration of the Earth's skin.

burrow. This picture causes us to wonder further about the origins of mammalian bonding and family ties. Here we have what appears to be the oldest fossil evidence of a couple that died at home during a natural disaster. These corkscrew burrows are the oldest evidence for burrowing by mammal-like reptiles, suggesting that these Permian bush pigs probably behaved similarly to modern rodents such as prairie dogs and beaver. Burrowing probably had several advantages. In addition to providing a safe refuge from large predators, a burrow provides a habitat with a constant air temperature. Such temperature regulation was probably advantageous to animals still in the process of evolving fully warm-blooded metabolisms.

One might ask why the burrows were spiral-shaped. Wolfgang Schad writes: "In a living organism, spiral forms always indicate the transition to total asymmetry."[50] The burrows of later underground mammals are therefore often more irregular than these Permian prototypes. I am also reminded of the form of the intestine, and the marine invertebrate burrows mentioned earlier. Nature reiterates again and again. Similar burrowing and digging traces in early Permian deposits of Utah are associated with footprints that represent digging and scratching activity by the same terminator track makers responsible for the alleged Dirty Devil murder.[51] All this evidence of digging activity shows that some of our Permian mammal-like ancestors were energetic creatures that had developed the burrowing habit very early in their evolutionary history.

It is amusing that there is so much macabre terminology associated with our early ancestors, all of it independently introduced by different students of the subject. Of course, paleontologists deal with dead things, and some colleagues goad us by calling us grave robbers. Though trackers deal with living animals, they still cannot always escape the dark side. In the two cases just discussed we have discovered that a certain type of protomammal that left its footprints in what is now the Dirty Devil River canyon has been labeled "the terminator." It also dug burrows labeled the devil's corkscrew, that end in a terminal chamber. All things must come to an end and so we find these devilish terminators, dead in the terminal chamber.

RUNNING FROM FLOODS IN THE DESERT

Fret not to roam the desert now, with all thy winged speed.

Caroline Norton

Anything that digs a lot must be fairly energetic, and trackways show that many Permian protomammals were no slouches. Indeed, trackway evidence from fossil sand-dune deposits suggest that some species may have run around quite happily on difficult terrain. Some of the first fossil footprints ever discovered and the first ever described were reported from Permian dune deposits near Annandale in the rain-drenched Scottish Lowlands.[52] Unlike today, during the Permian Scotland was a desert. But nineteenth-century geologists and paleontologists, including the famous Rev. William Buckland of Oxford, knew little of Paleozoic vertebrates other than fish and could assume only that these organisms were primitive. In a famous experiment to try and reproduce tracks similar to the fossil examples, Buckland and his friends put a tortoise on a sheet of pastry dough and were entertained by the creature getting its feet stuck. Consequently the fossil tracks were named *Chelichnus,* meaning tortoise traces; they do indeed resemble tortoise spoor.[53]

Buckland wrote that the "footsteps . . . alone suffice to assure us both of the existence and character of the animals by which they were made. . . . [I]t is to the . . . tortoises that we look, with the most probability of finding the species to which their origin is due." He continued: the tracks were "stamped upon the rock, distinct as the track of the passing animal on the recent snow; as if to show that thousands of years are but nothing amidst Eternity—and, as it were, in mockery of the fleeting perishable course of the mightiest potentates among mankind." Like many of his day, Buckland saw geological and paleontological wonders through the lens of the "power, wisdom and goodness of God."[54] Such flowery rhetoric may seem a little out of place in scientific literature today, but it is not much different from that of a number of modern scientists, especially physicists, who frequently bow down reverentially before the mind of God and the magnificence of the cosmos.

Buckland's words about mighty potentates are not without slight irony, for though he was one of the geological potentates of his day, the footsteps that convinced him his track maker was a tortoise appear to be those of entirely different creatures. When the tracks were examined by the great Thomas Henry Huxley, in 1877, he disagreed with the interpretations of Buckland and Owen, stating, "There is really no ground for ascribing these tracks to chelonians," and further doubted that they could "be safely referred to any known form of Reptile or Amphibian."[55]

Huxley was right, and we now consider that these are the tracks of protomammals. Between 1918 and 1928, "*Chelichnus*-like" tracks were also found in Permian rocks in the Grand Canyon by trackers from Yale and the Smithsonian Institution.[56] Although more or less identical to Scottish *Chelichnus* tracks, they were given the different name *Laoporus,* meaning "stone track." Uncertainty continued about the character of the track maker, and many museum displays still carry the label "amphibian tracks," reflecting the general notion that they are the tracks of "primitive" animals. What makes these tracks most interesting, however, is that many show unusual trackway patterns. Most were made on the lee slopes of sand dunes, and most are oriented heading up the dunes. Other animals moved obliquely, producing very odd trackway patterns.

Such patterns suggest intriguing behavior and call for some sort of explanation. The most recent chapter in this Permian saga was opened when the biologist Leonard Brand and his collaborators from Loma Linda University attempted to explain the oblique trackways as the result of the activity of swimming animals. Through a series of experiments with modern salamanders, they managed to produce tracks that resembled the dune tracks by drifting their animals along in currents in aquaria. Despite the obvious utility of experimental studies to help understand how tracks are made, this study was seriously flawed, because it failed to take into account the overwhelming evidence, such as wind-generated ripple marks, that the ancient environment was a desert. The suggestion that these classic examples of ancient desert sand dunes were made underwater is highly provocative to many generations of geologists and geology students who have been taught otherwise.

What makes the debate even more controversial is that it has spread into creationist circles, where it has been tied into the idea that the entire Paleozoic rock sequence in the Grand Canyon is the result of a catastrophic, biblical-style flood. This argument, popular among creationists, is geologically naïve and disingenuous, especially when one considers how it has been presented. Brand first published his swimming-salamander ideas in reputable paleontological journals, with no mention of the flood idea.[57] Then, in his recent book *Faith, Reason and Earth History* we find a whole chapter devoted to the flood hypothesis.[58] The track evidence is interpreted as evidence that "the flood. . . washed animals out of their natural habitat [and that] as the animals tried to get back where they came from, they were often going up the lee sides of the dunes, against the current."[59] We shall discuss the role of tracks in creation-evolution debates again in Chapter 6. In the meantime, however, we should state clearly that the evidence for flooding is nonexistent. The protomammal tracks are often found in association with countless trackways of spiders, scorpions, and other desert arthropods that could not have been walking around underwater.[60] The unusual trackway patterns have been interpreted as evidence of running behavior.[61] Though Brand is a biologist and can be excused for not fully understanding the geological implications of his hypothesis, it is disappointing to find that he omits any discussion of the objections published by geologists to his interpretations.[62] By leaving out the other side of the story, Brand is open to the criticism of bias and inconsistency. Geologists will surely question his motives as much as his conclusions.

Faith, Reason and Earth History is not without some merit, however, and I believe that Brand is genuine in his efforts to challenge certain institutionalized scientific paradigms. He honestly notes that creationists have made a lot of mistakes in the past, including wrong interpretations that call for humans' making tracks alongside dinosaurs (discussed further in Chapter 6). It is also pointed out that creationists are getting much better at doing science, and that science-bashing is a poor and counterproductive strategy for creationists. He stresses that "above all it is essential that we treat each other with respect, even if we disagree on fundamental issues."[63] When Brand discusses certain aspects of topics such as evolution and sociobiology, he raises some thought-provoking philosophical questions. These essentially come down to identifying the point at which we call on divine intervention (i.e., hitherto unknown or supernatural explanations) to account for the unknown areas that science cannot explain. If Nobel prize–winning physicists can talk about the mind of God and an intelligent or conscious universe, why should creationists not also speak of divine intervention?

The crux of the matter is that the universe is "run" by awe-inspiring powers beyond our feeble comprehension, so there is no reason not to attribute them to "God." It is just that we create confusion and belittle God if we suggest ridiculous supernatural causes for events that have already been adequately, even elegantly, explained by sound scientific theories. While the sophisticated scientist may cast aspersions on naïve hypotheses such as "big flood = Grand Canyon strata," we must remember who the audience is. Brand is writing for people of faith, who are unlikely to give up cherished creationist beliefs that God created the world simply because impatient scientists say that their explanation is utter nonsense. This is just the confrontational attitude that Brand is trying to avoid. If we are to educate creationists to a scientific way of thinking, a task that Brand is trying to do, we must, like Brand, be gentle and coaxing. It is no good going for the jugular and pointing out only what we perceive to be creationist errors. This only leads to polarized holy wars.

After we have explored all the scientific explanations, including the origin of scientific consciousness in the human mind, we can easily arrive at the conclusion that the universe is supernaturally intelligent, and, like our proverbial physicists, we may stand in awe of God's mind. It might be better for science to be more humble and say, "Look we do not have all the answers, but we shall try and write for persons of faith in such a way as to help validate their beliefs and point to the extraordinarily, fascinating way in which the universe is intelligently constructed." We should point to the convergence of worldviews between science and creationism and not only the divergences. Though I contest Brand's conclusions about tracks and find them inconsistent and confusing in relation to my worldview, I try to see his motives in their creation-culture context. Ironically Brand may do more to educate creationists in science than an irate anticreationist.

The creationists, stereotypically identified with right-wing, conservative, moralist institutions and forces, are ostensibly in awe of the power of God's thunderbolts and power to bring about a severe reckoning in a world gone astray. The stereotype of the other camp, labeled as left, liberal, sexually explicit, and humanistic, is that it is less God-fearing and ultimately has more faith in humanity. The former tribes pay homage to ascending, idealistic, heroic, masculine (yang) forces, looking to the heavens for redemption, while the latter clans pay homage to the descending (yin) feminist forces that remind us that we are an earthbound species wedded to corporeal reality. Surely humans are the quintessential expression of both ascending and descending forces, like the powerful Kundalini cycle, which rises from the sacred basal source (sacrum) to the crown of the head and back to the abdomen. We are simultaneously current and countercurrent—Mae-Wan Ho's coupled cycles. When in the ascending mode our orientation is outward and upward, toward our higher minds and nerve-sense pole. We point celibate prayer fingers to noosphere, stars, and God, and all beyond our ken. In the descending mode our orientation is inward and downward, toward lower, earthly reality, which tends to dismiss the notion that God or any higher power can intervene to help us. Is it any wonder that in our very language we cannot decide whether we "arose" or "descended" from the apes? Surely we did both. We are yin and yang, conservative liberals, immoral moralists, right lefties, celestial celibates, and sexually explicit anthropoid apes.

SHAPE SHIFTING

Most organisms are a kaleidoscope of parts [and] the number of potential life forms that can evolve . . . is enormous. . . . Developmental give-and-take generates profound morphological novelties that can open up entirely new evolutionary vistas—the tetrapod limb . . . was perhaps one of the greatest.

Ken McNamara, biologist

Whether it was Buckland in the 1830s or Brand in the 1990s, the study of Permian tracks does not proceed without the periodic appeal to God, faith, and even floods. In my story the track makers were protomammals. Their feet best fit the tracks, and besides, this was their heyday. They evidently adopted special gaits to negotiate steep dune slopes, as do modern lizards.[64] Trackways with long strides indicate that far from being like slow tortoises, these track makers were likely among the first runners. Moreover, our protomammal ancestors had a strong affinity for deserts, for their tracks are found there in abundance from the Permian through early Jurassic, a time span on the order of 100 million years.[65]

Let us revisit a holistic, Schadian perspective in considering the tracks and traces made by fish, early amphibians, reptiles, and protomammals.[66] Consider the difference between fish and early amphibians such as *Ichthyostega*: the difference between so-called "stem" reptiles—long-tailed ancestral, primitive, lizard-shaped forms such as *Hylonomus*—and some of the last of this line, the turtlelike pareiasaurs; and the difference between early protomammals such as *Dimetrodon* and more advanced protomammals such as *Diictodon*. The purpose of this exercise will be to show that their limbs and tracks reveal very interesting trends through time. Most notably, we see coupled cycles: limbs get longer while toes get shorter, and vice versa—limbs get shorter when toes get longer.

A fish is fundamentally a fast-moving, reactive, nerve-sense–dominated animal (essentially a skull and backbone), darting about with the aid of its a well-developed mobile tail. The fish in general holds its anterior pole static and uses its tail as its propulsive limb. This reminds us of our little spirochete developing directional motion. As a result of their movement, fish produce sinuous, snakelike traces on the substrate. The comparison reminds us that snake moves with a kind of swimming locomotion that seems out of place on land, but serves well in the water.

Just as the body changes shape as it grows, we must try to imagine the plasticity of form that arises in the evolutionary history of major groups. Fish, like tetrapods, have gone through multiple cycles of elongation and shortening with respect to their anterior/posterior axes, and many cycles of lateral compression and expansion with respect to their sagittal plane. Such multidimensional shape shifts are easily visualized by looking at the classic studies by d'Arcy Thompson, who showed how one may conceptually transform one animal shape into another by simply pulling, pushing, stretching or compressing different parts of its anatomy.[67] Like the proverbial pillow that bulges out on one side when pushed in on the other, there is always an equal and opposite reaction. As Ken McNamara explains these shape shifts have their origins in the dimension of developmental time, and are fundamental in opening up new evolutionary vistas.[68] What I wish to stress is that such shape shifts are reiterated in the evolutionary process, and so can be read in anatomy and tracks.

When salamander-shaped amphibians evolved limbs from fins, they moved their propulsive organs forward to the central region of the body, where they also developed lungs. In comparison with the anatomy of fish, the long bones in the limbs of amphibians became longer, but the toe bones became fewer and shorter, a coupled cycle—Goethe's compensation principle. Among long-tailed ancestral "stem" reptiles like *Hylonomus* the limb is short and the toes long, but this pattern reverses, as we shall see in their short-tailed, short-toed turtlelike descendants. Similarly on the protomammal line, long-tailed, long-toed, short-legged *Dimetrodon,* with such remarkable developments in the central, axial part of its body, is ancestral to long-legged, short-toed, short-tailed dune-running forms. These later protomammals also tend to have long necks and horns. Accompanying this physical shift to anterior emphasis there was a loss of posterior ribs, coupled with development of the diaphragm and improved protomammal lung. This all coincided with a shift of habitat, from watery habitats such as coal swamps to dry continental settings such as deserts. Just as coal swamps and amphibians reflect watery habitats, desert sand dunes and protomammals reflect the influence of air and wind. Indeed, there is an intimate relationship between increased activity (running), improved ventilation of the lungs, and improved thermoregulation/metabolism. Animal and habitat always coevolve, as we shall see in subsequent chapters. Such perspectives serve to emphasize how unlikely we are to find salamanderlike amphibians in the dunes.

PROPULSION

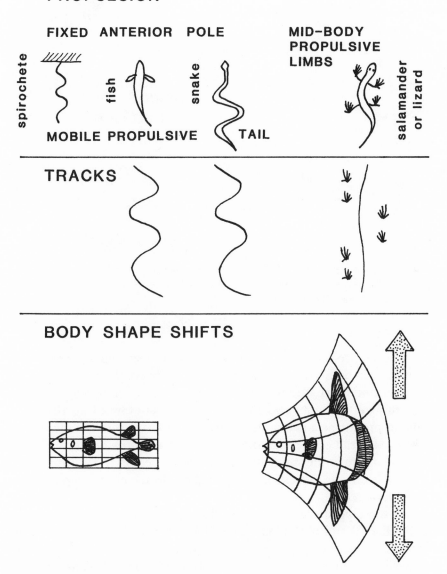

FIGURE 2.6   Fish, like microscopic spirochetes, use their posterior tails for locomotion, and produce sinuous snakelike trails, whereas tetrapods use limbs in the middle portion of their bodies for locomotion. All animals have undergone systematic shape-shift cycles of elongation and shortening, broadening and lateral compression (see also Figure 2.7).

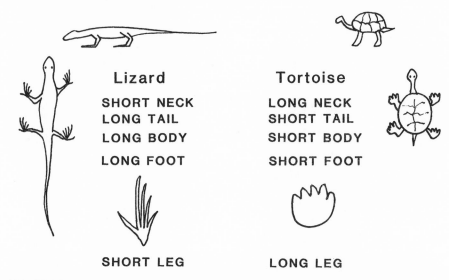

Lizard

SHORT NECK
LONG TAIL
LONG BODY
LONG FOOT

Tortoise

LONG NECK
SHORT TAIL
SHORT BODY
SHORT FOOT

SHORT LEG

LONG LEG

FIGURE 2.7   The shape-shifting pattern continues in tetrapod evolution, allowing us to see the effect of cycles of elongation and shortening, broadening and lateral compression on the shape of limbs, feet and tracks.

As trackers we can associate the widening and shortening of the pro-tomammalian body with shortening and widening of the foot and the development of short stubby toes. At the same time, however, there is a lengthening of the limbs and an inward movement toward the parasagittal plane, producing a narrower trackway. This shape shift is a measure of erectness. Instead of the high dorsal spines of *Dimetrodon*-like forms we see the elongate in the supporting limbs developed on the ventral side of the body. Is this the expression of another one of the many coupled cycles that we have seen?

From these perspectives we learn that clues to systematic patterns in morphological and physiological evolution can be discerned even from individual tracks. Let us remember the long track–long tail (i.e., posterior emphasis) relationship and the corresponding polar relationship, short track–short tail (i.e., anterior emphasis) and learn how to use it to good advantage in the interpretation of other groups that we encounter in our evolutionary journey.

SPIDERS IN THE DUNES

The spider's touch how exquisitely fine. . .
Alexander Pope, "An Essay on Man"

Along with the myriapods, spiders were among the first creatures on land. Among the least popular of God's creatures, at least among humans, spiders are a fascinating group. They range in size from microscopic species found floating in the stratosphere near the peaks of the Himalayas, to giant bird-eating creatures of the tropics and the Carboniferous *Megarachne*, a crab-sized creature a foot or more in length. Their evil reputation comes from the fact that they ensnare their prey in webs or through subterranean trap doors, and are all venomous. The latter characteristic is best known as a feature of the deadly black widow and the brown recluse, with their appropriately sinister-sounding names.

The *Tarantula,* also called the wolf spider, is a larger variety so named because it was thought to be the cause of a medieval dancing mania or malady known as "tarantism," from which we get the term "to rant." In one grotesque story of madness at an ancient outpost on the old Silk Road, an evil inn proprietor fed dried fruit to starving tarantulas, which they bit into voraciously, so releasing their venom. The fruit was then mixed with alcohol, producing a deadly brew known as tarantula vodka.[69] (Raw Central Asian vodka may produce paralytic effects even without the addition of spider venom!) This evil potion paralyzed unwary travelers, who were then thrown into a pit to be devoured by a starving bear. Those who like happy endings will be pleased to know that the insane proprietor was eventually brought to justice.

Spiders are not all evil creatures hell bent on scaring little Miss Muffett off her tuffet. One of Scotland's most famous historic figures, Robert Bruce, was inspired by watching a spider persistently climbing up its thread. Bruce had been beaten in his battle for the crown of Scotland six times, and had taken refuge on Rathlin Island near the famous Giant's Causeway—a giant "footprint" in Irish mythology. When he reputedly saw the spider succeed in fixing its web on the seventh attempt, his spirit rallied and he returned to Scotland to win the Battle of Bannockburn, in 1314, and establish Scottish independence. To this day in Scotland it is still regarded as wrong to kill a spider.

It might surprise many to learn that spider trackways are preserved at all in the fossil record. In Mystery Valley, a part of the megalithic Monument Valley area of southeastern Utah and northeastern Arizona, not far from a place appropriately named Spider Rock, we find dozens of Permian sand-dune surfaces with spider tracks.[70] Similar tracks are found in the Grand Canyon region, and have been given the appropriate name *Octopodichnus* (meaning "eight-foot trace"). The tracks are well preserved and up to four inches (10 cm) in width, indicating tarantula-sized creatures.

One is never far from the arthropod world, particularly in desert environments, where we also find unmistakable present-day trails of scorpions, myriapods, and insects that are virtually indistinguishable from those found in Permian rocks. Our studies also show that Permian spiders frequently clambered up the dunes following the contours of ripple-mark crests and troughs produced by swirling desert winds. Exactly what the spiders were doing is hard to tell, but they may have been catching insects and other arthropod prey blown from the windward side of the dunes onto the lee slopes. While spiders were among the largest arthropods to leave tracks in the Permian sands of time, they were probably not at the apex of the food chain. In desert settings, where vegetation is sparse, arthropods feed on each other. Spiders prey on flies and other insects, and small vertebrates prey on arthropods, as seen in the habits of lizards and kangaroo mice at the present time. The combination of spider, protomammal, and lizardlike tracks suggests that modern desert ecology has Permian roots.

Incidentally, whereas monster millipedes are long-bodied and short-legged and are associated with the dank forest-litter habitat of coal swamps (where we also find lizard-shaped *Hylonomus* and its kin), spiders are short-bodied and long-legged and their tracks are associated with protomammal footprints in the wide- open desert spaces. I am not suggesting that all arthropods always obey the rule of short legs with long bodies or short bodies with long legs just formulated for vertebrates, but it is worth thinking about as a general pattern. Spiders may have been among the first invertebrates on land, so have a longstanding affinity with the air, as seen in their spinning of webs in airy space to trap flying creatures. Even those that go underwater take air bubbles with them.

One of the most common questions ever asked about such clear and beautiful desert tracks is "How were they ever preserved?" For tracks in dry sands do not last. It appears that water is the key, making the sand more cohesive and likely to register footprints. A gentle and subtle mechanism is required, for heavy rains or catastrophic biblical floods would simply wash away delicate tracks of spiders and scorpions. One possibility is dew and the condensing of fog and mist onto track surfaces, as is common in coastal dunes in the present-day Namib Desert.[71] Dunes can also become saturated from below, given the right conditions of rising water table, rains, recharge of underground aquifers, rising sea level, high tides, and even changing air pressure. So in tracking spiders we may also be reading subtle changes in the desert water table and microclimate.

THE ITALIAN BIGFOOT

The story is extant and writ in very choice Italian.

Shakespeare, *Two Gentlemen of Verona*

To visit the late Dr. Piero Leonardi was a rare privilege. His home, a Venetian palace, is a wonderful museum arranged to reflect the stratigraphy of an archaeological site, with Greek artifacts on the ground floor, Roman artifacts on the second floor, and, on the third, a rich tapestry of history gleaned from a lifetime of discerning historic and prehistoric study among myriad Mediterranean treasures. Piero is the father of Father Giuseppe Leonardi, a Ph.D. in paleontology and a high-ranking priest, who formerly headed the Instituto Cavanis in Venice. Following in the footsteps of his father, Giuseppe became a tracker.

Among Dr. Piero's many accomplishments is a massive set of volumes on the geology of Italy's famous Dolomite Mountains, consisting of rock formations that yield fossil footprints. In 1975 father and son described a large footprint that they named *Pachypes dolomiticus* ("bigfoot from the Dolomites").[72] Because it is hard to find broad ledges and surfaces that yield footprints in this precipitous terrain, it took more than 20 years to piece together a continuous trackway sequence.[73] It appears that the track maker was a pareiasaur, a member of a rather ugly tribe of primitive reptiles described as clumsy, swamp-dwelling herbivores that may have been the ancestors of turtles and tortoises.[74] Our awareness of track-body relations should predict that this was an animal with a short, wide body and short tail. This turns out to be a good description of pareiasaur anatomy. The tortoise-turtle connection even converges somewhat with Buckland's discredited interpretation of short-toed, Permian tracks from Scotland, in the sense that the Scottish tracks are indeed somewhat pareiasaur or turtlelike in general appearance.

The hitherto poorly explored world of Carboniferous footprints and the world of "Permian perambulations"[75] is still not well known. But this offers us a challenge. In March of 1997, at the invitation of Hartmut Haubold, a world authority on Permian footprints, 19 trackers from all around the world descended on the museums and rocky outcrops of central Germany to pool their knowledge,[76] and compare stories of tracking in deep time. Here the reconstructed Italian bigfoot trackway was presented by Umberto Nicosia, whom I dub as the singing tracker.[77] In jubilant mood, Umberto hummed bars from favorite tunes in between ichnological pronouncements, ending his presentation with a picture of a Native American scout helping ichnologically illiterate white men to interpret the spoor of the animals they were tracking. It used to be that trackers performed a useful function in society, he concluded, and were even paid for their skills. Can trackers still be useful to paleontology?

The answer is surely yes! Trackers of prehistoric creatures have come a long way in a short time, as the tracking down of the little-known pareiasaur shows. We are already in a golden age of dinosaur tracking, and

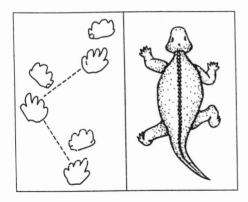

FIGURE 2.8 Short, wide bigfoot tracks (found in Italy) suggest a short, wide track maker—the pareiasaur, a probable ancestor to the turtles and tortoises.

the tracking fraternity looks on the Permian as one of several new ichnological frontiers. Tracking brings animals to life in their natural habitats and helps museums to reconstruct the ancient world realistically. Many lay persons have little knowledge of prehistoric animals other than dinosaurs, mammoths, and the occasional pterosaur or saber-toothed cat. Tracks can help introduce us to new menageries of obscure yet fascinating creatures such as pareiasaurs. In the next chapter we shall see how the track record adds much to paleontology because it reveals footprints whose makers are still unknown. Although we are about to leave the Permian, which marks the end of the Paleozoic era of ancient life, our study of trackways is just beginning.

chapter three

HAND SIGNALS

ACT III: CAST OF CHARACTERS

With the Mesozoic Era we are entering geological "midlife," and here we shall stay until the end of Chapter 7, concentrating on one period at a time: the Triassic, the Jurassic, and the Cretaceous. The Triassic period (about 250 million to 205 million years ago) is so called because, in Europe, it consists of three distinct sequences of strata. It has been called the Age of Archosaurs, archosaurs being a group of advanced reptiles that includes dinosaurs, pterosaurs, crocodiles, and various less familiar extinct forms. Many of these lesser celebrities existed only in the Triassic, and some remain so elusive as to be known only from their tracks. Herein lies an important theme, namely, that sometimes we may need to rely exclusively on tracks to fill gaps in the fossil record. As we learn to read an animal's morphology from its tracks we can do more than merely infer that it was large with a certain number of toes. We may use our growing holistic savvy to sketch out other body proportions, probable physiological tendencies, and so forth.

Most Triassic archosaurs are completely unknown to the layperson, and are not even well known to most paleontologists. As a result they have proved very difficult to track, as one of the most protracted and confusing footprint sagas—that of the "hand animal"—has proved. In recounting this story we shall witness some of Europe's finest geological and paleontological minds completely at sea in a miasma of confusion. But the saga has a happy resolution and there are signs of a promising future.

We shall also see the emergence of true dinosaurs from among the ranks of the elusive archosaurs, and discover evidence that they were among the first creatures ever to stand on their own two feet. The question of exactly when dinosaurs started making footprints is of great concern to paleontologists, and significant controversy has arisen as a result. The tracks of another dinosaurian group, the prosauropods, ancestors of the gigantic brontosaurs, were first brought to our attention by another tracker-priest, who has stirred up controversy as a result of overenthusiastic naming of far too many footprints. In this act we will also see a decrease in the spoor of protomammals, but find rare traces of what might be the first true mammals. The Triassic, then, was a time of transition. Archosaurs vied with protomammals for dominance and won, setting the stage for the ascendancy of the dinosaurs, and relegating mammals to playing second fiddle for the best part of the next 200 million years.

WORLDWIDE WANDERINGS

All the world's a stage and all the [creatures] merely players,
They have their exits and their entrances.
Shakespeare, *As You Like It*, II.vii

We speak of living in a global society, despite its fragmentation owing to wars and patriotism. But where does such ecological or holistic awareness come from? Even a cursory glance at Earth history tells us that global society is nothing new. The lithosphere, atmosphere, hydrosphere, and biosphere are integrated global systems. Together they form a dynamic living entity—Gaia, poetically described by Guy Murchie as "a spherical biofilm rotating in gravitational, electromagnetic and nuclear fields—a sort of gyrating bubble of evolving potency."[1] Fish that swam in Paleozoic seas swam in a global ocean. Continents, on the other hand, are islands in otherwise interconnected oceans, and these islands cause varying degrees of isolation. At the end of the Paleozoic Era, in the transition from the Permian to Triassic period, Earth's continents were coalesced together into a giant supercontinent known as Pangaea. The northern portion is referred to as Laurasia, and the southern portion as Gondwanaland. The supercontinent had a supercontinental climate to match its size, with intense seasonal and daily extremes of heat and cold. Many deposits from this time are the result of deserts and arid conditions. Sea level was also low, making the continent larger and climatic conditions harsher. Such conditions evidently favored the evolution of reptiles, which unlike their amphibian ancestors were capable of breeding and surviving in more arid conditions. It is in this context that we must look at the thriving reptilian track makers of the day.

The supercontinent made it easy for species to travel to and fro. In short, the world's fauna was then very cosmopolitan, or *pandemic* (meaning found everywhere), quite the opposite of the *endemic* (found only in a specific place) faunas that evolve when continents or islands are isolated from one another, as in the famous Galapagos archipelago. This was the first global society of land-based vertebrates. The cosmopolitan flavor of the world's Triassic fauna is reflected in the track record, where we often find the same footprints at widely separated localities on different continents. Such similarities allow us to use the footprints to compare the age of strata around the world. This specialized field of fossil footprint stratigraphy (the technical term is *palichnostratigraphy*—the longest word in this book) allows us to track the wanderings of entire groups of species.

In a sense, evolution spoke the same language at any given time, leaving similar tracks over large areas. The application of this principle allows paleontologists to identify exceptions. Where different types of fossils are found in strata of the same age, some factor such as geographic isolation or different environmental stresses is likely to have been at play. We should be careful about generalizing too much because the extent to which we speak the same track or paleontological language is not controlled only by whether the continents were joined or fragmented, and whether footprints were really the same or different. The human factor, and provincial pride, also come into play more than an objective scientist would like to admit. Pangaean tracks that once lay side by side are now located thousands of miles apart, where they are studied by trackers who speak different languages and work for different institutions. These trackers can only see a small part of the puzzle and often give the same tracks different names—a type of paleontological splitting. This Tower of Babel phenomenon is known as "provincial taxonomy," a taxonomy that reflects the present provinciality of science, instead of what it should express, the underlying unity of the ancient biota. Symbolically, once the similarity of the ancient biota is recognized, it brings about a convergence and unification of science and allows us to once again speak the same language—in this case the cosmopolitan language of the lost supercontinent of Pangaea.

The reverse situation also arises: tracks and other fossils that really are different may be given the same name. This process of excessive homogenization, what paleontologists call "lumping," can have the undesired effect of masking differences and diversity. In paleontology, splitting is usually more of a problem than lumping, because one is so often dealing with various bits of the puzzle that look different. One can point to examples where three paleontologists have found three different bones of the same animal, and like the three blind men touching the tail, trunk, and leg of an elephant, all declare that they have discovered an entirely different species.

THE HAND ANIMAL

This story . . . I am going to tell may sound . . . incredible. It has to do with . . . fossil footprints of an unknown animal. . . . The problem was to determine not only the size but the shape and type of animal that caused them, without any clue other than the footprints themselves. Just to make things a little more baffling, these footprints happened to have a highly misleading shape.

Willy Ley, *Willy Ley's Exotic Zoology*

The discovery of "petrified hands" sparked off one of the liveliest scientific arguments of the nineteenth century. The tracks in question bear an uncanny resemblance to the imprints of human hands, complete with thumb, and come in a variety of sizes ranging from those that would match the palm of an infant to those that could only be attributed to superhuman giants. Moreover, many of the footprints are very well preserved, in some cases revealing fine details of skin texture. Such tracks were probably first recognized in Europe during the Renaissance, when fossil collecting became popular, but it was not until 1824 that specimens found their way into the hands of paleontologists capable of describing them and attempting to pronounce on their origin. As we shall see, this was easier said than done.

The tracks were first reported in 1824 in Cheshire, England, in the Merseyside vicinity of Tarporley and Storeton, in strata known as the New Red Sandstone. They were later identified as Triassic in age. For a decade English trackers made little headway in describing the tracks or identifying possible track makers. Then in 1834, the year that the Triassic was named, new finds from Hildburghausen came to the attention of German geologists, who described "tracks of large, unknown primordial animals . . . in the . . . sandstone quarries." The track-bearing red rock is well known to geologists and builders as the Bunter Sandstone Series (*bunt* is German for "colorful"); it was used in the construction of the cathedral of Strassburg and the palace at Heidelberg.[2]

A number of fascinating theories were put forward to explain these petrified hands. Friedrich Voigt proposed that the track maker had been a great ape, which he named *Paleopithecus* (meaning "ancient ape"), then later suggested an alternate hypothesis, that the large tracks were those of a cave bear, and the small tracks probably those of a mandrill. These hypotheses, which we now regard as preposterous, were only the first in series of incredible interpretations put forward by otherwise capable scientists. Johann Jakob Kaup attributed them to a marsupial and so in 1835 coined the name *Chirotherium,* meaning "hand mammal." The famous explorer Alexander von Humboldt initially agreed with this hypothesis, though later, in collaboration with Kaup, he suggested that the track maker was a reptile. William Buckland also supported the mammal interpretation and went on to observe that kangaroos resembled *Chirotherium* in having considerable "disproportion between the fore and hind feet" and a first toe "set obliquely to the others."[3] Such unlikely hypotheses are a reflection of lack of knowledge early in the nineteenth century of fossil vertebrates and their evolutionary time frames. Such views also reflect a genuine interest in exotic animals such as marsupials, then regarded as very "primitive." We should remember here that the existence of kangaroos was only established in European science in the late 1700s, and the early reports of the duck-billed platypus, an egg-laying mammal, were still being greeted with absolute disbelief and derision as late as the 1880s!

The famous British zoologist Sir Richard Owen proposed that the footprints were probably made by a primitive amphibian or a reptile. The German geologist H. F. Link, among the first to advocate an amphibian, suggested that the tracks were those of a giant toad. The great geologist Charles Lyell went so far as to create a reconstruction of the *Chirotherium* track maker as an amphibian, but his is one of the most preposterous locomotor fantasies ever proposed. In order to explain the so-called thumb, whose print was situated on the outside of the trackway, he proposed that the track maker crossed its legs over as it walked. Lyell evidently believed that Triassic animals had been so "primitive" that they lacked the coordination to walk normally. A moment's reflection reminds us that there is no precedent for such a ludicrous and inefficient gait among animals with such well-developed feet and narrow trackways. It is a reflection of our self-imposed image of sophistication that allows us humans to equate ancient with primitive and inept. In Victorian eyes, primitive meant low on the scale of beings, and such creatures were regarded as pitiful; in this case they were given pitiful and hopeless gaits to match their lowly status. In reality, as we shall see, *Chirotherium* track makers represented the most advanced and sophisticated animals in existence at that time.

The first hint of a biological promotion for the *Chirotherium* track maker came in the 1860s through 1890s, when it was suggested that the track maker had been a crocodile, or a dinosaur. These interpretations put the elusive track maker higher on the evolutionary scale, and heralded the debates of the twentieth century. In short, the status of the track maker evolved with the improvement of paleontological knowledge. In 1917, the

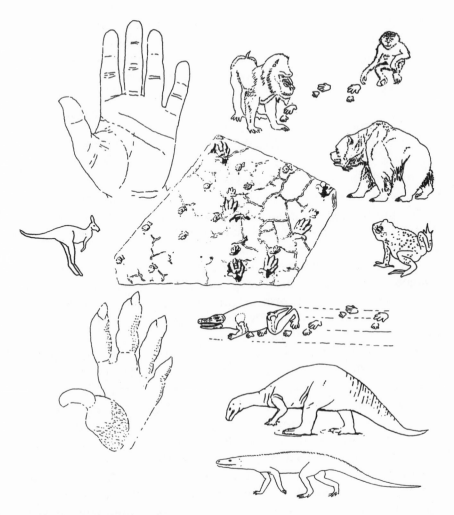

FIGURE 3.1   The mysterious petrified hand tracks shown in the center of the illustration have been attributed to apes, mandrills, toads, kangaroos, cross-legged amphibians, cave bears, and prosauropods.

young German paleontology student Karl Willruth concluded that *Chirotherium* had not walked cross-legged, as Lyell had proposed. Instead he interpreted the distinctive thumb mark as the impression of a fleshy appendage on the outside of the foot.

It is at this point that Baron Franz von Nopsca (1877–1933), one of the most interesting and eccentric characters in the history of paleontology, entered the debate.[4] Born of Hungarian-Transylvanian nobility, the brilliant Nopsca became interested in paleontology when Cretaceous dinosaur bones were discovered on the family estate. He quickly set about learning the field and soon became Hungary's leading geologist and a world-renowned expert on reptiles. Nopsca's adventurous spirit took him in many different directions, sometimes away from paleontology. He served as an intelligence agent in the Balkans and was adopted by the Albanian Mirdites tribe, whose language he learned to speak fluently. Disguised as a Skipetar (Albanian mountaineer) he aspired to become King of Albania when the country was liberated from Turkey in 1913. His hopes for high office were not realized, and he returned to paleontology in 1923, making a contribution on fossil footprints called by many a landmark in the European literature.[5] Nopsca argued that chirothere tracks could be attributed to dinosaurs, specifically to prosauropods, the forerunners of brontosaurs; prosauropods had recently been discovered in abundance in Germany. Although these creatures had four toes on their hind feet, not five, Nopsca was persuaded by Willruth's argument that the outer impression was that of a fleshy appendage.

Despite Nopsca's credible reputation, many had justifiable reservations about this claim. In a classic 1925 study entitled "The Tracks of the Chirotheria," Wolfgang Soergel argued that the so-called flesh appendage mark was the impression of a real digit, digit V, on the outside of the foot, not digit I on the inside.[6] He showed that small pseudosuchian (meaning "false crocodiles") reptiles from the Triassic of South Africa had the correct foot structure, even though their feet were only two inches (5 cm) long. Thus he predicted that large pseudocrocodilians up to 16 feet (5 m) in length with tracks at least 8 inches (20 cm) long must have existed. Everyone appears to have accepted Soergel's interpretation, even Nopsca, who was off on another espionage mission at the time, disguised as a Rumanian shepherd in the Balkans. This was to be one of the last of the passionate political pursuits that characterized his restive life, for not long after this he committed suicide.

As of 1925 the chirothere ("chirothere" is a generic term for *Chirotherium* tracks) case appeared solved after a century of speculation. In the late 1920s, Professor Friedrich von Huene, a colleague of Soergel's discovered a 16-foot-long (5 m) pseudosuchian in the Triassic of Brazil, named it *Prestosuchus* and noted that it would have made large chirotherelike tracks. Indeed it was not long before von Huene discovered such tracks in South America.

The 1930s was something of a golden decade for track discovery in North America. In 1925, the same year that Soergel published his seminal work on *Chirotherium*, a young American named Frank Peabody was just a boy with no inkling that he would become one of the world's leading authorities on petrified hands. The material that would preoccupy him for the better part of a decade was discovered by Major L. F. Brady at the Cameron Trading Post on the Little Colorado River in northeastern Arizona. According to local legend, Brady spotted the hand tracks in the wall of the stone fireplace where he was warming his hands one cold December day. He found that the rocks came from local exposures of the Triassic Moenkopi Formation, named after a small settlement near Tuba City. Peabody got the tracking bug and served his tracking apprenticeship in California, studying the tracks of modern salamanders, studies on which he wrote his master's thesis.

Through careful study Peabody identified "eight different species of *Chirotherium*" from the Moenkopi, and declared that many were the same as those found in Europe.[7] Peabody also established that there were two distinct chirothere groups, one with large front feet (large manus, or hand, group), the other with small front feet (small manus group). These differences can be of great interest in the interpretation of the biology of the track makers. He also concluded that "since skeletal remains of reptiles are fragmentary and scarce, the trackways provide the only clear picture we have of the early Triassic history of North American reptiles, particularly the Archosauria."

In the 1960s, the focus returned to Europe again, when another large pseudosuchian named *Tichniosuchus* was discovered in the Triassic of Switzerland. At last an animal had been found that was the right size to fit the large chirothere tracks. The track makers were not apes, toads, kangaroos, nor amphibians, but members of the advanced tribe of reptiles known as archosaurs that rose to prominence as the rulers of the Triassic.

A MESSAGE IN THE LADY'S HAND

Here lies a most beautiful lady, light of step and heart was she;
I think she was the most beautiful lady that ever was in the West Country.
But beauty vanishes; beauty passes; however rare—rare it be;
And when I crumble, who will remember this lady of the West Country.

Walter de la Mare

The mysteries that surround the famous "hand animal" are by no means fully resolved. The latest new puzzle posed by *Chirotherium* tracks arises from the recognition that both in Germany and England stout and slender tracks occur together in the same Middle Triassic strata. We are here reminded of the difference between the stout feet of the English and the slender feet of the Celts, and the alternate poles of narrowness and breadth expressed throughout human and vertebrate anatomy. In a cleverly titled article, "Sex in the Footprint Bed," Geoffrey Tresise, of the Liverpool Museum, proposed that the tracks reflect differences in gender.[8] As early as 1838 the slender tracks, named *Chirotherium stortonense,* after Storeton on Merseyside, were referred to as the "lady's hand," and the stouter tracks, dubbed *Chirotherium barthi,* after a German named Barth who had first dug them up, were described as resembling "a large man's hand in a thick fur glove."

"Vive la différence!" Certainly archosaur life did not proceed without a demarcation between the sexes. The question is whether footprints reveal such sexual dimorphism. There appears to be no reason to dismiss the possibility, nor is there evidence to confirm the suggestion. The supposed "female" tracks are longer than those of the stout "male" footprints. This rules out the possibility that the slender tracks are simply those of small individuals. The German paleontologist Hartmut Haubold has suggested that English *Chirotherium stortonense* is simply a variety of German *Chirotherium barthi* (i.e., the Brits can claim nothing new). To shed light on the question, perhaps we need to understand the difference between the male and female anatomy of modern reptiles. Such information as is available indicates that the size of the sexes is often different, though it is not always the females that are smaller.

We may ask what Tresise's paper implies for footprints in general. There are many examples of sexual dimorphism in the track record. Certainly among modern animals such as deer it is possible, with experience, to tell males from females by the size of tracks and the degree to which they toe in or out.[9] Among extinct reptiles such as dinosaurs, even the skeletal record does not yet provide definitive evidence to help us distinguish males from females in most cases. Some trackers hold that female dinosaurs were larger than males; others disagree or reserve judgment. The question remains open, but nonetheless very intriguing. Footprints again pose fascinating questions about the biology of extinct species. Could such differences be attributable to regional differences among populations, analogous to ethnic differences between the feet of Saxons and Celts? We now know that we can begin to learn a lot from footprints by looking at such simple measures as slenderness and breadth within groups. From this approach we can perhaps infer that the narrowness–width spectrum is not usually explained by sexual dimorphism, and is more likely to be a reflection of different species or races.

It is also interesting to consider the relative size of the front and back footprints of these animals: do they fall into the previously mentioned large or small manus groups. As we shall see in the following chapters, differences in the forefeet among the famous brontosaurs are the key to

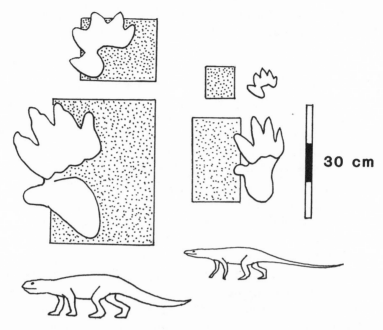

FIGURE 3.2   Large and small manus (hand) chirothere tracks also show the wide and narrow dichotomy seen in many animal groups.

distinguishing different families. These differences also tell us the extent to which animals favored their hind or front quarters, which in turn gives clues to overall body proportions, gait, and behavior. Those heavy at the rear end or on the rear axle were probably the "narrower" varieties. They had a smaller manus and a tendency toward being bipedal. They would have had longer tails, smaller heads, and physiological tendencies toward nerve-sense orientation. As we shall see they may have been closely related to dinosaur ancestors. By contrast, the large manus track groups indicate greater breadth of body, shorter tails, larger heads, and a physiological inclination toward the metabolic pole. Such varieties may well have been vegetarians, armored, and geographically widespread. When discussing narrowness and breadth, it is instructive to look simultaneously at the shape of the digits as well as the shape of the whole footprints.

On the subject of hands, it is worth mentioning Count Louis Hamon, a famous nineteenth-century palm reader nicknamed "Cheiro"—literally, "the hand." He was apparently able to read a palm with "wonderful accuracy." Mark Twain, a true cynic, is quoted as saying that "Cheiro has exposed my character to me with humiliating accuracy. I ought not to confess this accuracy, still I am moved to do it." The duke of Newcastle also praised his skill and insight as follows: "Cheiro has told my past and immediate future with wonderful accuracy, especially with regard to certain coming events which he could not possibly have known."[10]

As mentioned in Chapter 1, although palm reading is not considered a science, fingerprinting is considered objective and scientific. We also noted the evidence that fingerprint types and foot shapes are related to genetic and ethnic differences in the human gene pool. But in a holistic system we cannot acknowledge only a few useful connections without implying that all anatomical features are related. Could it be, therefore, that we have not paid enough serious attention to the whole form of hands and feet? Serious study of this field might allow a rational penetration of the occult world of mystics such as "Cheiro" and begin to make a "science" of such ancient practices as palmistry. In support of this statement, which scientifically minded skeptics might find difficult, I cite an article in *Clinical Pediatrics*: "To the cheiromancer, the lines which traverse the palm write the indisputable but secret history of an individual's past, present and future. To the knowledgeable physician, these same hand lines, until recently, have been little more than a curiosity. We now know that an abnormality in one hand line, the single palmar crease, is suggestive of an early insult to the developing fetus, which may have an ominous bearing on its future."[11] The clinical and anthropological literature on this subject can be very revealing, indicating that in some human hands certain lines are clear clues to "simian" developmental traits, which correlate with developmental abnormalities. Like Victorian ladies who were shocked to hear Darwin's views on our descent (or is that ascent?) from apes, we may not want to acknowledge that our hands still reveal close relationships to our anthropoid relatives. But in some cases they clearly do, and in such cases, primitive mental characteristics evidently accompany the primitive physical traits. Before there are any cries of political incorrectness, note that the holism implicit in self-organizing biological systems implies an advanced, or "nonprimitive," correlate to balance the "primitive" characteristics, maintaining the coherence of the whole. Such persons may be savants in areas were we are ignoramuses. Indeed this line of reasoning suggests that the savant in palmistry might have this skill precisely because of a blind spot in the analytical mind, while the accomplished analytical scientist is blind to the palmist's sight because of his or her own blockage to mystical sight. Such a view celebrates diversity, not uniformity, and suggests that we might need each other's skills more than we know.

NOW YOU SEE 'EM, NOW YOU DON'T

If a step should sound or a word be spoken, would a ghost not rise. . .
Algernon, Lord Swinburne, "A Forsaken Garden"

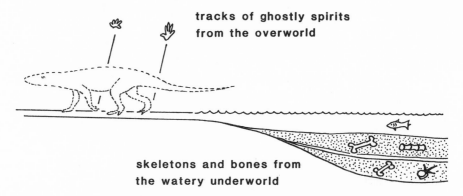

FIGURE 3.3   Bones of aquatic creatures from the underworld and tracks from elusive terrestrial travelers above tell two different stories about Triassic life.

It has been said that paleontologists are nothing but a bunch of grave robbers. This uncharitable allegation also applies, with even greater force, to archaeologists, who may sometimes be accused of violating the sanctity of human burial grounds. Even the final resting places of dinosaurs and vertebrates far removed from the hominid line are sacred in the eyes of many, and only those professionals who show the proper credentials and respect are permitted to break ground and meddle with our ancient heritage. Most human cultures have made a deliberate ritual of burying their dead, and in the case of kings, queens, and great rulers have gone to great lengths to ensure that they were unusually well preserved. Such rituals play into the hands of the archaeologist, by placing a disproportionate number of human remains of the rich and famous in well-protected tombs. It is generally harder to find the remains of beggars who passed away beside the monolithic mausoleums of the monarchy, for their remains crumbled to dust amidst the wear and tear of busy streets and thoroughfares.

In the fossil record similar principles apply. The best fossils are well preserved because they were well buried. Although Triassic animals did not bury their peers in elaborate graves, some lived closer to good burial sites than others, while some merely passed through such cemeteries, with still many a mile to wander before reaching their final resting places. This was one of the reasons why it was so hard to identify the chirothere track makers for so long. The animals were passing through like ghosts walking on the graves of others. In many areas a large proportion of skeletal remains are those of amphibians and aquatic reptiles that lived in the underworld of swamps and ponds. Upon death they were already more or less entombed. In the world above, however, reptiles roamed as free, restive spirits, evolved beyond the watery underworld to a realm of wind and atmosphere. In such terrestrial settings only their feet remained in contact with the Earth, and only their footprints show that they skirted the wet shores of the ancestral underworld.

Putting together the puzzle of ancient Triassic life is a game of "now you see 'em, now you don't." The bone record tells us of the amphibious (yin) inhabitants of the underworld, but tracks tell us of the world of peripatetic terrestrial wanderers (yang) moving to and fro. This is exactly why footprints assume great importance: in many cases they are the only record we have of these animals.

TRACKING THE CRIMSON CROCS

It depends on the adjustment of a hinge whether a door opens in or out.
Dr. T. Suzuki, psychologist, Columbia University

The Triassic Age of Archosaurs was a time of great evolutionary diversification among the reptiles. Archosaurs, characterized by upright posture, are among the best-known reptiles, for they include the dinosaurs, pterosaurs, and crocodiles. In comparison with lizards and their kin, many archosaurs were fully erect, as their trackways show. Many of the successful archosaur tribes bore a superficial resemblance to crocodiles, and so have been named accordingly. There are the Early Triassic erythrosuchids (meaning "crimson crocodiles"), the pseudosuchians ("pseudo crocs"), the ornithosuchids ("bird crocs), lagusuchids ("rabbit crocs"), and so on.

As we have seen, footprints are often much more abundant than skeletal remains. Such circumstances have prompted a number of European trackers to suggest that it is easier to track archosaur evolution through the footprint record than through the bone record.[12] This may at first seem like an unreasonable claim, but in fact, much of what we know about archosaur evolution is based on a study of ankle structure and locomotor characteristics, which can be inferred from tracks. But this should not be so surprising because some of the most fundamental differences between fish and amphibians, reptiles and birds, humans and other primates have to do with the specialization of the limbs.[13]

The story of archosaur success in the Triassic is the story of reptiles taking locomotor sophistication to new levels. Far from being the pitiful shambling creatures postulated by Lyell and his contemporaries, chirothere track makers were up and running early in the Triassic. Careful study of their footprints allows us to see how some groups had the axis of their feet turned outward, whereas others had their feet directed forward in the direction in which they were moving. The former direction is the primitive condition, retaining something of the sprawling posture. The latter orientation helps keep the limbs erect, maximizes propulsion in the direction of movement, and facilitates walking on one's toes rather than flat-footed.

If we compare the track record of archosaurs with that of protomammals (discussed in Chapter 2), we find notable similarities. Like the protomammals, the most successful archosaurs developed fully erect posture. This can be seen clearly in the narrow trackways of chirotheres. As their legs moved under their bodies, we also see significant changes in the shape of their feet. In the primitive condition, seen in "stem," or ancestral, reptiles, the foot is lizardlike in shape with elongated outer digits. This condition persisted in crocodiles, which never learned to walk fully erect. As a result, the ankle remained primitive, leading paleontologists to talk explicitly of crocodilian ankles. This ankle structure can be inferred from tracks in a number of ways. The simplest clue is that the individual track is not symmetrical or well aligned in relation to the midline or direction progression (sagittal plane). The foot is asymmetrical, with long outer digits, or it points outward, or both. By contrast, the tracks of fully erect archosaurs develop a much greater symmetry, with the middle digit (III) clearly the longest. This symmetry is seen in the tracks of advanced protomammals, with the notable difference that the tracks of protomammals were short and wide whereas those of the advanced archosaurs were elongate. Such differences are not by mere chance. They reflect corresponding differences in the length and breadth of the two groups' bodies.

These observations indicate that primitive archosaurs, like primitive protomammals, were sprawlers, with wide trackways, and that advanced archosaurs, like advanced protomammals, produced narrow trackways. In both groups the primitive condition indicates a primitive ankle, whereas the advanced condition indicates an advanced ankle, demonstrated by the ability to keep the foot in line and not have it turn outward. We can test these fundamental foot, ankle, and limb connections for ourselves quite easily. If we turn our feet outward, in imitation of Charlie Chaplin, and try to walk along we will find that our knees (limbs) turn out awkwardly, that we tend to squat and bend forward a little (i.e., become less erect), and that we shuffle with inefficient, short steps, making a wider trackway, rather than stride purposefully in a straight line. Here in a nutshell is the holistic locomotion system in action. There is a ripple effect throughout the entire system when any shift in foot ankle or limb orientation is introduced. It is precisely because this is true that we can read such a comprehensive account of ankles, locomotion, and posture from the trackways.

The advanced archosaur ankle is a simple flexible hinge mechanism, functionally like a simple door hinge. The most successful Mesozoic archosaurs—mainly the ancestors of dinosaurs and birds—developed such advanced ankles. This opened the door to fully erect posture and more than 150 million years of Mesozoic locomotor sophistication achieved by most dinosaurs. It is amazing how much may hinge on a hinge.

THE FIRST BIPED

We are risen and stand upright.
Psalms 20:7.

We speak of taking a stand or standing on our own two feet as if it were a great accomplishment. For a tetrapod with a long history of reliance on four feet, the switch to bipedal locomotion was indeed a significant step. Nothing is more symbolic of a radical change in our own evolutionary history than our ability to stand and walk bipedally. Indeed, our own hominid family is differentiated from other primates by being bipedal, and the verb "to stand" is synonymous with the verb "to be" in many languages, for example, Spanish. We often view ourselves as standing above other four-footed creatures and, alas, too often look down on them. By standing and walking bipedally, a creature raises itself up from the Earth and holds its head high, freeing forelimbs and hands for other uses and even getting away from sweltering temperatures at ground level. This is to say nothing of the ramifications for blood circulation, temperature regulation, and evolution of the brain and consciousness that have been attributed to the simple act of holding up one's head. In *The Evolution of Consciousness*, Robert Ornstein notes the heat-reducing properties of bipedalism and suggests that solving the problem of shedding the heat produced in the brain may have been a major factor in brain growth.[14] Bipedal dinosaurs and birds, both descendants of Early Triassic archosaurs, developed bipedal locomotion and perhaps benefited from the cooling effects of holding their heads high. They did not, however, bring their liquid crystal/spinal axes into fully vertical alignment in the same way as hominids (see Figure 1.5).

Trackway evidence suggests that archosaurs may have been standing on their own two feet as early as 240 million years ago. In 1982, two European trackers, Georges Demathieu and Hartmut Haubold, reported a Triassic trackway of a possible biped which they named *Isochirotherium archaeum,* which loosely translates as "ancient chirothere track."[15] When we consider that tetrapods had been walking on all fours since the Devonian (almost 400 million years ago), it was a considerable accomplishment on the part of archosaurs to become bipedal. It took 160 million years to achieve this radical new posture. What is important is not so much to determine which unpronounceable label fits the "first" biped, but to note that from among these "hand animals" the first bipeds arose. And it is extraordinary that in the process of raising their hands from permanent contact with the earth, their feet should leave such distinctive handlike traces.

A PROTOMAMMAL WITH HAIRY FEET?

All theory, dear friend, is gray, but the golden tree of actual life springs
ever green.

Goethe

It was not from the archosaurs that our ancestors arose, even though we may identify with their bipedal abilities and their extraordinarily hand-like tracks. Rather, our ancestry lies with the protomammals, which like their mammalian descendants were probably hairy creatures. Insulation helps regulate body temperature (or achieve a warm-blooded metabolism) and helps the organism adapt to life in colder, higher latitudes outside the tropics. The distribution of protomammal remains suggests that they were among the first vertebrates to achieve such geographical emancipation.

According to the French priest-ichnologist Paul Ellenberger, an incomplete footprint from southern France reveals the footprint of "a large but still unknown hairy vertebrate." He named the track *Cynodontipus polythrix* ("hairy footprint of a cynodont"; "cynodont" means "wolf teeth"), claiming that it showed traces of hair.[16] The cynodonts were advanced protomammals. Knowing that few mammals have hairy soles to their feet, he compared the track with one of the few groups that do—namely the rabbits—but obviously, a track maker with feet the size of a cart horse has little in common with a rabbit. The possibility of a large protomammal track maker without hairy feet has been confirmed by the discovery of a different complete trackway of similar age in Arizona. (Ellenberger has to be credited with considerable "imagination" to propose a hairy protomammal on the basis of a single, incomplete footprint, and it is fair to say that his interpretation is very much open to question.)

Paul Ellenberger is often in the thick of ichnological controversy, owing to his unconventional, some would say overenthusiastic, approach to tracking. He came by his perspectives honestly, however. Born in 1919 into a family of missionaries in Lesotho, Paul spent most of his early years as a free spirit roaming the countryside with his father, Victor, who was a serious student of Bushman culture and in fact had been born in a

Bushman cave. Victor Ellenberger witnessed the tragic end of Bushman culture, an experience he described in his book *La Fin Tragique des Bushmen* (*The Tragic End of the Bushmen*).[17] So Paul can justly describe himself as the son of a caveman. To a significant extent Paul Ellenberger learned tracking from the Bushmen, who could pick out any one of more than 40 species of gazelle by their footprints. It is perhaps this background that gave Ellenberger his reverential habit of celebrating all life by naming almost every different type of track that he has studied.

This overzealous naming of fossil footprints is called "oversplitting" by the paleontological fraternity, of whom few discern the many subtle differences apparently seen by Ellenberger. But when one meets Ellenberger one begins to understand his perspective. His worldview reflects his close affinity to nature and the land in his formative years. He is one of that rare breed of naturalists who communed with nature so wholeheartedly that he grew to love it not only with the intellectual passion of an inquisitive and active mind but also with a depth of heart and soul feeling that comes from being immersed in it throughout childhood. To this day his spirit and joie de vivre are a joy to behold, for every footprint is a natural treasure, a means to uplift the spirit and a cause for enthusiastic celebration.

LIZARDS GALORE

Early lizard fossils might be expected in rocks of Triassic or even Late Permian age.

Michael Benton, *Vertebrate Paleontology*

Despite the abundance of *Chirotherium* hand tracks, according to Frank Peabody footprints of small lizardlike reptiles that are virtually indistinguishable from the tracks of living lizards are more abundant than any other kind. There is some debate as to when true lizards first evolved, but there is little doubt that close relatives of true lizards were well established by the Triassic. One such group were close relatives of the famous but endangered New Zealand's tuatara, genus *Sphenodon*, which has a well-developed pineal gland, or third eye, situated beneath a transparent scale in the center of its skull. In addition to its scientific rarity value, it also, unfortunately, fetches a high price as an exotic animal on the black market.

Triassic lizard tracks, so abundant that "they provide an embarrassment of riches, [a] cluttered mass [that] is generally unintelligible,"[18] are so ubiquitous that we rather uncharitably referred to these track makers as vermin of the Triassic.[19] This derogatory label is not an indication of

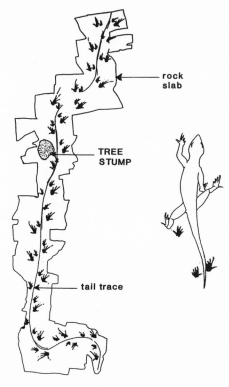

rock slab

TREE STUMP

tail trace

FIGURE 3.4   Meandering through the woods. The trackway of a small lizardlike species swerves to avoid a small tree stump.

prejudice toward lizards and their kin, but rather stresses that the tracks are so abundant that they often confuse and frustrate trackers.

Finding a complete trackway of these lizardlike creatures is like searching for a grain of sand on the beach. Nevertheless, Paul Olsen of Columbia University has reported finding a single spectacular example from footprint-rich Late Triassic strata of the Newark Basin, in New Jersey. In a clearly illustrated trackway consisting of thirty consecutive steps we can follow the trail of a tiny lizard as it turns from left to right, then swerves right to avoid a small tree stump, before veering right again.[20] The footprints are less than half an inch (1 cm) long, indicating a track maker only a few inches long, but despite their diminutive size some show clear skin impressions. Persistence and patience have been rewarded. After a prolonged search for a clear picture of this type of trackway, trackers have finally found an almost perfect example.

THE FIRST GIANTS

Theirs was the giant race before the flood.
John Dryden, *Epistles*

Among the tracks described by Ellenberger were tracks of truly elephantine proportions from southern Africa. Warning! Three of several dozen names are necessary! Ellenberger christened *Tetrasauropus* and *Pseudotetrasauropus* (tracks of "four-footed" and "pseudo-four-footed" reptiles), and attributed them to prosauropods, the ancestors of the well-known sauropods, or to brontosaurs, which had the largest tracks of any creatures known to have walked on Earth, up to three feet (1 m) in diameter.[21] Some prosauropods also grew quite large, and left tracks up to 24 inches (60 cm) long but were mostly bipedal.

Generally the quadrupedal track makers (*Tetrasauropus*) were large, but not all; some were quite small with a strong tendency to be bipedal. Ellenberger named nine types in this latter, mainly bipedal, category (*Pseudotetrasauropus*). Such observations are generally consistent with reconstructions of many prosauropods as bipedal. As we have seen, Ellenberger is considered overenthusiastic and inclined to shower us with dozens of new names. He is a "splitter," whereas those inclined to include many similar types under a few labels are called "lumpers." Ellenberger elevated splitting to a fine art. The tracker Don Baird has commented that splitters toss names around "like confetti at Mardi Gras." As a result of Ellenberger's confetti sprees, Baird's colleagues Paul Olsen and Peter Galton staged a rebellion: in a paper published in 1984 they gave a long list of Ellenberger dubbings they regarded as duplicative and invalid.[22] This they managed to do without actually going to southern Africa to look at the tracks he had described! Although one can see where their frustration arose, and can intuitively accept that many of Ellenberger's names may be invalid, this type of shotgun approach can create as many problems as it solves. In short, aside from the potential strain placed on Franco-American relations, the whole subject of southern African footprints is in a state of serious semantic confusion.

In an attempt to make sense of this confusion, I teamed up with my Swiss tracker colleague Christian Meyer, whom I have long known affectionately as Jacques Cousteau II, owing to his excellent French accent and widespread international expedition experience. With him at "zee helm" of Calypso II, the rent-a-car chartered from Marseilles airport, we set a westward course for Montpellier to conduct an investigation of French driving habits and an analysis of Ellenberger's collection. We found that most of the tracks assigned to *Pseudotetrasauropus* are quite similar to *Otozoum* (meaning "giant animal"), a track type from the Lower Jurassic of New England named by Edward Hitchcock, and discussed further in the next chapter.[23] In recognizing their tracks, Ellenberger pushed back the track record of giant prosauropods (brontosaur ancestors) into the Triassic. These "giant animal" tracks are pivotal in helping us understand the evolution of the whole tribe of lizard-hipped dinosaurs (saurischians) of the Mesozoic era, which include the carnivores (or theropods) and brontosaurs (or sauropods).

Splitter or not, Ellenberger found wonderful tracks and brought back a fine set of replicas for examination by the Calypso expedition without our having to stray from French territorial waters.

NOW WE SEE 'EM: A NEW LOOK AT
DINOSAUR NATIONAL MONUMENT

Now you don't see them, now you do.
Hoimar von Ditfurth,
The Origins of Life

The spectacular quarry at Dinosaur National Monument in northeastern Utah, discovered back in 1909, yields giant brontosaurs, stegosaurs, and contemporaries such as *Allosaurus* from the Late Jurassic (from about 150 million years ago), the middle of the Age of Dinosaurs. But the quarry occupies only a tiny geographical area at the extreme western end of a huge national reserve that covers thousands of square miles. This vast area remains largely unexplored by paleontologists. Consequently, the National Park Service decided to fund a series of small projects to search for tracks. At the start of the effort, in 1990, only one track had ever been found and nobody knew exactly where it had come from. After a cumulative total of only a few weeks of fieldwork in 1991 and '92, however, our University of Colorado Dinosaur Trackers Research Group had recorded approximately 250 trackways from half a dozen formations.[24] In comparison with the rate of excavation of skeletons in the quarry—one per year during a period of about 80 years—in less than three years the trackway documentation project proceeded to record about 80 trackways per year, mostly of chicken-sized bipedal dinosaurs. Here in a national monument whose raison d'être was the presence of dinosaur resources, no one had recorded any sign of tracks because no one had looked.

By convenient coincidence, some of the best examples of early dinosaur tracks come from in and around Dinosaur National Monument. It is still a well-kept secret that this world-famous graveyard for Jurassic giants is surrounded by numerous stomping grounds of Triassic chickens and the last of the "hand animals." Skeletons and bones provide us the key to one avenue and epoch of paleontological investigation while tracks take us into quite another. We shall see as we conclude this chapter that in the Late Triassic epoch we have reached a time when dozens of small bipedal dinosaurs are scurrying around our feet at almost any track site we care to investigate, especially in the wild west of Utah and Colorado. These tracks lead us to the gateway to the Age of Dinosaurs.

GATEWAY TO THE WORLD OF MINIATURE MAMMALS

Particularly dramatic biological events are sometimes responsible for
critical instances of punctuated equilibrium. . . breakthroughs or gateway
events that open up whole new realms of possibility.
Murray Gell-Mann, *The Quark and the Jaguar*

In the remote canyonlands of western Colorado in the Dolores River valley, near the tiny village of Gateway, a remarkable track story has been unfolding. The story began early in the 1980s, when a local resident, Chester
Jennings, began mining alabaster from the "hand animal" layers of Early
Triassic epoch. One day he hiked up the cliffs above his mine, to look at
the later Triassic strata above, and came across three-toed dinosaur tracks
and peculiar trails made by crayfish. As the dinosaur footprint renaissance
of the 1980s got underway, new Triassic track-site reports began to trickle
in, and by 1995 at least 60 track sites had been identified in the Rocky
Mountain region. Nowhere, not even around Dinosaur National Monument, are the tracks as abundant, well preserved, and diverse as in the
Gateway area of Colorado.

Most significant are what appear to be the first known mammal tracks.
Less than a half inch in length—a pair of them are easily covered by a single penny—to the casual observer they look like the tracks of mice. This is
as far back as we—the species that has walked on the moon—can look
and recognize the spoor of our own class, Mammalia. These most distant
of our 210-million-year-old relatives are often compared with the abundant rodents that populate the world today. Thus, the history of mammals
could be called the evolution of mice into men.

Although diminutive, our Triassic ancestors developed certain novelties
that allowed the mammal lineages to gain a foothold and survive the long
reign of dinosaurs before eventually flourishing. It is hard to say with certainty exactly what these novelties were, when the only available evidence
is a few teeth and a few tracks. Nevertheless, most paleontologists agree
that the early mammals' success had much to do with improved reproductive strategies, leading to enhanced care of their offspring.

As we know, ultimate evolutionary success depends on successful pro-
creation and care of the next generation. Some of the most fundamental
breakthroughs in biological evolution have revolved around improved reproductive strategies. Classic examples include the evolution of sexually
reproducing eukaryotes (cells with a nucleus) from prokaryotes (cells
without a nucleus); the evolution of a hard-shelled egg, allowing reptiles
to reproduce on land; and the development of mammary glands, pouches,

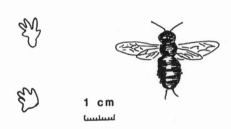

FIGURE 3.5   The world's oldest known mammal tracks, from Gateway, Colorado. Compare to size of a honey bee..

1 cm

and placentas for prolonged nurturing of mammalian offspring. Murray Gell-Mann calls these biological breakthroughs "gateway events."[25] How apt that the oldest known mammal tracks, the footprint evidence for the breakthrough of a new class, should be found at a town called Gateway.

The famous behaviorist Konrad Lorenz introduced the term *fulguration* to describe evolution's periodic sudden leaps (*fulgar* is Latin for "lightning").[26] Evolutionary novelty can strike with lightning speed, at least from the geological perspective. The punctuated equilibrium terminology of Niles Eldredge and Steven Jay Gould describes how a long period of stasis or equilibrium is "punctuated" by a sudden evolutionary event or leap forward.[27] When the evidence for this gateway event or fulguration is in the form of minuscule mammal tracks made more than 200 million years ago, it is easy to miss. It is exhilarating to realize that we can now track the lightning bolt that shot the first true mammals through the evolutionary stargate of the early Mesozoic. Among Ellenberger's papers is a real gem entitled (in abbreviated translation) "The Demographic Explosion of Small Quadrupeds of Mammal Affinity."[28] In it he documented large numbers of diminutive mammal and mammal-like footprints near the Triassic-Jurassic boundary in southern Africa. These are the same age as footprints subsequently found at Gateway and others found in South America.[29] Thus we must give Ellenberger credit for tracking down the "explosive" origin of mammals, one of the most significant gateway events in the history of evolution.

DINOSAUROIDS AND THE FIRST DINOSAURS

There is now a substantial mismatch in the age of the earliest skeletal evidence of dinosaurs and . . . supposed dinosaur footprints

Michael King and Michael Benton, paleontologists

In the last 15 years a large amount of effort, ink, and scientific-journal space has been devoted to speculating on the cause or causes of the demise, or "extinction," of the dinosaurs. The origin of dinosaurs is just as intriguing a topic, and it too has been the subject of extensive investigation during the last decade.[30] Our desire to understand our origins has driven paleontologists to delve deeply into the origin of all major animal groups. Anyone finding significant evidence of the "first" or "oldest known" tetrapod, mammal, bird, or dinosaur is assured of their 15 minutes of paleontological fame.

Currently the diminutive dinosaur (or dinosauroid) *Eoraptor* ("dawn predator") is one of the leading candidates for the world's first dinosaur. *Eoraptor* fossils have been found in the Late Triassic Ischigualasto Formation in South America, and the creature has been dated at around 220 million years. These fossils would lead us to expect footprints of *Eoraptor* to be small three-toed tracks of a bipedal turkey-sized creature. But this is not what we find at all. The first reports of footprints from this region were of a large four-toed quadruped with feet up to 14 inches (35 cm) in length.[31] Such tracks attracted enough attention to result in the naming of the formation as Los Rastros ("The Trackways"). These tracks have also been variously attributed to armored dinosaurs (ankylosaurs), plated dinosaurs (stegosaurs), ornithopods (early duckbills), and non-dinosaurs like the "hand animals," *Chirotherium*.[32]

Subsequent discoveries have added further spice to the tracking of South America's first dinosaurs by providing evidence of a large bipedal creature that made footprints closely resembling the tracks of large carnivorous dinosaurs.[33] There are also claims of dinosaur footprints from both England and France that, if they hold up, have significant implications for the origin of dinosaurs. Georges Demathieu speaks of a very dinosaur-like trackway from France's Massif Central region, in rocks that are Middle Triassic.[34] Demathieu sensibly used the term *dinosauroid* to describe the affinity of the tracks, suggesting either that dinosaurs existed in the Middle Triassic—before their bones are known—or that very dinosaurlike ("dinosauroid") archosaurs had developed as the forerunners of true dinosaurs.

The latest chapter in the oldest dinosaur tracks saga was published in 1996, and revolves around purported footprint evidence from the Early and Middle Triassic of England, originally reported in 1970 by Leonard Wills and Bill Sarjeant from boreholes in Worcestershire.[35] The concept of finding footprints 1,000 feet below the surface is intriguing, as the chances of finding two or more dinosaur tracks when drilling a narrow borehole to depths of between 230 and 1,100 feet seems slim. To cut a long and controversial story short, some of the purported dinosaur footprints have recently been reinterpreted as sedimentary artifacts (i.e., fossilized mud flakes and rock texture patterns) by the British trackers Michael King and

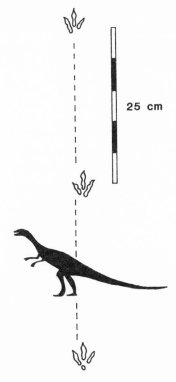

25 cm

FIGURE 3.6 A possible contender for the world's oldest dinosaur tracks, from the Middle Triassic of France.

Michael Benton.[36] Their conclusions suggest that there is no Early or Middle Triassic footprint evidence for dinosaurs in England. Sarjeant, however, does not accept the King-Benton reinterpretation and instead has proposed that the prints may represent other reptiles.[37] The role played by tracks in shedding light on the origin of dinosaurs is clearly controversial, but it is an indispensable piece of the puzzle.

THE STRESS OF MISSING AND BROKEN PIECES

Humpty Dumpty sat on a wall,
Humpty Dumpty had a great fall;
All the king's horses, and all the
King's men,
Couldn't put Humpty together
again.
Traditional nursery rhyme

Countless authors have likened the fossil record to a jigsaw puzzle that must be painstakingly put back together. But it is a puzzle from which many pieces are missing. Although we continue to find lost pieces, we can surely never find them all. It is also hard to find pieces that fit together cleanly, that "match." Though we find an assortment of bones, tracks, eggs, and other remains, they rarely interlock perfectly. Rarely—but there are exceptions. In 1954, Donald Baird, while engaged in a study of Late Triassic tracks from New Jersey, ran across a trackway specimen purported to originate from Saratoga, Pennsylvania. He recognized it as what paleontologists call the counterpart of an important chirothere track unearthed in 1885 in Milford, New Jersey. Baird was able to use the combined mold and cast (part and counterpart) to write an improved description of one of the last (youngest) known chirothere tracks. The pair— the "male" cast and the "female" mold—were reunited after a separation of three score and ten years.[38]

Another case of fitting jigsaw puzzle pieces recently came to light in England, where many *Chirotherium* trackway sequences had been broken up

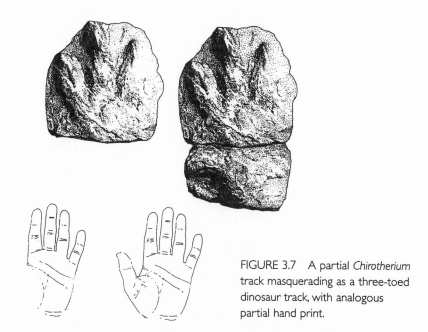

FIGURE 3.7 A partial *Chirotherium* track masquerading as a three-toed dinosaur track, with analogous partial hand print.

and put in different museum collections. William Sarjeant had described a large three-toed track, which suggested a very early origin for large bipedal dinosaurs. Subsequent studies by the King-Benton team suggested that such tracks are incomplete five-toed chirothere tracks.[39] In one case such an apparent three-toed track was reunited with its missing "thumb" (digit V). Symbolically the track had split along a line equivalent to its lifeline, which had literally fragmented its true identity.

I have introduced a new way of looking at tracks, and whole animal anatomy, as a dynamic evolving system of coupled cycles of flow and counterflow, but as always it is hard to improve on an example directly from nature. Like the distorted ancient myriapod trails from England that required a computer to be restored to their original shape, we can find dinosaur tracks that have been compressed and elongated by tectonic forces. In examples from the Eastern Seaboard we can see how the same forces acted to elongate one track and compress another because they originally had different orientations. These tectonic forces are analogous to the formative biological forces that shape vertebrate anatomy. They are uncannily like the examples of animal shape shifting presented by d'Arcy Thompson in his classic study *On Growth and Form*.[40]

Despite the challenge of missing, distorted, and broken pieces of the puzzle, remarkable strides have been made in tracking in the century and a half since the footprints of the enigmatic "hand animal" first came to

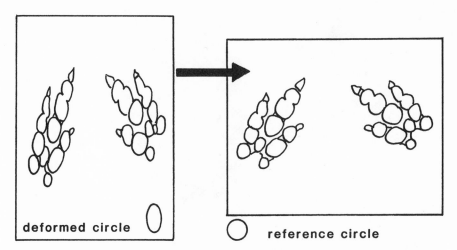

deformed circle   reference circle

FIGURE 3.8   Rock deformation can cause dinosaur tracks to succumb to stress and be distorted into stout or slender forms. Similar distortions can be achieved graphically by morphologists experimenting with animal shape.

light. We have among other things zeroed in on the origin of dinosaurs and mammals. The optimistic view, which I shall develop in the forthcoming chapters, is that we are now just beginning to see how the pieces of the puzzle really fall into place. Just as DNA analysis or the lines on our hands may reveal more about an individual than we might ever have imagined at first sight, the study of tracks continues to give us new and subtle insight into the biology, behavior, and evolutionary history of a diverse menagerie of extinct animals.

chapter four

NOAH'S RAVEN

ACT IV: CAST OF CHARACTERS

In this chapter we shall spend the whole time in the Lower and Middle Jurassic epochs, when the dinosaurs rose to dominance. We will review the classic work of the American theologist-tracker Rev. Edward Hitchcock (1793–1864), the author of *The Religion of Geology* and the father of the science of fossil footprints. We shall see how fossil spoor found in central Massachusetts were first attributed to races of extinct giant flightless birds, and dubbed Noah's Raven by local residents. Later the tracks were correctly identified as the first dinosaur footprints ever discovered. We shall continue to follow the exploits of yet another tracker-priest, Father Giuseppe Leonardi, as he bends the rules of municipal authority to "steal" track specimens for science.

We will also examine three miles of rock core, spanning 30 million years, or some 3 million dinosaur generations. We will meet large dinosaurs christened "true thunder" and the "giant animal," follow the tracks of some of the worlds smallest, sparrow-sized dinosaurs, and hop around in an ancient desert on the trail of the Jurassic equivalent of a gerbil. We shall enter the age of brontosaurs, and visit Portugal, where several world track records have been set.

We shall also continue our efforts to read the whole animal from its tracks. Having established that this holistic approach helps us understand lesser known predinosaurian species, we shall see just how effective it is for tracking dinosaurs and establishing a thoroughly intelligible relationship between their footprints and their entire anatomy.

FIBONACCI'S FINGERS:
THE SACRED GEOMETRY OF HANDS AND FEET

All science requires faith in the inner harmony of the world. Our longing
for understanding is eternal.

Albert Einstein

We have already introduced the mystery of fingerprints as potential clues
to a deeper understanding of the biological and psychological depths of
individuals of our own species. Now let us turn to the "deep structure" of
hands and feet and look at the bone structure in terms of mathematical
proportions.

The numerical series 1,1,2,3,5,8,13,21,34,55,89,144 . . . in which each
number is the sum of the two that immediately precede it, is called the
Fibonacci series. Many mathematicians independently noted this series,
actually a ratio, or proportion. The thirteenth-century mathematician
Leonardo Fibonacci (ca. 1170–1240) is remembered for defining it. In the
Fibonacci series the ratio between consecutive numbers becomes constant
at the value of .618 (known as phi: ∅) as the series progresses. In nature
this mathematical relationship shows up in the anatomical design of
many species ranging from sunflowers and pine cones to spiral shells,
pentagonal sea urchins and starfish, and even human hands.

The ratio of the length of one side of the pentagon to any diagonal is
also ∅. Euclid's "divine" or "golden section," also known as the golden
mean, is based on so-called "golden" rectangles, those with sides in the
ratio 1:0.618 (or 1:1.618). If the short side is 1, the long side is 1.618, and
if the long side is 1, the short side is .618. (One can approximate such a
rectangle oneself by placing the tips of one's index fingers on the tips of
one's thumbs.) These golden proportions manifest in nature have been
considered sacred since the time of Pythagoras. This same ratio corresponds to a musical interval of a fifth.

It is quite extraordinary that such a unique mathematical relationship
keeps cropping up in nature and in geometric figures such as rectangles
and pentagons, and it is no wonder that mathematicians have been fascinated by it for thousands of years. What interests us in relation to hands
and feet is that the ratio between our finger bones, called phalanges by paleontologists, follows the series 1, ∅, ∅2, ∅3. Thus, there is a hidden mathematical order to the anatomy of our pentadactyl ("five-digit") hands that
is related to the five-point geometry of the pentagon. Mere coincidence or
a manifestation of the deep structure of the universe? Perhaps the seventeenth-century German astronomer Johannes Kepler said it best in *Mys-*

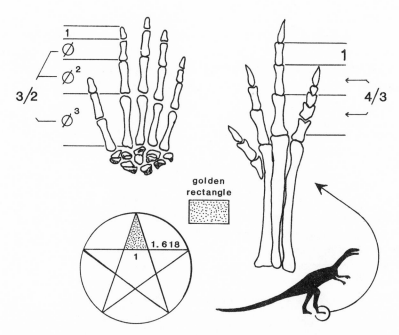

FIGURE 4.1   The phalanges (bones) of human fingers reveal a connection between pentagonal geometry and pentadactyl hands. Similar geometric relationships can been found in the feet of dinosaurs.

terium Cosmographicum: "Geometry has two great treasures: one the theorem of Pythagoras, and the other the line divided into mean and extreme ratio. The first may be compared to a measure of gold; the second to a jewel."[1]

If, as Mae-Wan Ho has suggested in Chapter 1, all animals are indeed ordered, liquid crystalline wholes, then the hidden geometry of human hands might be reiterated throughout the vertebrate kingdom in many forms. We now know that number and length of digits and phalanges changes in the process of evolution. For example, in the feet of some carnivorous dinosaurs (theropods) the average ratio between adjacent phalanges is consistently 1.33:1, or 4/3 (not 3/2, as one might expect). The theropod foot is well known for having four toes, but typically leaving only a three-toed track. One can go further: the 4:3 ratio, also known to the Greeks as the *diatessaron*, relates to the classic 3–4–5 triangle, and a musical fourth. By further coincidence the phalangeal formula for theropod toes II, III, and IV is 3–4–5 if we include the claws, and, of course, the track is three-toed, with the toe tips forming a triangle. The Great Designer is no slouch when it comes to mathematics.[2]

I shall not enter into mathematical analysis of tracks, but I do note that at least one analysis of the proportions of an *Allosaurus* skeleton reveals numerous Fibonacci ratios.[3] The point is that we can appreciate biological form through the lens of mathematics and geometry. Feet are expressions of the coherence of parts with higher orders of wholeness, and we can appreciate such order and harmony without having to be accomplished mathematicians.

THE FIRST DINOSAUR TRACKER

What a wonderful menagerie! Who would believe that such a register lay buried beneath the strata? To open the leaves, to unroll the papyrus, has been intensely interesting.... [Y]et the volume is only partly read. Many a page I fancy will yet be opened, and many a new key obtained to the hieroglyphic record. I am thankful I have been allowed to see so much.

**Edward Hitchcock, *A Report on
the Sandstone of the Connecticut Valley
Especially Its Fossil Footmarks***

In 1802, when a local farm boy, Pliny Moody, unearthed both large and small tracks at a field in South Hadley, Massachusetts, many of which looked like bird tracks, onlookers joked that the Moodys must have had very heavy poultry if their tracks could be impressed in stone. Local residents soon dubbed the maker of the birdlike tracks Noah's Raven. Hitchcock also regarded the tracks as those of birds; his first scientific paper, published in 1836, was titled "Ornithichnology-Description of the Foot marks of Birds (*Ornithichnites*) on new Red Sandstone in Massachusetts."[4] In it he said he was "much gratified by some unexpected disclosures . . . precisely resembling the impressions of the feet of a large bird in mud." Although often labeled Triassic by later researchers, in fact the tracks Hitchcock studied were all from the Lower Jurassic.

From the beginning we see Hitchcock puzzling over the best way to understand these classic footprints. In a marvelous yet apparently inadvertent pun Hitchcock cautions that "the geologist should be the last of all men to trust to first impressions."[5] Alas, geologists have too often gone to the other extreme of ignoring tracks completely. Hitchcock's work was considered flawed because of his interpretation of the track makers as birds rather than dinosaurs. But this judgment is too harsh. In the 1830s, 1840s, and 1850s giant fossil birds, including the New Zealand moa (*Dinornis*) and the Madagascar elephant bird (*Aepyornis*), were discovered, leading everyone to believe that giant birds had been abundant in pre-historic times. The fact is that all Hitchcock's most eminent contemporaries, including William Buckland, Richard Owen, Charles Lyell, and even Darwin himself, supported his interpretations of the tracks as those of birds.

Some discussions of Hitchcock's work also imply that his scientific conclusions were somehow compromised because he was a deeply religious man, troubled by a perceived atheism or pantheism among scientists, and too much influenced by the message of the theologian William Paley's book *Natural Theology: or evidence of the existence and attributes of the Deity collected from the appearances of nature.*[6] We have seen that contemporaries such as Buckland wrote on mineralogy and geology with reference to the power, wisdom, and goodness of God. Far from being archaic, such a perspective is in line with the views of later cosmologists who, like Albert Einstein or Paul Davies, want to know "the Mind of God."[7]

So Hitchcock should not be singled out for mixing theology and science. On the contrary, an argument can be made that the habit of separating science from theology is a bigger problem because it leads to fragmentation of thinking. The writer-philosopher Arthur Koestler, in *The Sleepwalkers,* suggests that thanks to Plato and his followers, the Pythagorean unity of science and divinity was undone in favor of a formal separation. While rationalists profess that the separation of science and theology is absolutely necessary, holistic philosophers warn of the dangers of dualism.

There was nothing wrong with Hitchcock's science. His *The Religion of Geology and Its Connected Sciences* was published quite separately from his papers on tracks.[8] He wrote of "our imaginations . . . carried by these relics, to that immensely distant period when feathered tenants . . . once occupied the now delightful valley of the Connecticut," noting correctly that "the scriptures are silent [on] a period of indefinite duration between the beginning and the creation of man; and geological monuments . . . furnish us with few chronological dates." Such observations neither deny the geological evidence nor try to mold it to fit a narrow biblical interpretation. On the contrary, they simply point out what is not yet known. He continued objectively: "If birds lived during the deposition of the new red sandstone, they doubtless existed during the formation of each successive group of rocks to the highest. . . . [S]urely the geologist will be led to inquire whether he has been too hasty in inferring the non-existence of . . . animals and plants, in earlier times of our globe."[9] With the hindsight of countless discoveries since Hitchcock's day, these comments are sensible geological prophecy uncomplicated by religious dogma.

BIG BLUE BANANAS IN THE GREAT RIFT VALLEY

There is not in the whole wide world a valley so sweet
As that vale in whose bosom the bright waters meet.

Thomas Moore,
"The Meeting of the Waters"

If the chronology of Earth's history cannot be found in theology, how can scientists find a reliable guide for dating events that took place millions of years ago? One of the best places in the Earth's crust to track time is in areas called rift valleys. Rift valleys are places where forces on the Earth's crust make it open up, thus creating new geological territory. The creation of a rift valley is like the Earth's giving birth. Rift valleys are fertile and vibrant terrain, alternately shaken by volcanic eruptions, faults, and fracturing and soothed by the flow of water into new lakes that give birth to new life. East Africa's great rift valley—often simply called the Rift Valley—is considered the cradle of humanity, for this was the home of our earliest hominid ancestors. It was created when powerful plate tectonic forces split East Africa from the Red Sea, almost down to Cape Horn in South Africa.

In late Triassic and early Jurassic time just such a giant rift was opening up all along the eastern margins of the North American continent from what is now the Carolinas to Nova Scotia. At that time the entire region was in contact with the continent of Eurasia, but the Earth was already programmed by tectonic forces to open inexorably, first creating the Atlantic Rift, which flooded like the Red Sea, and then widening steadily to become the Atlantic Ocean.

In much the same way that hominids later emerged in the East African rift, so dinosaurs emerged from the great Triassic-Jurassic rift system in North America to make their first appearances. As well, the science of vertebrate ichnology was born in this great rift complex, providing evidence of dinosaurs long before any bones were unearthed.

The importance of the rift basins owes much to a happy combination of geological and biological circumstances. Using drilling techniques unheard of in Hitchcock's day, a remarkable geological story has been wrung from the rocks of the rift in the Newark Basin along the present-day Eastern Seaboard. A team from Columbia University made an intensive study of this very complete rock sequence that revealed a remarkably detailed and continuous record of earth's climatic and magnetic history over a span of approximately 30 million years.[10] Much of this record accumulated in large lakes in the Newark Basin, like Lake Tanganyika and others that are found in the East African rift today. Some of these lakes were shaped like a banana, so the Columbia team christened one in the Newark Rift Basin the Big Blue Banana, a fitting fruit for a rich rift then situated in the tropics.[11]

This ambitious drilling project extracted a sequence of rock core some three-miles deep, representing 30 million years of climatic and magnetic history, or half the time since the dinosaurs died out. As the planet rotated, regular changes in Earth's orbital geometry, known as the Milankovitch climatic cycles, affected the amount of solar radiation received at the Earth's surface.[12] These in turn affected the climate and caused lake levels to rise and fall in a cyclical pattern. These cycles are well known and include the precession of the equinoxes (cycle duration about 26,000 years), the eccentricity of the ecliptical (about 109,000 years), and longer 413,000- and 2-million-year cycles.

Using our "24-hour" clock that represents the time since the Big Bang, the 30-million-year-history of the Newark Basin spans the interval from about 10:45 to 10:55 P.M. There were more than seventy 413,000-year cycles in the history of the Big Blue Banana lake. Each one represented only about eight seconds on our 24-hour timescale, and each of the 1,500 shorter (26,000-year) precessional cycles represents a mere .4 second. This is what geologists call "high-resolution" measuring, or discrimination, of time intervals. It is analogous to a dendrochronologist's finding a 1,500-year-old tree and taking a core sample to use the tree rings to distinguish wet and dry years. Likewise, geologists can read wet or dry cycles in the core from the Newark Basin.

The Columbia team was also able to use the Newark core sequence to reconstruct Earth's magnetic history. As sediments were laid down as individual grains, especially those containing magnetic mineral material, they aligned themselves with the Earth's magnetic field, either "normally," as they would do at the present time, or in reversed alignment, reflecting a "flip" of the field. This has allowed the rocks to be calibrated with an accuracy that would have amazed Hitchcock. The sequence is the best standard available for measuring and subdividing Late Triassic and Early Jurassic time.

All that now remains is to fit our tracks into this very precise time frame. In so doing we should be able to establish how long particular track makers were around before being displaced by other groups. Some estimates put the average longevity of dinosaur species at about 7 million years, equivalent to 350 precessional cycles. (The data of Paul Olsen, on the range of various tracks through this rock sequence should help confirm or adjust this estimate of the dinosaur species' longevity.) The range of particular track types helps establish track zones. If a dinosaur generation lasts about ten years, so some 2,600 dinosaur generations might be represented in 26,000 years, the length of a precessional cycle. Thus, we can estimate that 1,500 such cycles represent more than 3 million dinosaur generations.

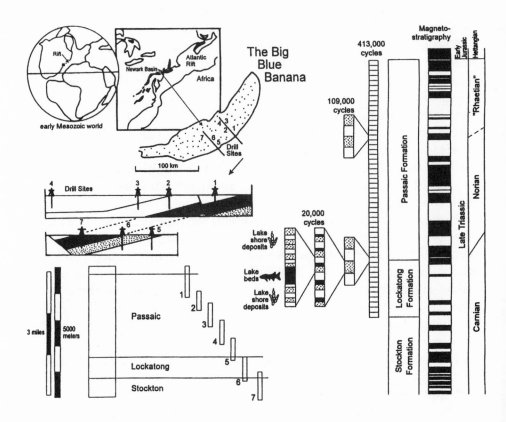

FIGURE 4.2   Peeling the Big Blue Banana: An ambitious drilling project unravels the 30-million-year history of a great rift valley, down to details of each magnetic reversal and 20,000-, 109,000-, 413,000-, and 2-million-year climatic cycles. Rock sequences are replete with tracks and other fossils at literally thousands of different levels.

GRALLATOR:
THE "EARLY BIRD" THAT WENT ON STILTS

The Grallae above all other birds, might be expected to frequent the muddy shores of the estuary or lake where doubtless the fossil footmarks were formed.

Edward Hitchcock, A Report on the Sandstone of
the Connecticut Valley Especially Its Fossil Footmarks

Hitchcock gave the name *Grallator* to tracks bearing a close relationship with water birds of the family Grallae (literally, "one who goes on stilts"). Anyone who has seen a well-preserved example of an Early Jurassic *Grallator* track will agree that "[t]he fine-grained muddy sediments appear to be particularly well suited to preserving the finest details of footprints with exquisite clarity." The tracks "exhibit two phalanges on the inner toe, three on the middle, and four on the outer . . . as (in) the feet of living birds."[13] *Grallator* tracks are narrow and elongate, more like those of songbirds than those made by the widely splayed feet of water birds. Trackways are also extremely narrow with long steps. Such a pattern, which indicates a fast-moving and very erect animal, prompted Hitchcock to name one species *Grallator cursorius*, indicating a running, or "cursorial," ability.

Let us adopt our holistic Schadian perspective to examine the theme of narrowness.[14] Small tracks of the *Grallator cursorius* type are typically ascribed to the well-known carnivorous dinosaur (theropod) *Coelophysis* (the New Mexico State fossil) or a close relative. Standard depictions show *Coelophysis* to be an extraordinarily slender, elongate, narrow-bodied animal with narrow skull and narrow teeth.[15] Looked at from above it is almost a spinal chord and not much else. Add to this the narrowness of both the track and trackway and we have an integrated theme of "narrowness." The narrowness of the foot and footprint is accentuated by having only three toes in contact with the ground, unlike the tracks of many other dinosaurs. The *Grallator* track maker sets the standard for ultra-narrowness. Assuming that *Coelophysis* is indeed the track maker, the correlation is qualitatively pleasing because the narrowness in track and track maker have been observed independently. As noted previously, elongate narrowness has also been reported as a distinctive quality in the feet of humans and in the tracks of the "hand animal." We shall see whether this line of reasoning holds up when we examine other tracks and correlate them with their probable makers.

TRUE THUNDER

... deep scars of thunder ...
John Milton, *Paradise Lost*

Eubrontes giganteus ("gigantic true thunder") is the name of the print of a much larger theropod first found in the Connecticut river valley region. In contrast to slender *Grallator cursorius,* which measures only 3 inches (7–8 cm) long, *Eubrontes* is large—up to 16 inches (35–40 cm) long—and comes from an animal that is not only bigger but broader. Hitchcock

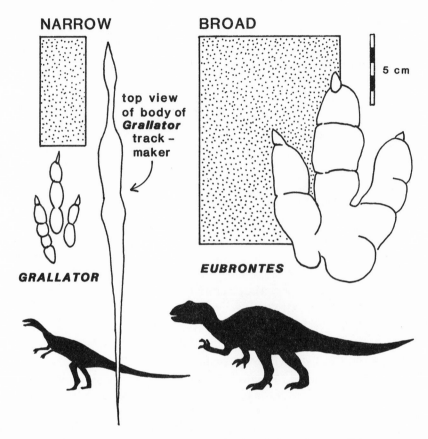

FIGURE 4.3 *Grallator cursorius* tracks suggest a slender, stilt-legged runner. *Eubrontes giganteus,* by contrast, represents a thunderous giant.

regarded the giant track with awe, reporting that "I threw [it] aside at first, because I could not believe that an impression three or four times larger than that of the great African ostrich's foot, could be a track.... For a time I regarded it as the giant ruler of the valley. But subsequently I have found that the tracks of others might have successfully disputed the palm of superiority with it."[16] Hitchcock might have been interested to learn that the Navajo people use the term "*yetso beta*" (meaning "God's tracks") to describe the same tracks in the western United States.[17] He would also be pleased that *Eubrontes* has been adopted as the emblem on the Connecticut state flag flying over Dinosaur State Park in Rocky Hill, Connecticut, a track site now open to the public.[18] "God's truly thunderous track" carries the flag for the little state where the science of dinosaur tracking was born.

When we compare *Eubrontes* with *Grallator* we find that it is not just a scaled-up version of its smaller contemporary—it is proportionally much wider, the width being 65 percent of the length, as compared with 42 percent for *Grallator*. Thus, *Eubrontes* is at least a third wider than *Grallator*. What do such differences mean, and can they help us identify differences between the track maker that go beyond mere size? Some say that as animals get bigger they generally grow wider at a faster rate than they grow narrower. This is used as an argument that *Eubrontes* is just the spoor of an adult of the species that made *Grallator* tracks.[19] But this seems unlikely because we often find lots of small *Grallator*, a few large *Eubrontes*, and nothing in between. If they were all growing up together we would expect intermediate sizes. The alternative explanation, seen in the "hand animal," in humans and, as we shall soon see, in brontosaurs, is that species tend to fall into the "broad" or the "narrow" category.

THE GIANT ANIMAL

It is interesting to observe what progress has been made in tracing out the character of the tracks of this species.

Edward Hitchcock, *A Report on the Sandstone of the Connecticut Valley Especially Its Fossil Footmarks*

In the previous chapter we learned of the footprints of the "first giants," described by the delightful French tracker Paul Ellenberger. These giant footprints mostly represent species of prosauropods, forerunners of the brontosaurs that were at least partly bipedal. Similar tracks from Lower Jurassic strata in New England, named *Otozoum* ("giant animal"), are among the most famous of this epoch. In Hitchcock's original paper we see his reaction to the discovery of footprints that were entirely new to science, and were, moreover, attributable to animals that were completely unknown and undreamed of.[20] We also see how these extraordinary finds prompted Hitchcock to ponder the very foundations of his new science and develop a coherent tracker's philosophy that justifies the interpretation of the prehistoric world through the fossil footprint record.

He described *Otozoum* as the "most extraordinary track yet brought to light in this valley," representing a "bipedal animal . . . distinguished from all others. . . in the sandstone of New England" by the impressions of four thick toes directed forward.[21] He also noted the excellent quality of preservation of some of the tracks, which reveal details of the skin, pads, and even raindrop impressions beautifully exhibited on the slab. "So peculiar is the shape of the track . . . and so different from those of the feet of any living

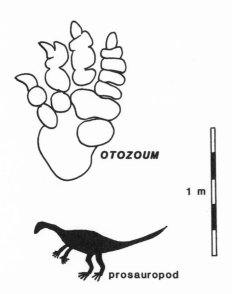

OTOZOUM

1 m

prosauropod

FIGURE 4.4 *Otozoum*, the giant animal track, provided Hitchcock with evidence that some Connecticut River valley track makers were large reptiles, not birds.

animal ... that I should hardly have dared describe it from a single specimen ... had I not found its essential features exhibited in other tracks." We see clearly that Hitchcock realized that these tracks were not in the least birdlike. He realized "more forcibly than ever [that] these animals ... may ... have ... characters ... found in some of the Saurians." Here Hitchcock anticipated the modern view that the track maker was a dinosaur.

The magnificence and audacity of *Otozoum*'s very existence helped Hitchcock formulate a tracking philosophy that led to the founding of the science of vertebrate ichnology as we know it. His 1845 article is full of gems of tracking wisdom that provide food for thought and questions that vertebrate anatomists might find a little provocative such as "Why is it not as desirable, and as consonant to the laws of zoology and comparative anatomy, to derive the name of an extinct animal from its tracks, as from a fragment of a skeleton?" Hitchcock reveals that he was "surprised ... to learn that some object to giving scientific names ... to ... footmarks ... or the animals that impressed them." He disagreed and gave eight "reasons for a contrary opinion," which remain valid in principle and practice today:[22]

1. Tracks demonstrate the existence of certain extinct animals.
2. Facts concerning the skeletal remains render it certain that the track-making animals have not been described.
3. All who have seen the tracks are convinced that they represent a diversity of species. Such diversity can and should be expressed in words, thus names for distinct track types must be given.
4. "Comparative anatomy teaches us that some of the surest and most constant characters by which animals are distinguished, are derived from their feet."
5. "Living animals could to a great extent be divided into families, genera and species, by their tracks." (Our Bushmen friends would surely agree wholeheartedly).

6. If no fossil animal were to be named until we had a complete skeleton and description of it, we would hardly name any, and most of those already named would have to be stricken from the list.
7. There is precedent for naming animals based on tracks, as in the case of *Chirotherium*.
8. Scientific names are necessary for communication.

It is not necessary to be an experienced tracker to appreciate the subtleties of Hitchcock's arguments. However, it is another matter to clearly establish the identity of the *Otozoum* track maker—or indeed any animal known only from its tracks. The paleontologist Richard Swann Lull of Yale revised Hitchcock's work in 1953, suggesting that the track maker might have been a prosauropod, a position consistent with the views of Ellenberger and the global proliferation of prosauropods at this time.[23] However, in recent years it has been suggested that the *Otozoum* was a crocodile, an ornithopod dinosaur, or an armored dinosaur.[24]

Obviously, the maker of *Otozoum* tracks, however versatile, could not assume all these identities. I prefer the prosauropod interpretation, not least because the *Otozoum*'s giant maker appears almost always to have been bipedal, as the prosauropods are generally thought to have been. But let us look at the problem using Schad's approach.[25] The theropod described previously was small and narrow. The tail is long in comparison to the body and the posterior limbs are specialized for bipedal locomotion (posterior-pole emphasis). Much debate surrounds the high abundance of tracks, and it has been suggested that the track makers were very active. Theropods belong to a group known as the saurischians ("lizard hips"), which includes the sauropods, or brontosaurs, at the large end of the size range. They are called "lizard hips" because of their typical reptilian pelvic anatomy. Sauropods were all large and quadrupedal herbivores with very long necks (anterior-pole emphasis). Their limbs were well developed, but they generally moved slowly, and much debate surrounds the efficiency of their digestion and metabolism. Whereas the carnivorous theropods made narrow trackways with their narrow, skinny feet, herbivorous sauropods made broad trackways with broad feet and had footprints with five toes (instead of three, like the theropod) enclosed in broad fleshy pads. Sauropods even had round eggs, whereas those of theropods were long and narrow!

If this Schadian analysis is to be taken seriously, an intermediate group (i.e., the putative prosauropod makers of *Otozoum* tracks) can be inferred to have intermediate, alternating, or rhythmic characteristics. It appears that the prosauropods fit this bill. They are intermediate in size and many aspects of morphology. There is much debate about whether they were quadrupeds or bipeds, and trackway evidence suggests they switched back and forth, acting like theropods at one time and like true sauropods at another time. There is a comparable history of debate as to whether they were carnivores or herbivores. Their teeth had pronounced broad serrations intermediate in character between those of theropods, which are finely serrated, and the smooth surfaces of sauropods' teeth. They also have four toes, between three (theropods) and five (sauropods). Thus, it appears that an evolutionary sequence is recorded for theropods, prosauropods, and sauropods that is similar to the one we saw in the development of small rodents, carnivores, and then large ungulates.

ODD FOOT

Those with unequal feet, as the Kangaroo, have five toes in the fore foot, which are placed, when the animal brings them to the ground, as in the *Anomoepus.*

Edward Hitchcock, *A Report on the Sandstone of the Connecticut Valley Especially Its Fossil Footmarks*

It was not only *Otozoum* that presented a puzzle for Hitchcock by seeming to originate with track makers that were clearly not birds. *Anomoepus* (literally, "unlike feet") has hind feet like a bird, but small front feet that left five-toe impressions.[26] Understandably, Hitchcock reported that he had "been more perplexed to determine the . . . mode of progression of the animals of this genus, . . . than . . . almost any other."[27] Let us try to understand this odd animal.

In Hitchcock's day, marsupials were viewed as primitive in comparison to placental animals. Hitchcock thought that the footprints might have been made by a kangaroolike animal, and so ascribed the *Anomoepus* track maker to a group he characterized as "ornithoid marsupialoids." Here we have, at least nominally, a truly odd and entirely hybrid creature—half bird, half mammal. But we should remember that mammals and birds both arose from reptilian ancestors, so why could not a third group have arisen from the same ancestors? Hitchcock's interpretation is quite reasonable from some perspectives, because we now know that the *Anomoepus* track maker is considered an ornithopod, a dinosaurian group with both bird and some mammalian characteristics. As we shall see in later chapters, it is not unusual to find zoological groups that exhibit a blend of characteristics normally associated with entirely different groups. For example, bats are mammals with pterosaur-like wings, and dolphins are mammals that resemble fish.

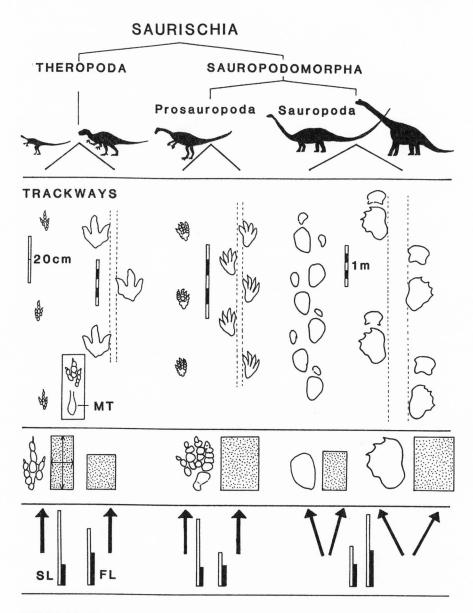

FIGURE 4.5 The principle of narrowness and breadth illustrated through the spectrum of theropod, prosauropod, and sauropod tracks. (Compare with Figure 7.5.)

If we look through our holistic lens we will see that Hitchcock was quite astute in describing what he saw. *Anomoepus* footprints are similar in size to *Grallator* tracks, but they generally have more widely splayed toes on the hind foot and five toes on the front foot. Both *Anomoepus* and *Grallator* indicate track makers that had the tendency to squat down and make

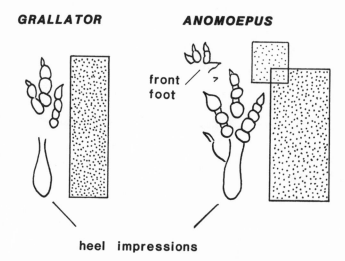

FIGURE 4.6   A comparison between Anomoepus and Grallator tracks that show heel impressions.

long heel (metatarsal) impressions, so in a sense the tracks sometimes have a tail. But in the case of *Anomoepus*, the manus print is a very significant anterior component. As we know, this anterior component in quadrupeds suggests greater general width, so it is interesting that we find that the toes are widely splayed. We also find similar patterns reiterated among birds.

THE DIMINUTIVE DUNE RUNNERS

He prayeth best, who loveth best
All things both great and small:
For the dear god who loveth us,
He made and loveth all.

Samuel Taylor Coleridge,
"Answer to a Child's Question"

We tend to think of dinosaurs as big, but in fact many of Hitchcock's poultry-sized track makers were obviously small creatures. There is also a stereotype of Mesozoic mammals as small, rodent-sized creatures, which seems well supported by the fossil evidence. This explains why their tracks are small and hard to find. The size of animals in a given area can also depend on ecological factors. In modern deserts, most vertebrates are small mammals such as gerbils, jerboas, or kangaroo rats and small reptiles such as lizards; they share their world with spiders, scorpions, and other arthropods.[28] Footprint evidence from Mesozoic desert deposits indicates similar ecologies.[29] We saw that Paleozoic desert deposits contain many cat-sized protomammal tracks and spider trails. We find almost identical track assemblages in Jurassic sand-dune deposits in both North and South America.

Father Giuseppe Leonardi became one of the first to study such tracks when he discovered a bonanza of footprints during his ecclesiastical wanderings through Brazil's São Paulo province in the 1970s. In the town of Araraquara he came upon numerous footprints in flagstones quarried for use as paving stones (flagstone is a type of slabby sandstone). Intrigued, he soon found the source of the tracks in quarries preserving ancient desert sand-dune environments. The sandstone, named the Botucatu Formation after the town of Botucatu, 60 miles (100 km) to the south of Araraquara, is found in a large region on the order of a half million square miles.

Intent on obtaining samples for his research, Leonardi suggested removing track-bearing pavement and replacing it with barren rock of the same type, but was warned that any meddling with the level of the pavements could be dangerous. Unsuccessful with the honest approach, he decided to resort to subterfuge: he traced out the shapes of coveted slabs on newspaper and prepared substitute pieces in the same shape so that he could make the switch when a convenient opportunity arose.[30] During subsequent fiestas he masqueraded as a city worker and switched the slabs. He explained to curious onlookers that he had to replace the slabs that had irregularities (tracks) because they were a hazard to public safety. Before he had finished Leonardi even resorted to operating at night with a map of the town on which he had marked the actual street addresses of all the important specimens. He was not stealing the tracks, but "saving them for science." After the inauguration of a new mayor, he finally obtained official permission to continue his slab substitutions legally.

Leonardi named the tracks *Brasilichnium*, meaning "Brazilian trace," and identified the track maker as one of the last of the protomammals. When first discovered in the Botucatu Formation, they were thought to be from the Permian, the first so-called Age of Reptiles, because they so closely resembled the much older protomammal tracks, *Chelichnus* and *Laoporus* (discussed in Chapter 2). In South America and North America Permian, Triassic, and Jurassic sand-dune deposits look almost identical and are very hard to date. Therefore, to assess the ages of the formations Leonardi looked at the declining proportions of protomammal tracks and the growing number of dinosaur footprints. Eventually the age of the Botucatu was established as Lower Jurassic, and tracks found there could be compared with footprints from the Lower Jurassic Navajo Sandstone of the American Southwest. Named after the Navajo people and their nation in the Four Corners region of Arizona, Utah, Colorado and New Mexico, this formation is regarded by geologists as a classic example of fossilized sand dunes. It is now abundantly clear that mammal-like tracks found there are very similar to *Brasilichnium*.

Leonardi and another South American tracker, Rodolfo Casamiquela, also recorded tracks attributable to small dinosaurs and to small, gerbil-sized hopping mammals.[31] Theropod dinosaur tracks no more than 1.5 inches (3.5 cm) long in the Navajo Sandstone of Arizona indicate an animal with a foot the size of a starling or sparrow and a hip no more than 6 inches (15–16 cm) above the ground. Long steps suggest that they were moving at four to six miles per hour (6–9 kph), a considerable speed for a diminutive animal. Casamiquela also reported diminutive dinosaur tracks from Middle Jurassic deposits in southern Argentina. Like some tracks from the Navajo Sandstone, these are also tiny, starling-sized footprints. They tell of a world of diminutive dinosaurs living among sand dunes and perhaps feeding on spiders, scorpions, and other arthropods, and perhaps even dining on the last of the protomammals. Such reports show ecological adaptations and behaviors reminiscent of vertebrates in modern desert environments.

HOP TO IT

It remains possible that some bipedal dinosaurs used a hopping gait, but there is no very convincing evidence that they actually did so.

Tony Thulborn, *Dinosaur Tracks*

Talk of sparrows and starlings allows us to imagine tiny creatures weighing a few ounces that sometimes hopped, rather than walked, through the Jurassic deserts. Leonardi's report from Botucatu of three-toed footprints only .75 inches (2 cm) in length suggests a hopping creature with feet half the size of the diminutive dinosaurs just described.[32] Leonardi thought it was probably a dinosaur, but there is little evidence that dinosaurs hopped, so it might have been a mammal.[33] Even smaller tracks, from Argentina, that Casamiquela discovered and named *Ameghinichnus* (in honor of the paleontologist Carlos Ameghino) measure less than .5 inch (1 cm) and clearly indicate five digits, which is highly suggestive of mammal affinity.[34]

Mesozoic trackway evidence for hopping vertebrates raises fascinating questions about the locomotion of reptiles, protomammals, mammals,

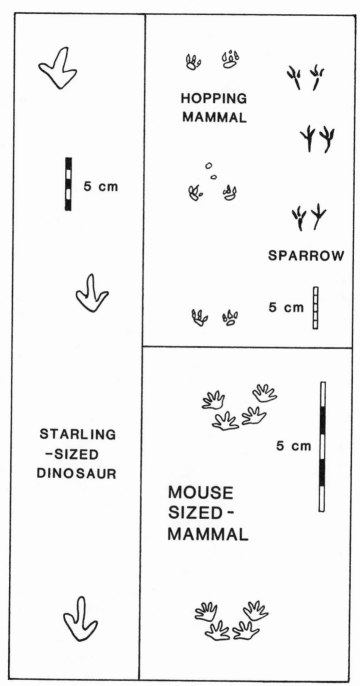

FIGURE 4.7 Tiny sparrow-sized tracks (left) from North and South America, provide evidence that diminutive theropod dinosaurs inhabited desert sand dunes. Tracks of tiny hopping vertebrates (right) suggest that some mammals adopted a jerboa-like locomotion early in the Jurassic.

and even birds. Paul Ellenberger first reported footprint evidence of hopping mammals from the earliest Jurassic of Southern Africa.[35] In light of the South American discoveries of such well-preserved tracks one might have anticipated a flurry of interest among mammal paleontologists. Only one study, in 1994, by the Polish paleontologists Z. Kielan-Jaworowska and P. P. Gambaryan, noted that the tracks resemble those of the modern field vole (genus *Microtus*) and that the subject of hopping locomotion among small rodentlike Mesozoic mammals is of more than passing interest.[36]

In each group that exhibits hopping locomotion, we can ask when did it originate? Tracks appear to answer the question at least for mammals, and may eventually answer the question for frogs and birds. Mammals, frogs, and birds all represent small animals that began hopping early in their evolutionary history, during the Triassic–Jurassic transition or somewhat later in the case of birds (probably in the Early Cretaceous). This novel behavior also appeared in three groups that had been around for a long time. The mammal is a new expression of a protomammalian form, the frog a new type of amphibian, and the bird a new type of bipedal dinosaur (or reptile). Does the fact that these new evolutionary products of previously well-established groups show such novel yet convergent behavior mean that hopping is an inherent "emergent" tendency at some deep level of biological organization? There are a few larger hoppers such as kangaroos and rock-hopper penguins, but they are not widely distributed today, nor is there much compelling fossil or track evidence for large hoppers. As with birds, mammals, and frogs, the reason these larger mammals began hopping is a mystery.

The idea of hopping dinosaurs was first suggested by M. A. Raath, who inferred a kangaroolike gait for a theropod based on a pair of tracks found side by side in Lower Jurassic deposits of Rhodesia.[37] As he did not illustrate a trackway with consecutive pairs of left and right feet impressed simultaneously, the possibility exists that the tracks represent two different animals' trackways, impressed side by side.

The debate about hopping dinosaurs deepened in 1984 when French paleontologists reported tracks of a hopping dinosaur from Late Jurassic limestone deposits in the region of Cerin, near Lyons.[38] They even named the trackway *Saltasauropus* ("track of a jumping reptile"). The trackways consist of pairs of large, widely spaced, three-toed tracks quite different from normal trackways of walking dinosaurs. At about the same time these same authors also described the trackway of a giant turtle. The Australian tracker Tony Thulborn very sensibly suggested that the tracks had been made by a swimming sea turtle, touching the substrate with synchronous strokes of its flippers.[39] If we accept Thulborn's compelling interpretation, we are left with little convincing evidence that dinosaurs ever developed hopping gaits, except among their avian descendants known as birds.

LEFT LIMP

Stepping . . . off on your right foot . . . was the practice of Western military man until King Frederick William I of Prussia . . . decided something drastic was needed to instill the kind of discipline that wins wars. What better tactic than to make recruits put not their best foot forward, but, in the eyes of the Western world their worst [i.e., the left]? The Prussian style swiftly spread, giving rise in one European country to the legend that its soldiers were so dense that they could only march if hay were tied to one foot and straw to the other, and the command given, Hayfoot, strawfoot!

Jack Fincher, *Lefties: The Origins and Consequences of Being Left-Handed*

Statistics show that left-handed people are more accident-prone than right-handed people and that they also have a slightly shorter life expectancy. Jimi Hendrix is perhaps an example of shorter life expectancy; other famous successful lefties Leonardo da Vinci, Paul McCartney, Lord Nelson (who was sufficiently accident-prone to lose his right arm), Pablo Picasso, Babe Ruth, and Queen Victoria lived much longer.[40] If we asked any of these people to kick a ball we might find that they were also left-footed. When we look into the whole question of handedness or footedness, we find that we live in a fascinatingly complex, ambidextrous universe. Everything from the rotation of stars and planets to the spiraling of seashells and the coiling of molecules displays a preference for left or right, depending on environmental and developmental conditions and a host of other influences, many of which we don't yet understand.

Perhaps it is therefore not surprising to find that many dinosaur trackways show subtle or not-so-subtle signs of animals that were favoring one foot over the other. Such trackways are characterized by alternating long and short steps, indicating a pronounced limp or perhaps something more subtle, such as left- or right-footedness. In a few cases there is clear evidence that a track maker limped because of some accident or congenital abnormality. The best example is the trackway of a Lower Jurassic theropod from Morocco that had an injured right foot, shown by the abnormal impression made by the inner toe on the right foot.[41] As a result it took short steps when it transferred its weight from the left foot to the right—i.e., it stepped gingerly onto its right foot—but took normal steps when moving off its right foot onto its left. Anytime we suffer a foot injury we also adjust our locomotion—the amount of weight we may put on the injured foot—to avoid pain, and we may do this without thinking consciously of the mechanics involved. Imagine then how much less we think about subtle differences in our trackways that might result from unconsciously favoring one side more than another, or carrying an object.

We have found evidence of irregular, or "abnormal," gaits among theropods, sauropods, and ornithopods of both the Jurassic and Cretaceous ages, but it is impossible to say why such individuals limped, unless there is obvious evidence of a foot injury.[42] It is possible that the animal had suffered an injury to another part of its body, or perhaps had one leg a little shorter than the other. The possibility that limpers were carrying something or looking over their shoulders while walking can also not be ruled out.

Let us assume that, as hypothesized, the short step led to, rather than away from, the injured foot; of the nine reports of lop-sided trackways so far documented, six suggest an inclination to favor the left side. Could this mean a greater proportion of injuries to left footed animals? Perhaps dinosaurs were like people in this regard, and lefties were more accident-prone. Such a suggestion is highly speculative, but it does give food for thought. I suspect that trackways indicating footedness are far more common that previously supposed, and that much more data on left- and right-footed dinosaurs and other extinct vertebrates is readily available.

THE WORLD'S LONGEST BRONTOSAUR TRACKWAYS

A newly discovered dinosaur track site in a quarry near Fatima, Portugal, represents the largest and most important Middle Jurassic tracksite currently known. . . characterized by the longest sauropod trackways known anywhere (147 m and 142 m). . . . Due to the large size of the site, [and] the long and visually spectacular trackways . . . the site has great potential as an outdoor interpretive center. With optimal illumination the site resembles a mudflat only recently crossed by dinosaurs.

> **V. F. Santos et al., *A New Sauropod Tracksite from the Middle Jurassic of Portugal***

Portugal is noted for the performance of its long-distance track and field stars. Men and women athletes have excelled at track events from the 5,000-meter race to the marathon. Dinosaur trackways from Portugal have also been getting into the record books. In 1992 Portugal claimed the world record for the world's longest trackway, when a 417-foot (127 m) segment was reported from a site in a suburb of Lisbon. Excavation later extended the segment to 462 feet (141 m). But records are made to be broken, and in 1994 reports emerged of brontosaur trackways measuring 465 and 482 feet (142 and 147 m).

These Middle Jurassic tracks from near the sacred site of Fatima represent our first encounter with true sauropod tracks. The trackways are wide-gauge, with large front footprints and an inwardly turned claw impression on the front foot (manus). We established earlier in this chapter that the sauropods are the largest lizard-hipped dinosaurs (saurischians) at the "wide," "metabolic" end of the morphological spectrum. Our Schadian perspective would lead us to predict—on the basis of a wide trackway and large front feet, for which we have track evidence—a short body and tail to match the breadth.[43] It appears that this is the case, as wide-gauge trackways are associated with short-bodied forms with larger front legs (cetiosaurs and brachiosaurs) and narrow-gauge trackways are associated with diplodocids and apatosaurs, which were narrow forms with narrow heads, smaller front feet, very long tails, longer hind legs, and probably the capability of rearing up on their hind legs. The increasing "width" corresponding to a shift from posterior to anterior emphasis not just in the sauropods but throughout the saurischians (theropod, prosauropods, and sauropods) is nicely expressed not only in the increase in the number of digits but also in the shortening of toes and the forward movement of more flesh to enclose all but the anterior extremities of the digits. The pattern is very similar to that seen in the Triassic "hand animal," where width matches with increased manus size and greater fleshiness.

Portugal has suddenly burst onto the paleontological map, and since 1990 has designated five protected or "sacred" track sites, including the Fatima site. Fatima is already a sacred pilgrimage site because it is situated close to a field where many witnessed visions of the virgin Mary between May and October of 1917. The crowds of tens of thousands of religious pilgrims have recently been swelled by additional tens of thousands who visit to see the trail trodden by Iberia's oldest brontosaurs.

GLIMPSE OF A DINOSAUR FROM THE DARK AGES

Deep into that darkness peering,
Long I stood there wondering.

**Edgar Allan Poe,
"The Raven"**

Just as we can tell the story of a giant Middle Jurassic sauropod from Portugal, and describe some of its characteristics without having any skeletal remains of the actual animal in hand, so we can also tell the story of a small carnivore from Utah and England. In the entire United States there are no known Middle Jurassic dinosaur remains. It is a dark age in terms of yielding dinosaur evidence. Near Dinosaur National Monument in eastern Utah, however, at least 60 trackways of a small dinosaur have been reported.[44] These indicate a very distinctive foot anatomy with a shortened ankle bone on digit IV. So we can predict that the skeletons of individuals of this species of track maker—if we ever find any—would clearly show this characteristic. We can also predict that we might find the skeletons in Great Britain or in Wyoming, because that is where more of the same type of tracks have been found. Not only can we describe the feet of this unknown dinosaur from the dark ages in precise detail, but we may be able to make some surprisingly interesting inferences.

Why does it mean that this animal had a shorter ankle bone on digit IV? The holistic way of getting an answer to that would involve first simply describing the phenomenon: a posterior shortening of the outer digit. The track is shorter and wider overall than a typical small theropod track, with less outer and less posterior emphasis. The coherent laws of coupled biological form might suggest a corresponding posterior deemphasis and anterior emphasis of other skeletal elements. In birds the distal (foot) end of both inner and outer ankle bones is shortened in relation to the middle bone. Perhaps we are seeing the first subtle hints of a shift toward a bird-like anatomy, for as birds evolved from theropods their ankles became more elongate and erect than those of theropods. The shift made by birds away from the ground could be expressed in something as subtle as the first upward shift of the outer ankle bone. From the holistic perspective, no such subtle sign in a track is without potential significance.

chapter five
_____

LOST SOULS

ACT V: CAST OF CHARACTERS

This act takes place entirely in the Late Jurassic epoch (about 160 million to 140 million years ago). This epoch is known as the Age of Brontosaurs, but pterosaurs also fly prominently into the limelight and leave their footprints center stage. We shall also find Marco Polo tracking crocodiles in China, examine the tracks of the giant mule ridden by the Virgin Mary, and meet another great tracker-priest, Pierre Teilhard de Chardin, a man whose vision of the future of the Earth and mankind has been the subject of high praise and scornful ridicule. Those who are keeping count will note that we have met five tracker-priests so far. Perhaps it is a trend.

By the Late Jurassic there is considerable evidence of social behavior among dinosaurs, and we will find that tracks have played no small part in revolutionizing our perceptions of dinosaurian community life. In this chapter I shall also begin to broaden my holistic theme to suggest that we can see nature's coherence and integration at all levels: not just in the morphological resonance between foot and animal, nor just between gregarious individuals traveling in unison in a herd, but in the relationship between different species and their intimate relationships to the environment. As we shall see, this process of reading the ecology of tracks can be done on many different levels.

MEGALOSAURS AND MEGATRACKSITES

The original was probably carnivorous, and nearly thirty feet in length. As it was contemporary with . . . several . . . large saurians, it may have preyed on their young and on the lesser reptiles.

Gideon Mantell, *Geological Excursions Around the Isle of Wight and Along the Adjacent Coast of Dorsetshire*, 1847

One must think big in the Late Jurassic. The concept of a "mega-lizard" is older than the concept of dinosaurs or "terrible lizards" and dates from 1824, when William Buckland published the first description of a dinosaur, found near Oxford, England, which he called the "Stonesfield lizard." *Megalosaurus*, he noted, was to "a high degree carnivorous," a formidable predator, on the order of 40 feet in length. Its teeth were a "combination of mechanical contrivances analogous to those which are adopted in the construction of the knife, the saber and the saw" and were "admirably adapted to the destructive office for which they were designed."[1] Despite the historical importance of the Stonesfield lizard in helping shape our concept of dinosaurs, additional *Megalosaurus* bones have proved hard to find, though a few relatives have been unearthed in North America.[2] Presently the skeletal remains of megalosaurs are restricted to Middle and Late Jurassic strata.

The concept of megalosaur footprints evolved in the 1950s, when French paleontologists, working on Late Jurassic finds from Portugal, suggested the name *Megalosauripus* (meaning "megalosaur footprint").[3] Before long the term was being used quite indiscriminately and had been reduced to a catchall or "dustbin" category. Discoveries of similar tracks in North America and Asia have encouraged trackers to take another look at the megalosaur tracks problem.[4] In doing so we find we can recognize distinctive tracks larger than any known Jurassic theropod track, up to 30 inches (75 cm) in length.

Perhaps most striking is the remarkably wide-gauge configuration of the trackway, quite different from most narrow theropod trackways. By a remarkable convergence of circumstances, the famous American paleontologist Robert Bakker reported the presence of large megalosaurs in Wyoming that were evidently bow-legged with thigh bones (femora) angled outward, with shorter hind limbs and longer torsos than most contemporary theropods.[5] They are clearly good candidates for producing the wide trackways we report.[6]

Just as small, narrow theropod trackways suggest nerve-sense–dominated animals so the wide trackways of megalosaurs indicate large animals more likely to be at the "metabolic" end of the spectrum that exists within the theropods. Such tracks occur in conjunction with wide-gauge sauropod trackways at one Middle Jurassic site, and may suggest an ecological relationship in which large theropods preyed on or scavenged large sauropods. In general a relationship between large theropods and sauropod tracks is well established for many sites and many geological epochs.

At present, purported Late Jurassic megalosaur tracks are known from Uzbekistan, Turkmenistan, Tadjikistan, Portugal, Spain, Utah, Arizona, Oklahoma, and New Mexico.[7] In Utah they form part of a huge complex

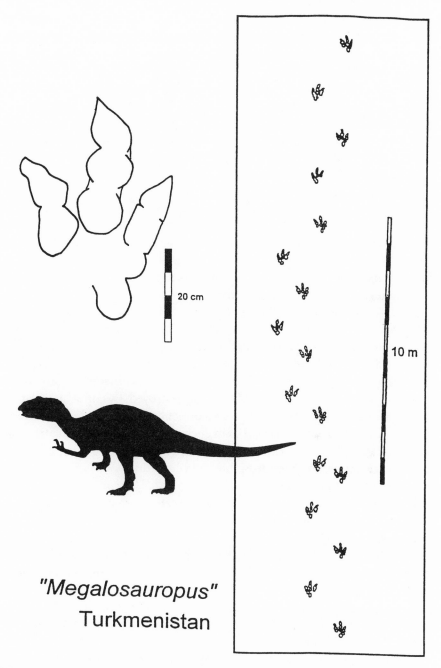

FIGURE 5.1 *Megalosauropus* tracks are found in peculiarly wide, irregular trackways, indicating a track maker with a primitive, rather waddling gait.

of tracksites spread out over an area of approximately 800 square miles (2,000 sq. km), which encompasses all of Arches National Monument, near Moab, and much of surrounding vicinity and has been dubbed the Moab Megatracksite.[8] A megatracksite is defined as any large area in which many tracksites are associated with a single surface, or discrete package of strata that accumulated during a relatively short interval of geologic time.[9] The term refers to the size of the site, not of the tracks, and has nothing to do with giant tracks or exceptionally long individual trackways. Confusion is understandable, however, because, by coincidence, large tracks, named *Megalosauripus*, were found there—after the Moab Megatracksite was named.

Present evidence suggests that megatracksites formed and were preserved as the sea level was rising. This conclusion may appear counterintuitive, for we expect tracks to be made when the tide goes out and be destroyed when the tide comes in. But sea-level rise and fall is a long-term phenomenon not at all like the ebb and flood of tides.

Changes in sea level are measured relative to arbitrary fixed points on land. Not all land is equally "fixed." Land can subside: it can happen that sea level appears to rise, and so flood homes built too close to the shore, but it could be that the land is subsiding. When there is a drop in sea level, new regions of the low-lying continental shelf are exposed, and rivers begin to erode the newly emerged land. For this reason, tracks made at a time when the sea level is at "low stand" are likely to be destroyed. By contrast, when sea level is at "high stand," rivers carrying sediment to the sea are dammed and held back on the coastal plain. It is in such sedimentary accumulations on coastal plains that track-rich megatracksites form and are preserved. Often the bone record is also good from high-sea-level times.[10]

To get the whole picture we must think big. The completeness of the track record is ultimately affected by large-scale cycles of rising and falling sea level. Geologists have labeled cycles first-order (30 million to 50 million years), second-order (6 million to 10 million), third-order (1 million to 3 million), fourth-order (400,000 years), fifth-order (100,000), sixth-order (42,000), and seventh-order (21,000).[11] Third- through seventh-order cycles are the same Milankovich climatic cycles identified by Olsen and his colleagues from the Triassic rift lakes. When such cycles are superimposed one upon another, the result is an intricate fractal pattern, one that repeats at many different scales.

First- and second-order cycles are probably the result of changes in the rate of plate tectonic activity, which affects the topography of the ocean basins. The Late Jurassic sea-level high stand was the culmination of a second-order cycle. Thus, in an endlessly connected chain of events, the convection within the Earth's interior mantle affects the rate of produc-

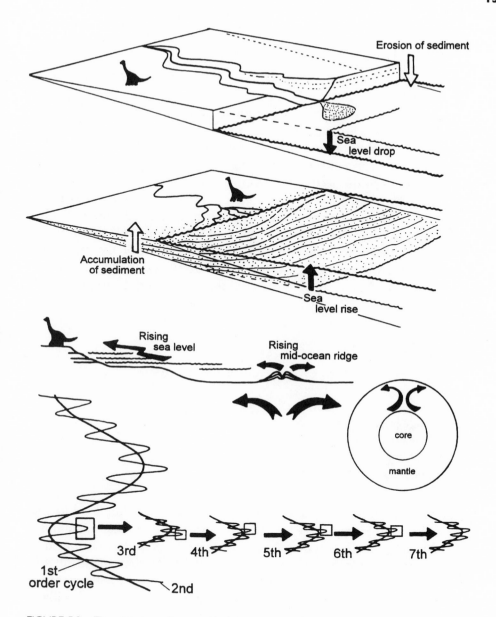

FIGURE 5.2 The structure of a megatracksite owes much to the influence of rising sea level, which causes accelerated accumulation of sedimentary layers in coastal regions. Such layers contain many tracks. The Moab Megatracksite probably owes its existence to a second-order sea-level cycle.

tion of new ocean crust and the mid-ocean ridge. This process, known as "sea-floor spreading," affects long-term—first- and second-order—sea level changes. As sea level rises to embrace the land, the coastal plains, rivers, and wetlands become pregnant with their rich accumulations of sediment and well-watered fauna.

This embrace of sea and land produces fertile paleontological terrain through the rapid growth of the very sedimentary layers that make up so much of the geological record. From pulses in the mantle to ripples in the newly hardened lava of the ocean floor, the swell of the ocean encroaches on the shore and the accumulated ripple affect gives birth to an enlarged sedimentary record rich in tracks, bones, and all manner of fossil remains. With each ebb and flow of the tides, with each eccentricity and obliquity cycle, the sea caresses the land with changing tempo. All the while as rhythm interleaves with rhythm, Earth's creatures dance on the ephemeral shoreline, leaving an interwoven and intricate trail, which ultimately owes its detailed configuration to the orbit of the planets and the deep pulse of the mantle.

HOLY FATHER OF THE ELEPHANTS:
THE WORLD'S LONGEST DINOSAUR TRACKWAYS

So geographers, in Africa maps,
With savage-pictures fill their gaps;
and o'er unhabitable downs
Place elephants for want of towns.

Jonathan Swift, "On Poetry"

For many centuries large and exotic creatures trod the Silk Road to and from Xanadu. From Venice to Beijing, a distance of almost 6,000 miles (10,000 km), inhabitants of Asia and the Middle East were accustomed to seeing camels, elephants, and other large beasts that made up the transcontinental caravans. In Turkmenistan, traditional carpet designs are centered on elephant footprints. The name of the remote village of Khodja Pil Ata, in eastern Turkmenistan, means "Holy Father of the Elephants."

But not all creatures observed by the local inhabitants were identified and understood. Not all were living! In Mongolia, for example, the complete skeletal remains of the beaked dinosaur *Protoceratops* gave rise to the reconstruction of the "mythical" griffon, guardian of the gold mines.[12] Fossil footprints also gave rise to speculation about the existence of beings that inhabited remote regions and about spirits of animals that walked through mountains. Throughout history, mapmakers have often regarded wilderness areas as the habitat of fantastic animals. In many cases their designation of these regions as the home of lions, elephants, and other large or ferocious wild beasts has inadvertently contained more than a smidgeon of truth.

Sometime in the last millennium or two, in the foothills of the Kugitang Mountains of Turkmenistan, near Khodja Pil Ata, a large landslide left a stark scar of light gray Jurassic bedrock standing out against the green slopes. These surfaces, exposed to daylight for the first time in 155 million years, are the same age as the strata of the Moab Megatracksite in Utah. Following leads and rumors of an extensive tracksite near Kodja Pil Ata, reported in obscure Russian literature, our University of Colorado Dinosaur Trackers Group obtained a grant from the National Geographic Society to travel to Uzbekistan and Turkmenistan to investigate this and other sites.[13] All our group knew was that several thousand tracks had been reported from this intriguing locality. Pictures and sketch maps indicated an unusual and wide gauge trackways, some up to about 900 feet (275 m) in length.

In the spring of 1995, when our motley crew of three Russians, four Uzbeks, four Americans, a Swiss (Christian Meyer, whom I call Jacques Cousteau II), and a Welshman arrived at the site, we recorded 31 trackways at the site, including 26 that we labeled *Megalosauropus*. The largest footprint was 28 inches (72 cm) long, as large as any theropod footprint known from the Jurassic. On May 23, 1995—155 million years since the days of Jurassic megalosaurs—we were about to obtain accurate measurements on the longest visible trackways anywhere in the world. After measuring several segments shorter than the previous record of 482 feet (147 m), we bagged trackway number 14 at 604 feet (184 m). A whoop of triumph went up from Servere, the Uzbek archaeologist who entered most gleefully into the euphoria of new discovery and the exploits of fellow trackers. Then came trackway 17 at 640 feet (195 m), trackway 18 at 741 feet (226 m), trackway 22 at 873 feet (266 m), and the crowning glory, trackway 21 at 1,020 feet (311 m). As each measurement entered the record, a joyous cry reverberated around the foothills of the Kugitangs. The world record for the longest dinosaur trackways passed from Portugal to the remote Silk Road outpost of Khodja Pil Ata.

Some simple arithmetic allows us to calculate the speed of the large megalosaurid track makers that wandered across the limy mudflats of Central Asia 155 million years ago. Animal 21 was moving at about 6 feet (2 m) per second, or four miles per hour (7 km per) hour—a typical walking speed for a large theropod. At this rate the animal would have taken between two and a half and three minutes to make the 1,020-foot (311 m) segment

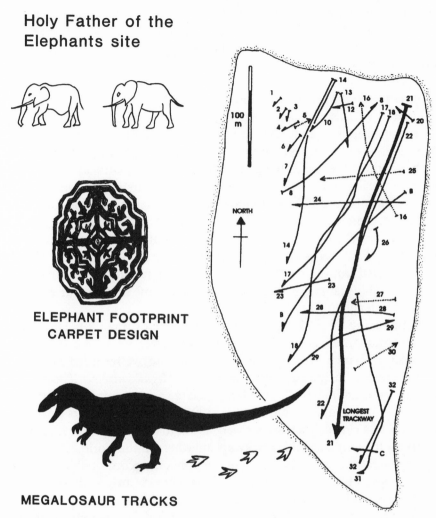

Holy Father of the Elephants site

ELEPHANT FOOTPRINT CARPET DESIGN

MEGALOSAUR TRACKS

100 m

NORTH

14

LONGEST TRACKWAY

FIGURE 5.3   The Kodja Pil Ata ("Holy Father of the Elephants") tracksite in Turkmenistan reveals the world's longest trackways. Tracks from this site include *Megalosauropus*, which is also found in strata of the same age in Europe and North America.

of trackway now preserved in the Kugitang hills. The four other long track-way segments record strolls of between one and a half and two minutes. Put another way, had we been transported back to the Late Jurassic for only two and a half minutes we could have observed five megalosaurs make track-ways with a cumulative length of about two thirds of a mile.

Such statistics underscore the immensity of the track record in terms of number of footprints created. Had our five megalosaurs been equally active for an hour on that day they would have created more than 12 miles (20 km) of trackway. Twenty-six animals could have made more than 60 miles (100 km) of cumulative trackway just by being active for a single hour in any day. At the rate of 400 miles (700 km) per week or about 1,750 miles (2,800 km) per lunar month, a relay of two dozen megalosaurs could walk a cumulative distance equivalent to the entire length of the Silk Road, from Venice to Beijing, in less than four months. In this business we count ourselves lucky to have a couple of miles of trackway still preserved for the record books.

PADDED OUT: OF FLESH AND BONES

. . . in time and place, since flesh must live,
needs spirit lack all life behind.
Robert Browning

Tracks found alongside *Megalosauropus* in Utah and Turkmenistan tell of a smaller, more erect animal that was probably more nimble than its giant contemporary.[14] It had a very fleshy or well-padded foot in comparison with the bony, birdlike feet that we associate with forms like the *Grallator* track maker. There has been very little effort to understand why tracks of the same size sometimes differ so much in the thickness of the toe impressions. Robert Bakker has recently suggested that dinosaur "species with small foot area for their body weight should have preferred harder soils than species with wide spreading feet"; this relationship between foot type and soil conditions he finds "quite satisfying."[15] Bakker may have a point worth exploring. Certainly it is well known that the differences between the feet of a water rail, which walks on lily pads, the hooves of a rock-climbing mountain sheep, and the hairy paws of the snow-bound Canadian lynx reflect the substrates on which they walk. So why not expect similar differences in the world of dinosaur podiatry?

Surprisingly, small animals such as rodents have very large feet relative to their body size and place their feet very flat on the ground (*plantigrade*), whereas large ungulates have very small feet relative to body size and stand on tiptoes (*ungulatigrade*). Carnivores occupy the intermediate central position (*digitigrade*) and may flatten themselves to the ground like nerve-sense, alert rodents or, alternately, stand tall like ungulates. From rodents to ungulates we also see a reduction in the number of toes in contact with the ground. In Schadian vocabulary the outwardly focused nerve-sense rodent is fully in contact with the surface of the ground, whereas the inwardly focused ungulate makes minimal contact with the Earth. *Foot size therefore is not an obvious indication of an animal's need for structural or mechanical support!* Try telling that to a biomechanics expert! Rather than strictly obeying laws of mechanics, the size and configuration of the foot is to a large degree a physiological and morphological expression of inherent properties in the animal itself. So is Bakker on the right track when he says that species with small feet prefer harder soils? Not really, because ungulates really don't care too much where they walk. Mountain goats negotiate rocky crags and moose trample soft lake beds. Trackers and soil experts have noted that sauropods, which had the largest feet of any dinosaurs, generally preferred harder, drier soils.

Let us avoid simple cause and effect "explanations" such as small feet = hard soils, and simply acknowledge that characteristics of the fleshy, soft-tissue anatomy of feet may be inherently variable. For example, many tracks of carnivorous dinosaurs indicate series of distinct individual pads clearly separated by prominent creases. But through time, in some cases as dinosaurs got bigger, many tracks show a lack of creases. This might be caused by poor preservation, but when we never see creases in large samples, another, simpler explanation may be necessary. In some dinosaurs there were evidently no footpad creases and the flesh of the foot coalesced into long cigar-shaped bundles. While spectacular paleontological discoveries of animal embryos, internal organs, molecules, and even congealed egg yolks make the news, we should not forget that tracks are the most abundant expressions of soft-tissue preservation. As noted in reference to brontosaur tracks, all such increases in flesh are an indication of a shift toward the wider, "metabolic" end of the spectrum.

PTEROSAUR RUNWAYS

Who . . . walketh on the wings of the wind . . . maketh his angels
spirits. . . . He laid the foundations of the earth.

Psalms

Pterosaurs, or "flying dinosaurs," have a remarkable pedigree in the annals of paleontology. They were first unearthed in the 1780s, long before the discovery of dinosaurs, but despite having been known to science for more than two centuries, there is still much we do not know about their biology. Paleontologists still debate the design of their wings and how they attached to the body or hind legs. Even more controversial has been the debate as to how they moved on land. Were they cumbersome quadrupeds like bats, or erect bipeds like birds? Older interpretations show them hanging upside down from trees and cliffs and suggest that when on the

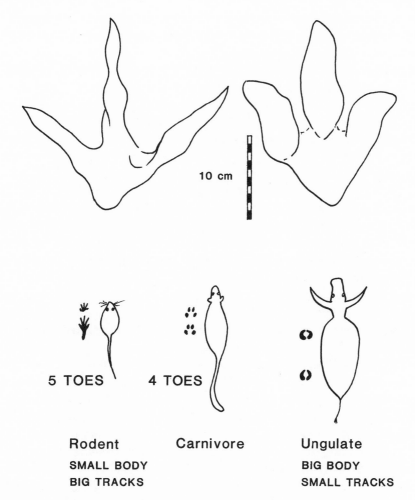

10 cm

5 TOES   4 TOES

Rodent   **Carnivore**   **Ungulate**

SMALL BODY   **BIG BODY**

BIG TRACKS   **SMALL TRACKS**

FIGURE 5.4 Comparisons between tracks of the same size show that there is considerable difference in the amount of flesh on the foot. Furthermore, feet are not necessarily designed in consistent proportion to an animal's size. Among mammals, for example, the size of the feet of rodents, carnivores, and ungulates is not determined by the animal's body size.

ground, like bats they would be unable to take off and would be at the mercy of passing predators.

Of course, if pterosaurs spent all their time in trees, on cliffs, or in the air, they would never make any footprints, and until quite recently it was claimed by some that pterosaur tracks simply did not exist. Reports in 1862 and '63 of pterosaur tracks from the famous Solenhofen Limestone Formation in Bavaria—which contains well-preserved complete skeletons of pterosaurs and *Archaeopterix,* the first birds—were debunked when they were shown to be traces of horseshoe crabs, some of which had quite literally died in their tracks. Such an inauspicious start did little to promote confidence in pterosaur tracking.

A century passed before Utah geologist William Stokes discovered footprints that he named *Pteraichnus*, meaning "pterosaur trace."[16] By an interesting coincidence, he found the tracks not far from the sacred rock monolith known to Americans as Ship Rock. This famous geological landmark of the Four Corners region is known to the Navajo people as *tse be dahi*, meaning "rock with wings"; in their lore it represents the site at which a giant bird brought the Navajo people (the *Dineh*) down to Earth.

At first no one bothered to dispute Stokes's claim, but in 1984, the pterosaur paleontologist Kevin Padian, of the University of California–Berkeley, and the Columbia University paleontologist Paul Olsen suggested that the tracks had been made by crocodiles.[17] They even went to the zoo to get a small caiman to make tracks for them. Despite the fact that the crocodilian tracks were different and included a tail drag impression, Padian and Olsen asserted that the similarities were sufficient to support their interpretation. Thus influenced by the pronouncements of so-called experts, most trackers accepted that the fossil record contained no known pterosaur tracks.

In the mid–1990s, however, *Pteraichnus* tracks were found at many localities in five western states, over an area of tens of thousands of square miles. The debate resurfaced with new vigor.[18] So we must ask again: pterosaurian or crocodilian? The trackway is highly distinctive. The hind footprint is long and slender and the front footprint is very unusual, with three toes making an E-shaped configuration quite unlike anything else known in the fossil footprint record. This anatomy fits very well with what we know of pterosaurs. The footprints are also all very small, in the range of one to four inches (2.5–10 cm)—a size more consistent with pterosaurs than a hypothetical tribe of diminutive crocodiles.

The track layers are all associated with strata that represent a shallow marine setting comparable to coastal lagoons of Europe from which pterosaur remains are well known. The tracks typically occur in association with worm burrows—which is what geologists call many worm-shaped burrows probably made by many invertebrate organisms, including worms. This association suggests that the pterosaurs were probably dabbling in the shallows as they foraged for food and then walked out on emergent surfaces. The discovery of surfaces with many parallel scratch marks made by pterosaur hind feet seems to suggest that they floated in the shallows like birds, perhaps raking the bottom with their feet, to stir up worms and other potential food items. We even find what we consider to be beak traces consisting of pairs of prod marks in the sediment.

The case for *Pteraichnus* tracks being pterosaurian seems compelling. After two centuries of exclusion from the debate about pterosaur locomotion, culminating in four decades of controversy and denial, tracks are now at the forefront of discussion. As if to underscore their importance and abundance, they are now popping up in France, England, Spain, and Korea, as we shall see in later chapters. It seems clear that pterosaurs were capable of landing and taking off from level mud and sand flats. Indeed it seems an insult to the gods of adaptation to suggest otherwise. Several dozen sites show that pterosaur landing grounds were flat and open, away from unnecessary obstacles. The nearest trees and cliffs were probably many miles away.

DINEHICHNUS: THE PEOPLE'S TRACK

On they pushed, north beyond Shiprock. . . . Some of them moved as if they never intended to stop, northwestward toward Kayenta and Monument Valley and Navajo Mountain. . . Into a thousand canyons, big and little, into a thousand valleys, hidden and open."

John Terrell, *The Navajos*

The folks at the hogan wondered what was going on about a mile to the west, over toward Boundary Butte. A couple of pickup trucks had been there all day, and whatever they were doing they were making a lot of noise. Perhaps it was road construction, but that hardly seemed likely, as there was nothing over there but a dirt track, a couple of old sheep corrals, and some rock outcrops. No one does much construction out here halfway between Ship Rock and Kayenta. This is the heart of the Navajo Nation, ten miles from Mexican Water, the nearest wide spot in the road.

Boundary Butte is an impressive black monolith on the Utah side of the state line, not a stone's throw from Arizona. The huge volcanic plug is perhaps not quite as magnificent as Ship Rock, off on the eastern horizon, but it is very inspiring nonetheless and well worth a visit. There is a handy water pump on the north side, and even a dinosaur track not a stone's throw from the well. But whoever was there that Thursday afternoon was not pumping water or admiring the rock. They were busy with a noisy project and could be heard miles away. On Friday they continued, and made even more of a racket. In addition to the drone of a machine, they punctuated their activity with blasting. The folks at the hogan decided it was time to find out what was going on in their backyard.

When they got to the scene of the activity they found a gang of young men who had already loaded several 500-pound rock slabs containing dinosaur tracks onto their large pickups. All the best tracks had been neatly sliced from the outcrop with a powerful rock saw. Another hour or two and they would have been gone. This was an inside job. One of the vandals was apparently recognized as coming from a camp just a few miles away.

Usually the FBI and the Navajo tribal police have to deal with burned-out cars, the occasional stabbing or murder, and other intrigues appropriate for Joe Leaphorn's casebook—not theft of 150-million-year-old rocks. (Joe Leaphorn is the fictional detective hero of many of Tony Hillerman's novels, which are set in Navajo country.) But each case is new, and the Boundary Butte incident added the crime of vandalism and attempted theft of dinosaur tracks to the season's caseload. Much of the Wild West is public land, and until recently, many people have regarded dinosaur bones, arrowheads, pottery, and any or all fossils and artifacts as fair game. Though laws protect such resources, violators are rarely prosecuted, and usually only the most blatant repeat offenders are actively pursued. One sobering statistic puts the conviction rate for infractions ranging from picking up an arrowhead to bulldozing or dynamiting a ruin, at one in ten thousand.

But little by little perceptions are changing, and the would-be offender has to be careful. In this neck of the woods, in and around the tribal lands of the Navajo nation, a few villains have been brought to justice for stealing fossils and have done time. In a much publicized case from the Dakotas involving excavation of a *Tyrannosaurus rex* skeleton and other commercial exploitation of fossils on Native American lands, the defendant faced indictment on 40 charges. Although he beat most of them he still received a two-year jail sentence.

In the former Navajo nation incident, to make their case, the FBI needed to collect evidence and so they went to the crime scene to make plaster casts of the tire tracks. They also recovered the stolen fossil tracks, dumped not far from the scene of the crime when the thieves were run off by the locals. These were impounded at the FBI offices as evidence. But notwithstanding their ability to track suspects in the great western desert, FBI agents and Navajo tribal police have no expertise in tracking dinosaurs. So they called us in as paleontological experts to determine what kind of material they were dealing with, and what it was worth.

The tracks turned out to be a new type never previously reported in North America. Moreover, they originated from the Morrison Formation (named after a little village nestled in the foothills west of Denver), one of the world's most famous dinosaur graveyards, but until recently not well known for footprints. The uniqueness of the tracks, not to mention their fine aesthetic quality, markedly increased their scientific and commercial value. The whole collection could easily fetch $5,000 or $10,000 at the right outlet. Fortunately for science, the thieves had been caught in the act, and the neatly trimmed slabs could be matched with saw cuts at the site, so most of the original configuration of the outcrop could be reconstructed. The Navajo gods had intervened in time to prevent complete desecration of the site.

In the course of mapping out the site, it became evident that the tracks were those of a small turkey- to emu-sized creature that evidently traveled in social groups. There are at least 16 trackways of this creature, clumped together in two groups in which the individuals were all going in more or less the same directions. We named the tracks *Dinehichnus socialis* (to emphasize ownership by the Navajo people, or Dineh, and to emphasize the evidence of the animals' social behavior).[19] The tracks were probably made by small ornithopod dinosaurs such as dryosaurs or hypsilophodotids, both small bipedal relatives of the duck-billed dinosaurs. Our general knowledge of ornithopods from around the world suggests that they were often gregarious. The 150-million-year-old evidence from the land of the Dineh, however, is the oldest to confirm this social tendency, so it's fortunate that the grouping was saved from being forcibly split up.

SOCIAL COMMENTARY

Some dinosaurs were herding by the Early Jurassic, leaving open the possibility for development of complex social systems over the next 140 million years.

Walter Coombs, *The Dinosauria*

We live in a social universe. Hydrogen atoms congregate together in deep space, and stars cluster together in galaxies. In the animal world we have shoals of fish, gaggles of geese, charms of canaries, exaltations of larks, flocks of sheep, and conglomerates of geologists. Lions gather in prides and whales swim in schools, even though, according to *The Yellow Submarine,* they are big enough to go to university. Our social vocabulary is so extensive, specific, and rich that perhaps we should extend the tradition to prehistoric animals and speak of a drove of dinosaurs, a preen of pterodactyls, an orchestration of ornithopods, and a bevy of brontosaurs.

Although the idea that dinosaurs are social creatures was once highly unorthodox, since the 1970s the idea has caught on in a big way. Excellent evidence for social dinosaurs comes from nest sites that reveal a colonial existence, at least during breeding season. Track evidence can be just as

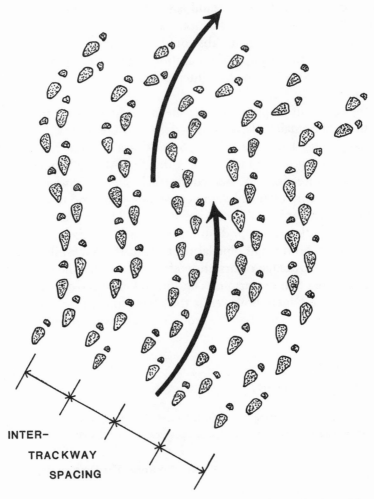

FIGURE 5.5   Parallel, regularly spaced trackways suggest animals traveling in social groups, especially where individuals show synchronized changes of direction. Illustration based on brontosaur trackways from the Purgatoire site in Colorado.

compelling. In recent years dozens of sites have come to light revealing multiple parallel trackways attributable to particular types of dinosaur. When trackers find only a few such trackways trending in the same direction, the possibility that they represent the random wanderings of a few individuals cannot be ruled out, but when we find dozens of the same type together, and often clustered in groups of similar-sized individuals, the probability that they represent a gregarious group is very high.

We most frequently find that it was herbivorous ornithopod and sauropod dinosaurs that left large numbers of parallel trackways.[20] In addition to the orchestration of small ornithopods suggested by the trackways at Boundary Butte, we now know of dozens of additional ornithopod track-sites from Colorado, New Mexico, Canada, South Korea, England, Spain, and elsewhere. The same is true for brontosaurs from sites in Texas, Colorado, Utah, Bolivia, and Portugal.

In ideal situations large surfaces are exposed on which we find multiple parallel trackways that are equally spaced, as if recording the passage of a marching unit or the tight pattern of birds flying in formation. Each trackway is equidistant from the next, and in some cases, as one trackway drifts or curves in one direction, so too does the one adjacent to it and the one adjacent to that. Each animal leaves a well-demarcated space for the next, just as we do when we walk in a group. We like to leave room for our invisible auras and not have our space crowded. Such regular spacing or group structure suggests herds moving as a coherent entity, a sort of superorganism, rather than a collection of individuals. Whether at the molecular or herd level, nature tends toward coherence and rhythmic entrainment. A good example of this is the tendency animals have to fall into step with one another when traveling together. Birds may fly in such a way as to synchronize their wing beats, even their heartbeats. No one has demonstrated statistical proof of rhythm entrainment from analysis of dinosaur footprints, though it appears that adjacent trackways are sometimes in step with each other. There are, however, enough herd tracksites to prompt us to pursue this intriguing problem further, and to wonder whether we might not one day speak of an oscillation of ornithopods or a synchronicity of sauropods.

THE RISE AND FALL OF
THUNDER BEINGS AND THUNDER LIZARDS

In a sacred manner I am walking
With visible tracks I am walking
In a sacred manner I walk.
>   **Black Elk (Oglala Sioux holy man),**
>   ***Black Elk Speaks***

We noted in a previous section that *Dinehichnus* originates from the Morrison Formation, or "the Morrison," as geologists call the fossil lode. It is legendary in the annals of paleontology as one of the richest dinosaur-bearing deposits in the world. It is the source of *Stegosaurus, Brontosaurus* (*Brontosaurus* is a popular but not scientifically correct name for *Apatosaurus*), *Diplodocus, Allosaurus,* and at least 20 other dinosaur genera, not to mention mammals, crocodiles, turtles, amphibians, and fish. These discoveries, from 1877 onward, have been described as the single most important series of contributions ever made to paleontology, save perhaps for the publication of Darwin's *Origin of Species.*

These fabulous discoveries, which now stock the world's major natural history museums, were made in an atmosphere of intense, selfish, and often acrimonious scientific rivalry that had developed between Professor Othniel Marsh, of Yale, and Edward Cope, from the Philadelphia Academy of Sciences. Both men resorted to subterfuge to wrest the best bones from the fossil beds and beat each other in getting their reports into print. Add to this the backdrop of a lawless Wild West, at the time of the battle of Little Bighorn (1876), and we have the stuff of which legends are made. This chapter in American paleontological history has been characterized alternately as a dinosaur gold rush, the bone wars, and the "fossil feud," but it has never been given the full epic treatment it deserves.[21]

Rather than attempt such a treatment, I prefer to place these heady, acquisitive days of nineteenth-century frontier paleontology in their historical context, and ask: Why such dinosaur fever? Though sometimes regarded as a time of triumphant scientific discovery, it was also an era of unmitigated anthropological disaster. The American Indians and their culture was being driven to the brink of extinction by the monstrosity of the United States' Manifest Destiny ideology and unbridled greed and hostility. In one of history's great ironies, as the world of buffalo, passenger pigeon, and native "thunder beings" was systematically slaughtered, the ever-hungry white man feverishly extracted the bones of extinct tribes of "thunder lizards" (i.e., dinosaurs). Dinosaur hunters like Cope literally prospected for thunder lizards in one valley while the U.S. cavalry battled "Indians" in another. We do not think enough of these connections. Had Takatanka-ohitika (Brave Buffalo) of the Sioux been asked his opinion, he might have echoed his belief that "The earth contains many thousand stones. . . . The thunderbird is said to be related to these." Chief Dan George predicted that "the thunderbird of old shall rise again,"[22] and Chief Seattle remarked that there is no death, only a change of worlds. As the scientific concept of evolution impressed itself on the minds of paleontologists like Cope and Marsh, so too they became aware of extinction. At this time we first see conscious awareness of the previous existence of extinct species and the extinction of living species. This awareness of turnover, of the change of worlds, is intimately related to evolutionary time consciousness. William Irwin Thompson put it succinctly when he stated that "the mischievous thunderbird is consciousness."[23]

The dinosaur is surely a symbol of the great thunder beings and of paleontological time consciousness. Had a nineteenth-century medicine man dreamed of a fantastic beast like *Stegosaurus*, or the giant tracks of a thunder beast in the rocky bed of the Spirit River, his vision would probably be dismissed as fanciful myth by those who espouse white rationalism. Yet Edward Hitchcock named his track maker *Eubrontes* "true thunder," and we, at the close of the twentieth century, spend millions of dollars to create the thunder beings of Jurassic Park, using bones plucked from the sacred lands of Sioux, Cheyenne, Ute, and Lakota. Our dinosaur idols are merely thunder beings in a new guise. There has been a change of worlds, perhaps, but thunder beings still dwell in human consciousness. Unlike aboriginal cultures whose members are still conscious of the psychic reality of this world, we prefer to keep thunder beings at a safe distance inside the fences of Jurassic Park.

Paleontologists are like medicine men who go to the wilderness and return with new images of monsters from the borderland of fantasy and reality. They can even materialize bits and pieces as tangible proof that these creatures are more than ephemeral visions or shadowy spirits. Tracks, however, seem always to retain that shadowy quality that links the tangible and intangible realms. *Stegosaurus* tracks vividly illustrate the elusiveness of the intangible paleontological spirit world. Though *Stegosaurus's* bones are prized material trophies in great museums, the tracks that tell us where *Stegosaurus* roamed free have proved very hard to find. We described one, *Stegopodus,* near Moab, that matches the skeletal anatomy of the front foot, and Robert Bakker identified what he believed to be hind footprints from another site in Utah.[24] *Stegosaurus* is being tracked down literally one footprint at a time! Precisely because of the lack of tracks there is considerable debate about its ecology and habitat preference, as I shall soon explain. The complete picture—the imaginative vision of *Stegosaurus* making tracks across late Jurassic landscapes—is still held partly in the domain of the thunder beings. Perhaps only they know when the spirit of *Stegosaurus* will finally be released from the mountains and leave its trail in the clear light of day.

DINOSAUR TRAILS IN PURGATORY

I can call spirits from the vasty deep.
> **William Shakespeare,**
> *Henry IV,* **Part I**

The Indians called the river in southeast Colorado the Spirit River, reminiscent of River of Spirits, or Spirit Trail, the name given by some tribes to the Milky Way. The Spanish called it Rio de las Animas Perdidas—river of lost souls—and the French called it Purgatoire, and so it appears on maps. But these days the French is simply translated as Purgatory, or corrupted to Picketwire. The Spirit River is a beautiful and still remote corner of the West. Long before it became associated with the sad purgatory of lost souls, this spirit trail was an artery of turbulent snow melt that flowed out of the Rocky Mountains to water the High Plains. As the white man came west, hungry for direction and thirsty for a muddled destiny, the spirit trail showed the way to Santa Fe. But long before pilgrims plodded the Purgatory's northern banks, blazoning the Santa Fe Trail into cartographic history, the thunder lizards had passed the same way, their high gaze set on a western horizon then unobscured by the high barrier of the Rocky Mountains.

We do not know when the tracks were first noticed by white settlers, but by the turn of the century the river valley was well populated, and not long thereafter the site was dubbed Elephant Crossing. The broad expanses of gray limestone in the riverbed provided a convenient place to ford the river, and the site also became known as Rock Crossing. By the thirties, drought and the Dust Bowl had driven most settlers from Purgatory, leaving only a few inhabitants in some of the larger settlements. One of these local residents was Betty Jo Riddenoure, a 15-year-old schoolgirl, who lived few miles downstream from the crossing. She had heard about the tracks from her father and was eager to see them.[25]

So it was that in 1935, she instigated an expedition to the site to show the footprints to her schoolteachers and local journalists. From this simple outing to Elephant Crossing came the first pictures of "Dinosaur Trails in Purgatory." The news was covered in the local press but it was a remarkable individual by the name of John Stewart MacClary, who adopted the site and crusaded zealously to bring it to the attention of the world at large. MacClary suffered from a debilitating disease that confined him to a wheelchair, but despite his handicap he had a great love of natural history and was an enthusiastic amateur writer. He struck up a correspondence with Betty Jo, and though he had no scientific credentials, he alerted professional paleontologists to the importance of the site by publishing short notes in *Life* magazine, *Scientific American,* and *Natural History.*[26] Barnum Brown of the American Museum of Natural History in New York responded and in 1938 sent his trusted assistant, Roland T. Bird, to Colorado to visit the site.[27]

Brown and Bird had been on the lookout for tracks to enhance their dinosaur exhibits. But the Purgatory valley was too remote. Destiny had put Bird en route for another tracksite in Texas, where, within a month of his Colorado visit, he discovered brontosaur tracks that were to prove better preserved and more accessible than those in Purgatory. These Texas footprints quickly stole the ichnological limelight, as we shall see in the next chapter. The forgotten and aptly named dinosaur trails of Purgatory were destined to languish in scientific obscurity until 1986. On this date, exactly 50 years after MacClary's first report, I and my tracker colleagues at the University of Colorado–Denver published a full description of the site documenting more than a thousand footprints exposed, along a quarter-mile stretch of ancient Jurassic lake shoreline. We dubbed this ancient watering hole Dinosaur Lake.

What had been a popular stomping ground for dinosaurs was to become a popular hangout for paleontologists measuring tracks and collecting fish, snail, clam, crustacean, and plant fossils. Because parts of the limestone surface were underwater, particularly in spring snow-melt and summer irrigation seasons, the trackers worked in winter to map this large area section by section. Camped out between October and March for 20 weekends over four winters (40 days and 40 nights in Purgatory), we pursued a slow and respectful ichnological courtship with the muds and waters of Purgatory and the elusive Jurassic spoor hidden therein.

At the end of the day we entitled our scientific contribution "North America's Largest Dinosaur Trackway Site: Implications for Morrison Paleoecology."[28] Suddenly a formation known almost exclusively for its bones had yielded a huge tracksite that could be placed squarely in the context of its ancient lake-shore ecology. We reported that about 40 brontosaurs had left their tracks on the shores of Dinosaur Lake, including one group of a dozen that had left parallel trackways as it walked west—the first reported Jurassic evidence of gregarious behavior among brontosaurs. This Jurassic Santa Fe Trail has been immortalized on celluloid by the former *National Geographic* photographer Louis Psihoyos, who shot the whole Purgatory panorama by night from atop a 30-foot scaffold. His breathtaking photo shows river, skyline and the footprints of five parallel brontosaur trackways filled with water reflecting the light of a magical moon.[29] Did the shores of Dinosaur Lake look much different almost 6,000 precessional cycles ago?

From a scientific viewpoint this set of trackways indicates subadult individuals that were all the same size, evidently representing a particular age group. The track makers walked for a considerable distance without converging or diverging. The subtle curvature of one trackway is reflected in the fluid parallel curves of the one next to it. The trackway cluster has a coherent, organic feel to it, reflecting the movement of an integrated group, if not in rhythm entrainment, then at least in social entrainment. Some form of sauropod collective consciousness or group mentality imprinted itself into the overall trackway picture. A synchronicity of sauropods.

Unlike the wide-gauge trackways from Fatima, those from the Purgatory are narrow-gauge with small front feet and narrow heels to the hind foot. I suspect they were made by diplodocids, or their relatives the apatosaurids (which include so-called *Brontosaurus*). A drove of diplodocids or was it an appearance of apatosaurids. These animals had very long, narrow bodies, narrow heads, and narrow teeth, as well as larger hind quarters than front limbs—a suite of characteristics that, along with the trackways, suggests a gracile sauropod more inclined to its nerve-sense pole than to the broad metabolic pole of wide-gauge track makers. According to conventional wisdom, these animals were much lighter than brachiosaurs, though just as long or longer; moreover, they are often depicted rearing on their hind limbs.

TRAMPLED TO DEATH

These dead bones have . . . quietly rested under the . . . tramplings of three conquests.

Sir Thomas Browne

On the basis of both bone and track evidence, there is no doubt that brontosaurs were the largest of all dinosaurs. Add to this the evidence that they were gregarious and traveled in large herds, and there is little doubt that they had considerable impact on vegetation and small animals in the habitats they frequented. Today, wildebeest and other large mammals plow, churn up, or otherwise trample large areas as they migrate. Hippos, too, may trample frogs and other creatures around their water holes and wallows.[30]

The trampling of invertebrates by horses and cows around fords, streams, and lakes shows that freshwater clams often fall victim to trauma. But these luckless victims often survive, regenerating their broken shells into bizarre shapes referred to in one report as "molluscan monstrosities."[31] At the Purgatory site more than two dozen clams suffered worse fates: they were trampled by brontosaurs and apparently were killed instantly as one individual dispatched as many as six clams per footstep. Such clam carnage adds up quickly when multiplied by many footsteps and many brontosaurs. The footfalls of the thunder lizards also squeezed the water from the limy sediment, trapping clams in mud rims around the tracks that effectively acted as instant quick-drying cement.

One wonders how clam populations in the shallows of Dinosaur Lake could survive the passage of a large herd of brontosaurs, which might kill thousands in mere minutes. But nature is remarkably resilient. Modern clam-eating birds such as oyster catchers can devour more than 300 clams per day and still not upset the balance of nature. It has even been suggested that trampling has a beneficial, pruning effect. A fascinating suggestion, explored later, is that dinosaur trampling, or "dinoturbation," would favor plants that regenerated rapidly.[32] The same pressure would be put on plants by browsing or cropping, often by the same animals responsible for the trampling. In fact, it was in the Late Mesozoic that the rapidly regenerating flowering plants first evolved. In the catchy words of the dinosaur celebrity Robert Bakker, "Dinosaur invented flowers."[33] In the world of footprints and trampling, this hypothesis would translate into the notion that the downtrodden will arise again with new vigor.

CROCODILES IN THE CHANNEL

The book of the Liang Dynasty (A.D. 502–556) says that in the Kingdom of *Lin Yi,* now southern Cochin-China, crocodiles (*Ngo*) are reared in moats of the capital. . . . Criminals are thrown to them. If during three days they are left unharmed by the animals, they are considered innocent and restored to liberty. These crocodiles measure twenty or more feet in length. . . . The Chinese book adds that they are very ferocious.

A. A. Fauvel, *Alligators in China,* 1879

Crocodiles, from the Greek *krokodilos* ("lizard"), are the last of the large living archosaurs (a group of advanced reptiles that includes dinosaurs and pterosaurs) and have long been venerated by human societies. The Egyptians regarded them as sacred, tamed them, adorned them with gold bracelets and earrings, and even embalmed and mummified them. According to Herodotus, the Egyptian word for crocodile is *champsa,* hence the paleontological term "champsosaur," used to describe a group of extinct, crocodilelike reptiles. Crocodiles are also the basis for many dragon myths, and indeed they are the living relatives of the pterosaurs and dinosaurs. Together these three groups constitute the archosaurs of air, land, and water.

Human fascination with crocodiles no doubt comes from the fact that they are large, dangerous, and demanding of healthy respect. They are also creatures of great ambiguity and mystery—motionless inhabitants of the watery underworld, yet able to move with lightening speed, and even run on their hind legs like lizards. They have strange habits, and may lie with their mouths wide open, or may close their third eyelids, known as nictitating membranes, giving themselves a glazed, sleepy appearance. Crocodiles incidentally are a great reservoir of trypanosomes, parasitic protozoans that cause sleeping sickness. These parasites can be transferred to victims by tse-tse flies that have bitten crocodiles. So they can cast their debilitating sleeping spell on victims without even stirring from their primordial limbic slumber. Such tidbits of information might enliven future Jurassic Park scripts if there is ever a need for variation on the theme of mosquitoes sucking DNA from dinosaur blood.

As stereotypic baskers on riverbanks, crocodiles make few wide-ranging terrestrial excursions and so produce few walking trackways. When they do they usually drag their tails like lizards. Marco Polo provided a description of a crocodile trackway observed in Yunnan Province, China, in A.D. 1280, describing the crocodiles as "great serpents of such vast size as to strike fear into those who see them." Their weight, he continued, "is so great that when they travel in search of food or drink, as they do by night, their tail makes a great furrow in the soil as if a full ton of liquor had been dragged along."[34]

Why Polo had liquor barrels on his mind when observing crocodile trackways it is hard to say; perhaps because the huntsmen he observed needed fortification to pluck up the courage to take on these creatures. This they did, as if snaring a rabbit, by setting a gyn "in the track over which the serpent has past, knowing it will come back the same way. They plant a stake deep in the ground and fix on the head a sharp blade of steel [which] they cover . . . with sand. . . . On coming to the spot the beast strikes against the iron blade with such force that it enters his breast and rives him up the navel, so that he dies on the spot." Polo goes on to tell that the flesh makes excellent eating, and reports that crocodile hunters extract the gall, which is highly prized as an instant cure for the bites of mad dogs and the pain of heavy labor.

But long before crocodiles swam in the sacred waters of the Nile or Chinese moats, they were inhabitants of countless prehistoric rivers, where their remains are common. Their tracks, however, are quite rare. Obviously, when in the water they did not produce many footprints. Though they would touch the bottom of river channels and ponds from time to time, tracks made when swimming are quite different from those made when walking, and often consist only of incomplete footprints and scrape marks, distributed with little sense of pattern or order. As a result, the interpretation of fossil crocodile tracks is controversial and footprints have been attributed to pterosaurs and various other creatures. Recently, an interesting example of a fossil crocodile trackway was described from the roof of an abandoned uranium mine in the backwoods of eastern Utah. The trackway, from the famous Morrison Formation, provides evidence of an animal with a hefty tail which it swished back and forth, leaving distinctive zigzag traces on the river bottom. It had feet about six inches (15 cm) long, which extrapolates into a creature at least six or seven feet (2m) in length.

The track, named *Hatcherichnus,* is similar to one from Colorado described by John Bell Hatcher, a renowned collector of horned dinosaurs

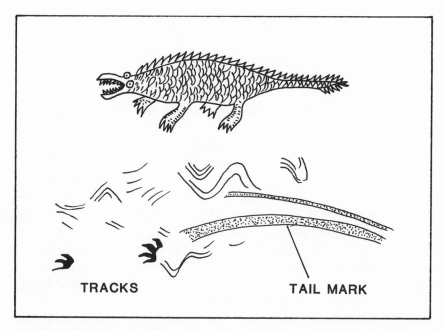

TRACKS

TAIL MARK

FIGURE 5.6   A crocodile track named *Hatcherichnus*, with tail mark, and an ancient Chinese depiction of a crocodile.

and a "legendary hero of vertebrate paleontology."[35] The tracksite's location in a radioactive uranium mine caused problems of access, but fortunately we gained entry before the mine was sealed up. The world's oldest, most distinctive, and most radioactive crocodile tracks are now locked into a remote hillside behind an impregnable cast-iron door, symbolically keeping the dragon in its den so that it cannot escape to terrorize the region with its fiery radioactive breath.

SACRED MULE TRACKS FROM LOBSTER BAY

Be ye not like to horse and mule, which have no understanding.
Book of Common Prayer

According to a local legend dating back at least to the thirteenth century, fossil footprints on the cliffs above Lagosteiros Bay, south of Lisbon, were made by the Virgin Mary's mule as she rode up from the sea. The tracks, referred to as "*pegadas de mula*"—"mule tracks"—are still revered by local fishermen, and are considered symbols of safe landfall for all mariners.[36] Beside the Cabo Espichel monastery, atop the spectacular sweep of strata that rises from the cove of Lagosteiros Bay some four hundred feet below, sits a small shrine to the tracks. Inside the simple onion-domed structure, an inconspicuous mosaic of blue and white tile tells the story of our lady's landfall at this sacred site.

Given that the tracks are brontosaur footprints that measure up to 3 feet (1 m) in diameter,[37] one wonders what sort of monstrous mule the Portuguese locals visualized as Mary's steed. Though it is easy to view such legends as childlike or naive, it is presumptuous to assume that our ancestors' interpretations of their surroundings are naive, or that their legends and myths lack substantive foundation. We assume that we understand the world better than they did, but in truth we only understand our world better, just as they understood their world better than we can. Many of the limestone surfaces around Cabo Espichel in fact are covered with semicircular traces that look exactly like the tracks of mules or horses wearing horseshoes. If we accept that Portuguese fishermen also saw such markings, then perhaps the legend is based on an interpretation of traces that really do resemble mule tracks.

In 1858, Edward Hitchcock described horseshoe-shaped markings from Jurassic strata in New England and even named them *Hoplichnus equus*, meaning "the hoof track of a horse."[38] By an odd coincidence, the markings originated from a locality called Horse Race. Many such enigmatic markings have been found in rocks that predate even the most remote of horse ancestors. For example, such traces were known to the Supai Indians of the Grand Canyon, who recognized the difference between the tracks of shod and unshod horses, and so apparently interpreted the markings as the spoor of a band of early invaders. Soviet geologists reported horseshoe-shaped traces from the Republic of Uzbekistan and named them *Gumatagichnus ungliformis*, meaning "hoof-shaped trace from Gumatag." Largely oblivious to any previous debate on the subject of such markings, they speculated that they were made by a dinosaur resembling the famous raptor *Deinonychus*.[39] In fact the conclusion is inescapable that geologists and paleontologists have contributed to more confusion than any Portuguese fishermen. But this is not the end of the story. Creationists have had their say also. Having heard of the Uzbekistan site, they claimed new evidence had emerged contradicting the geological antiquity of dinosaurs, and placing horses as dinosaurs' contemporaries.[40]

What then is the origin of these horseshoe-shaped markings? Many were probably the work of burrowing animals. Some were made by limulids, otherwise appropriately known as horseshoe crabs. Others are U-shaped burrows made by shrimplike species. Thus the *pegadas de mula* legend may conceivably have begun some 150 million years ago, not with dinosaurs but with Jurassic shrimp and lobster that burrowed away in the

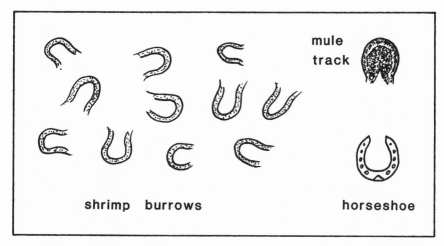

FIGURE 5.7 Traces made by shrimplike marine arthropods look uncannily like horseshoe tracks, and have been misinterpreted as such in some cases.

shallow waters of a young Atlantic. These marine arthropods were presumably oblivious to the dinosaurian perambulations nearby and to future directions in mammal evolution that would lead to horses, humans, and religious beliefs. Yet they, too, left enduring signatures, embodied in the very name of this sacred locality. *Lagosteiros* Bay means Lobster Bay.

TEILHARD'S ENIGMATIC CHINESE FOOTPRINT

The spirit of earth comes to explain to men the reason . . . it reveals itself as the force which is destined to set underway and organize . . . human . . . discovery.

Pierre Teilhard de Chardin

The French Jesuit priest, Father Pierre Teilhard de Chardin (1881–1955), affectionately known as Teilhard, was one of the most fascinating, and controversial of all paleontologists. A direct descendent of Voltaire, he is best known in the scientific community for his paleoanthropological work, excavating Peking Man at Choukoutien near Beijing in the 1920s. Prior to this Teilhard had also visited a famous site at Piltdown, Sussex, in England, where remains of early hominids had been reported.[41] But the fossils turned out to be frauds—the hoax was revealed in 1953—and so-called Piltdown Man has gone down in history as one of the greatest scientific hoaxes ever perpetrated. No one really knows "whodunnit"—who created and "planted" the false evidence, but Teilhard was one who came under suspicion, as did Sir Arthur Conan Doyle, who lived nearby and so presumably had the opportunity. As we shall see in the next chapter Conan Doyle's interests extended to paleontology, and of course he liked a good mystery.[42]

Teilhard has been accused of being the perpetrator by such influential persons as Louis Leakey, an Englishman, and Steven Jay Gould, an American.[43] Some people think their suspicions grew out of the chauvinistic belief that the French would stop at nothing to embarrass the English! Parochial as it may seem, a tradition of lively debate over the origins of the first "true hominid" has extended to claiming that particular nations represent the birthplace of humanity. The reader need only imagine the potential for "objective" debate between French and English anthropologists on the subject of which side of the Channel to search for the origins of civilization.

Many however, believe that Teilhard had nothing to do with the Piltdown hoax, and recent discoveries essentially prove that the perpetrator was an Englishman named Hinton.[44] In any event, such deceptive behavior on the part of Teilhard would have been out of character, for by all accounts he was a man of impeccable integrity. He is regarded by many as one of the greatest philosophers and mystics of the century. His study of evolution went far beyond the bones of Peking Man and took him into the realms of psychology and the analysis of consciousness and human nature. He introduced the concept of the *noosphere* as a sphere of the mind or collective consciousness. His books, including *The Phenomenon of Man, The Future of Man,* and *Building the Earth,*[45] are regarded as classic explorations of human nature by a man of great genius and vision, though some like Gould rail against his mysticism.[46] We shall review some of his ideas in the next chapter, when we discuss the evolution/creation debate and Teilhard's firm belief in the compatibility of science and divinity.

Teilhard wrote many scientific articles on the geology, paleontology, and anthropology of China. To this day China still remains virtually terra incognita as far as fossil footprints are concerned.[47] Whereas there are literally thousands of sites reported from North America, in the equivalent area of China only a few dozen are documented. Teilhard has the distinction of having coauthored the first article ever published about a dinosaur track from China.[48] The track in question, from Shansi Province, has since been named *Sinoichnites* ("Chinese trace"), and in truth we still know very little about it.

In the 1920s nothing was known of dinosaur tracks in China. Teilhard pointed out that the rocks in "Western Shansi and Northern Shensi are extraordinarily barren of animal fossils." At that time, in China, there were not even any dinosaurs known in pre-Cretaceous rocks, so this "Jurassic" track represented "the most ancient trace of reptile yet known in China." Like anyone interested in hominid origins, where a single tooth can be considered a major find, Teilhard was used to working with limited evidence. Those who knew him said he had an extraordinary ability in this regard, perhaps owing to a kind of cosmic consciousness.[49] His ability to recognize the significance of a single dinosaur track, and even identify the potential problems of correctly assessing its age, are clear signs of an incisive mind. His association with what became known as the "Chinese footprint" reflects his breadth of interest and love of all branches of Earth and life science. As we shall see in the next chapter, Teilhard's love of science transcended the ultimately selfish motive of publishing papers and establishing a reputation. He saw rational science, like faith and awareness of spirit, as a natural outgrowth of the evolution of consciousness and sought therefore to integrate and reconcile agnostic scientific rationalism with spiritual mysticism.

So we find ourselves in the middle of Mesozoic midlife, in a world of dinosaurs and mouse-sized mammals, already looking ahead to the potential realms of higher human consciousness. Despite detractors like Peter Medewar, who evidently regarded Teilhard's views as woolly and unscientific,[50] he has many scientific supporters, including "agnostics" like Vladimir Vernadsky, the originator of the concept of the *biosphere*;[51] the Nobel Prize–winning biologist Christian de Duve; and the astronomer-physicist team of John Barrow and Frank Tippler, who regard Teilhard's noosphere theories as "the only framework wherein the evolving cosmos of modern science can be combined with an ultimate meaningfulness to reality."[52]

Teilhard's thesis is that evolution is the rise of consciousness; since we cannot follow the eternal trail without pondering evolutionary progression, let us prepare to ruminate on such philosophical matters, for the trail leads quickly to a convergence of thought on issues surrounding evolution, creation, science, and spirituality.

chapter six

WITH GOD
ON OUR SIDE

ACT VI: CAST OF CHARACTERS

We shall spend this act in the Lower Cretaceous period (about 140 million to 100 million years ago). The act begins with the original story *The Lost World,* in which Sir Arthur Conan Doyle inadvertently transformed the science fiction suggestion that *Iguanodon* left its tracks in remote regions of South America into a serious scientific hypothesis. We shall continue to track quite a few brontosaurs and the occasional pterosaur, but this time we shall find lots of birds getting into the act. We shall also investigate the subtle realm of ghost footprints, phantom taxa (phantom scientific names), and footprints that seem to walk backward in time.

Brontosaurian antics will prove quite diverting, owing to controversial claims that they were irresistibly drawn into the sea, acted as magnets for predators, and displayed parental care when on the move. We shall have to learn to discriminate between scenarios that can be believed on the basis of evidence and those that just seem appealing or satisfying. Indeed, we shall even find scientists discussing what they would "like" to believe. In this vein and for light entertainment we meet one tracker who set a brontosaur trackway to music as part of a dramatic scenario.

We shall briefly visit the South Korean seashore to find evidence that there were far more baby brontosaurs running around than we had previously supposed. But for historical reasons we shall spend most of our time in Texas, where we will get to know Roland T. Bird, a celebrated amateur paleontologist who was the darling of tracker folklore, and will see how tracks led him to become embroiled in an evolution/creation debate that has persisted for 60 years. I shall review this debate to remind us that there is more to science and spirituality than partisan scientific or antiscientific dogmatism. In fact, I shall suggest that the debate may serve as a starting point for deeper and more fascinating questions about evolution, which we can investigate in the final chapters.

TRACKING DINOSAURS IN *THE LOST WORLD*

"Not a bird my dear Roxton—not a bird."
"A beast?"
"No; a reptile—a dinosaur. Nothing else could have left such a track."
Sir Arthur Conan Doyle, *The Lost World*

Had Sherlock Holmes turned his talents to track extinct animals, he would no doubt have made useful contributions to paleontology. His philosophy, like that of his creator Sir Arthur Conan Doyle, has been encapsulated in a favorite quote of trackers (from his story "A Study in Scarlet"): "There is no branch of detective science which is so important and so much neglected as the art of tracing footsteps." But Conan Doyle created not only the world's most famous criminal detective but also a paleontological hero, Professor Challenger, who led an expedition through the impenetrable jungles of South America to discover "the lost world."[1] In this classic tale in which Challenger and his expedition discover living dinosaurs, Conan Doyle created the prototype scientist-adventurer-explorer. The swashbuckling archaeologist Indiana Jones is a direct descendant of this heroic figure, as are the protagonists of the film *Jurassic Park* and its sequel *The Lost World*—with title recycled from Conan Doyle's 1912 yarn.

All three stories use dinosaur footprints to build suspense into the adventure. Conan Doyle was closely involved with the discovery of real *Iguanodon* footprints in his native Sussex. Dinosaur tracks were first reported from the Cretaceous of England in 1846 and soon after were attributed to *Iguanodon*. Moreover, the fortuitous discovery of *Iguanodon* tracks and bones in the same county resulted in a convincing correlation between a dinosaur and its tracks. According to Dana Batory and Bill Sarjeant, a British tracker and an expert on the work of Conan Doyle, a similar fortuitous convergence of circumstance may have inspired the writing of *The Lost World* in the first place; it was certainly instrumental in the creation of scenes featuring footprints.[2] Doyle lived near Crowborough in East Sussex, where dinosaur tracks were discovered in 1909. (A re-

FIGURE 6.1 *Iguanodon* tracks from England are among the first dinosaur footprints correctly identified. They were incorporated into Sir Arthur Conan Doyle's famous adventure story *The Lost World*, which was the first fictional tale to really popularize dinosaurs.

construction of the animal has been incorporated into the coat of arms of the town of Maidstone in nearby Kent. Long before Chaucer and the pilgrims plodded their way to Canterbury, this region was a thoroughfare for *Iguanodon*.) An illustration of one of these tracks in a later edition of *The Lost World* is the only published record of this particular find. In the story, when the *Lost World* expedition comes upon large three-toed tracks they are identified as the footprints of *Iguanodon*. Challenger's colleague, Professor Summerlee, even remarks that the south of England was alive with them back in the early Cretaceous "age of *Iguanodon*."

Although *Iguanodon* is exceptionally well known, its tracks have not been studied in detail, so until recently they had no formal name. Historically they were simply referred to as *Iguanodon* tracks, and as early as the 1860s were probably the first dinosaur tracks correctly matched with the track maker. For technical reasons they have been named *Iguanodontipus*, to convey the precise connotation "*Iguanodon* footprint."[3] The first name formally proposed, by the South American tracker Rodolfo Casamiquela, turned out to be rather a disaster, for he coined the name *Iguanodonichnus*, meaning "*Iguanodon* trace," for footprints from Chile that turned out to be the tracks of brontosaurs.[4] *Iguanodon* is in fact not known from South America. It is ironic that *Iguanodon* and its tracks were first discovered in England, and were only introduced to South America in a work of science fiction. Thanks to Conan Doyle, the global fame of

Iguanodon meant that its tracks were formally named from South America in a work of science fiction!

Many paleontological textbooks make much of the major transitions from one time period to the next: from Jurassic to Cretaceous and later from Cretaceous to Tertiary. The track record suggests that large *Iguanodon*-like ornithopods only began to leave abundant tracks after the beginning of the Cretaceous. The decline of brontosaurian faunas and the rise of *Iguanodon* and its duck-billed relatives has been linked to changes in the environment such as the evolution of fast-growing flowering plants. We must remember that holism extends to the ecosystem, not just to coherence in biological organization at the level of individuals, herds, and larger population groupings. Now we are starting to find that there are distinctive sauropod and duckbill soils, in other words, certain groups were closely tied to particular ecosystems and the soils and vegetation therein.[5] Such observations imply ultimately that all species are an integral part of the environment rather than mere inhabitants of geographic space. The sauropod-ornithopod transition appears to have been a reflection of a shift toward more humid conditions.

A PTEROSAURIAN BIGFOOT OR TWO

The creature was a fit companion for the kindred reptiles that swarmed in the seas, or crawled on the shores of a turbulent planet. . . . [W]ith flocks of such like creatures flying in the air . . . the shores of the primeval lakes and rivers, air, sea and land must have been strangely tenanted in these early periods of our infant world.
William Buckland, *Geology and Mineralogy*, 1858

Since the early days of *Iguanodon* track hunting in the fossil beds of East Sussex, toward the eastern end of the Pilgrim's Way (which winds from England's west country eastward to Canterbury in Kent), there have been many equally important discoveries farther to the west, in the famous Wealden and Purbeck fossil beds in Dorset. The Isle of Purbeck is a peninsula sticking eastwards from the Dorset coast into the English Channel. It is famous for the Purbeck limestone, or "Purbeck marble," quarried since the time of the Roman invasion, and distributed widely throughout Britain. Geologically the Purbeck limestone beds originated as an ancient freshwater lagoon deposit full of snails, fish, turtles, crocodiles, petrified tree trunks, and abundant dinosaur tracks. Some dinosaur tracks acted as traps for fossils, just like modern (and ancient) tracks do today. When footprints trap seeds and water they make excellent germination sites.

Trackways on the cliffs of the Portuguese coast are full of plants, making a delightful natural array of flowerpots. Modern tracks are sometimes used as nest sites by birds.[6]

Among the dinosaur tracks that have been found and identified in the Purbeck limestone beds are those of smallish *Iguanodon*-like ornithopods that sometimes moved in gregarious groups, theropods, armored dinosaurs, and brontosaurs.[7] Enigmatic nondinosaurian footprints named *Purbeckopus* ("track from the Purbeck") are the first tracks of our flying friends the pterosaurs recognized in the British Isles.[8] They represent a species larger than any known from bones anywhere else in the world at this time—a sort of unknown pterosaurian bigfoot that had a wingspan estimated at 12 to 15 feet (3.5–4.5 m). Possible beak prod marks indicate that it might have been feeding in the lagoons.

One wonders how many pterosaurs, iguanoditids, carnivores, and other dinosaurs have marched out of the Purbeck quarries in slabs of stone used for building since the Roman invasion some two millennia ago. Until now we have paid little attention to extinct animals' spoor, but this situation is changing. The true treasures from these ancient quarries are not the marble slabs used for paving courtyards and patios, but the footprints of some very ancient Britons.

A RARE COMMODITY:
THE TRACKS OF ARMORED DINOSAURS

[O]f particular importance to Phil was a section of cliff where a few lone armored dinosaurs, ankylosaurs, had wandered off by themselves . . . To get accurate measurements Phil would have to rappel down the cliff face.

Louis Psihoyos and John Knoebber, *Hunting Dinosaurs*

The armored dinosaurs, or ankylosaurs—"ankylo-" means stiffened, fused (or "ankylosed") armor—are sometimes referred to as the tanks, dreadnoughts, or armadillos of the Mesozoic. They are close relatives of the stegosaurs, or "plated" dinosaurs, but unlike these forms, which are elongate and narrow in body, neck, tail, limbs, and trackways, the ankylosaurs are generally short and broad by comparison. To date, all known discoveries—from sites in Canada, Bolivia, Germany, England, and Tadjikistan—are from the Cretaceous.

We did not arrive at our present understanding of ankylosaur tracks without the usual stumbling, controversy, and confusion. The first discovery, from the Peace River area of British Columbia, was described by Charles Sternberg, one of the most famous dinosaur collectors of his generation. In 1930 he named the tracks *Tetrapodosaurus borealis*, literally "four-footed reptile from the north."[9] For some 30 years they represented the most northerly dinosaur track occurrence known, until the discovery of tracks on the Arctic Ocean island of Spitsbergen, 360 miles north of Norway, astonished the paleontological world with evidence that dinosaurs, still regarded as cold-blooded, had existed north of the Arctic Circle.[10]

Sternberg inferred that his tracks had been made by a horned dinosaur, or *ceratopsian*. This turns out to be a chronological impossibility, however, because horned dinosaurs did not exist until 20 million years later. The mistake was rectified by the Colorado paleontologist Ken Carpenter, who pointed out that the tracks were the wrong age and shape and thus failed the Cinderella means and opportunity test.[11]

In the 1970s the provincial government of British Columbia decided to flood the Peace River Canyon as part of a large hydroelectric scheme, so a rescue mission was mounted by the Royal Ontario Museum to preserve as many tracks as possible. The two trackers involved in this mission were Bill Sarjeant, whom we met previously on the trail of controversial early Triassic dinosaurs, and Philip Currie, now one of the world's leading dinosaur experts. As waters rose around their feet, their team extracted almost a hundred footprints and mapped many more, but they never found Sternberg's original *Tetrapodosaurus* tracks.[12] As a consolation, however, the team discovered the oldest known bird tracks, described below.

It's not the only example of bad luck when it comes to documenting dinosaur tracks. In their lavish book *Hunting Dinosaurs*, Louis Psihoyos and John Knoebber tell of other dinosaur tracks that got away.[13] In 1991, Psihoyos, Knoebber, and Currie planned to shoot photographs of a spectacular array of Early Cretaceous tracks on a steep cliff face above the Narraway River in British Columbia, but when Currie went out by helicopter to investigate, he found that the entire site had fallen into the river.[14] Disappointed but undeterred, the team went off to another site at a vast open-pit coal-mining operation, near Grand Cache Alberta, where long segments of ankylosaur trackway had recently been unearthed. They had already been to the site, and Currie had dangled over the 200-foot cliff trying to get measurements. But they had failed to get the best light for photographs so they were now trying again, after their trip to the Narraway River. But the cliff at the Grand Cache had also disappeared. The surface on which Currie had dangled two days previously had collapsed just hours before. The world's longest ankylosaur trackways lay at the foot of the cliff as an indecipherable heap of rubble. Only a few consolation photographic prints remain from their earlier trip. There are two other consolations, however. Richard McCrea, a Canadian student, has found more important trackways by exploring other Canadian coal mines, where he also dangles on ropes and dodges falling rocks.[15] And in Bolivia a remarkable new site has also come to light, again on a very steep face in a large quarry. No earthquakes, please! For some unfathomable reason the tracks of almost all of the world's most heavily armored dinosaurs seem to be preserved on surfaces that are now nearly vertical and highly susceptible to being dismantled by gravity.

THE EARLY BIRDS:
NEW PERSPECTIVES ON BIRD EVOLUTION

And did god care? . . . A bird-
had stamped her foot
  Emily Dickinson

While evidently holding the power to crumble vast cliffs, Phil Currie has also demonstrated the ability to extract some of the finest and most elusive detail from the track record. The Peace River mission produced a dense cluster of 220 bird tracks that are remarkably similar to those of modern shorebirds. They measure barely two inches in length and are very slender, with toe impressions no wider than match sticks. . . . In 1981, Currie named them *Aquatilavipes*, meaning "track of a water bird."[16]

Until recently Mesozoic bird tracks were considered very rare. Before Currie's discovery only two sites were known, from Colorado and South Korea.[17] This all changed in the 1990s with spectacular discoveries of bird tracks too numerous to count in South Korea near Samcheonpo, the picturesque region of "three thousand bays." Two types of tracks, small *Koreanaornis* ("Korean bird") and large *Jindongornipes* ("bird track from the Jindong Formation") occur in dozens of layers alongside traces of minuscule flatworms and insect larvae no wider than a piece of thread. These invertebrate traces surely represent animals that provided a rich source of food for thousands of waders.

On the basis of skeletal remains it was thought that shorebirds evolved in the Late Cretaceous, perhaps 70 million to 80 million years ago. But the track record from Korea and Canada pushes the history of shorebirds back to around 115 million years, and we may go back further yet. New reports of much older and larger bird tracks, named *Archeornithipus* ("ancient bird track") have just emerged from Spain, and suggest a hitherto unrecognized species of large heron-sized shorebird in existence almost 140 million years ago.[18]

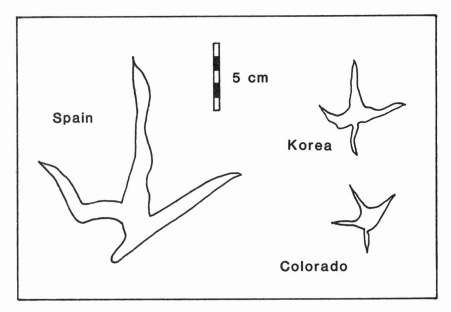

FIGURE 6.2   Tracks of early 140-million-year-old shorebirds tell us that some water birds evolved much earlier than paleontologists ever suspected.

Bird tracks provide a beautiful illustration of the narrowness/width polarity noted for other groups. Songbirds (also known as perching birds), which represent the small, nerve-sense end of the spectrum, have short limbs, and make long narrow tracks that show the rear claw (hallux) extended posteriorly. This morphology mirrors their long tails, another element of the posterior emphasis. In contrast, typical "metabolic-limb" water birds have long limbs, short wide tracks, short tails, and longer necks—obvious anterior emphasis. Again we see the coupled cycles of short limb–long foot or long limb–short foot. Among modern birds, raptors epitomize the central carnivorous group, but neither these nor the songbirds make many tracks. The reasons for this are obvious in relation to habitat, for songbirds live mainly in trees, while shorebirds are always active in ideal locations for making footprints.[19]

SHIPS THAT PASS IN THE NIGHT:
A MESOZOIC CROSSROADS

Ships that pass in the night, and speak to each other in passing;
Only a signal shown and a distant voice in the darkness
Henry Wadsworth Longfellow, "The Theologian's Tale"

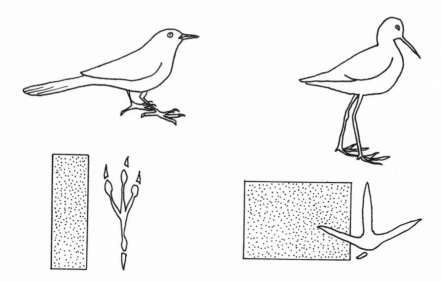

FIGURE 6.3   Tracks as clues to bird anatomy: Songbirds have long tails, short legs, and long, narrow tracks with "tails," whereas waders have short tails, long legs, and short tracks. This reminds us of the narrow-wide polarity seen in other groups.

South Korea provides a good example of the vast wealth of track evidence that still lies undiscovered in Asia. This particular bonanza of bird tracks was discovered serendipitously as a by-product of the study of dinosaur tracks on the rocky foreshore near Samcheonpo, in the picturesque Hallyo Haysang marine park, along Korea's south coast. These footprints represent "the most important paleontological resource currently known from the Korean peninsula."[20] They were found in rocks of the Jindong Formation, a shaley lake-basin deposit of Cretaceous age. In the January 1991 edition of *Seoul*, Korea's glossy monthly magazine, an article on footprints stole the limelight from a Moscow summit meeting, setting the scene poetically: "On a moonlit beach where the tide has ebbed out, one can see distinct trails of the dinosaur. . . . Seawater is pooled in each of the footprints . . . and tiny marine creatures that were swimming playfully . . . scurry away at the sound of human approach."[21]

This image reminds us vividly that dinosaur tracks are part of the modern landscape. Take a walk on a moonlit beach in Korea, and see the same water-filled tracks that once reflected the light of a Cretaceous moon along the shores of a 100-million-year-old lake. In the stillness of night with the waters of the same eternal hydrosphere lapping on the shore, it is hard to conceive that 1.3 billion lunar months have elapsed since a brontosaur made this trail. It is also a source of wonder that the moon, now 4.6

billion years old, was already 4.5 billion years old when the tracks were made.

Is this why Korea's dinosaur tracks are seen to be as important as the Moscow summit? They are likely more enduring than well intentioned political declarations or any of Korea's brief historic dynasties. As paleontologists we can look into deep time to find which dinosaur dynasties held sway when. Since our serious study of the Jindong Formation began just a few years ago, more than 500 dinosaur trackways have been documented. The tracks occur at literally hundreds of different levels, indicating the presence of a large lake basin well established for a prolonged period. During this time literally thousands of dinosaurs traversed the lake shoreline, returning season after season and generation after generation. We have recorded more than a hundred brontosaur trackways, the largest single sample in the world, and more than 250 ornithopod trackways, also a world record. The site also yields the world's largest known sample of Mesozoic bird tracks.

Yet more intriguing than the triple world record is the story behind it. The Jindong lake system is one of the few areas in the world where sauropod and ornithopod tracks occur in abundance in the same formation. Normally they are "mutually exclusive" in an ecological sense, meaning that they require different types of habitats. One group evidently took the high road, while the other took the low road. Sauropod tracks appear to be found only at low latitudes, mostly within 30 degrees of the equator, whereas large Cretaceous ornithopods are found mostly at higher latitudes, mainly in the northern hemisphere.[22] They were ships that passed in the paleontological night. This explanation makes intuitive sense, for both are large, gregarious herbivores that would be competing for the same food resources if they frequented the same habitats. Why then do these two groups occur together in Korea, but hardly anywhere else? The answer may be that Korea falls close to 30 degrees north latitude, and so is in the transition zone where the animals' ranges might have overlapped.

We have also observed that the Korean sauropod and ornithopod tracks generally do not occur together. That is they occur on *different* surfaces within the formation which appears to suggest that different herds of herbivores came into the area at different times, possibly at different seasons. Perhaps sauropods preferred drier, warmer climates, whereas ornithopods were associated with more humid seasons. Closer inspection of the track evidence lends further support to the idea that the two groups came in and out of the area on entirely different schedules, perhaps seasonally.

If the same biological and environmental influences were affecting both groups we might expect the track evidence to be similar in terms of numbers of individuals, size, direction of travel, degree of gregariousness, and so on. But this is not the case. The sauropods were mainly small, moving in random directions, with little or no evidence of gregarious behavior. By contrast, the ornithopod track makers were more numerous, mainly medium- to large-sized individuals, typically moving in groups, and with preferred directions of travel. At many successive levels these trackways are all oriented in the same direction, toward the southwest, as if they came into the area again and again and traveled in the same direction, perhaps along a well-established migration route. The tracks are all more or less the same size, suggesting subadult or adult dinosaurs on the move. Hints are that the track record encodes intriguing biological and ecological phenomena, hitherto unrecognized in ichnological studies. Whether these speak of seasonal migrations, population dynamics, or the partitioning of food resources among herbivore species, it is safe to say that the shores of the Jindong lake system were a Mesozoic crossroads for many species.

BABY BRONTOSAURS

And in such indexes, although small pricks to their subsequent volumes,
there is seen, the baby figure of the giant mass of things to come at large.
Shakespeare, *Troilus and Cressida*

And now for the Korean brontosaurs. According to *The Dinosauria*, the essential reference work for all dinosaur researchers (published in 1990): "[A]ll sauropods were large and ranged from 12 to 30 meters or more in total length."[23] This points to a simple conclusion: any sauropod less than about 40 feet (12 m) in length must not be fully grown. The whole issue of baby sauropods is quite controversial, and some experts, such as Robert Bakker, have claimed that some sauropods gave birth to live young.[24] Although it is hard to imagine a sauropod hatching out of an egg not much larger than a big grapefruit and then growing to be more than 100 feet (30 m) in length, nest-site evidence suggests that they probably did just this. In order to reach their large adult size they probably grew quite rapidly, at least when young.

The Dinosauria also states that "the overwhelming majority of sauropod specimens are at least 80 percent of adult size."[25] Since sauropod footprints range in size from less than 8 inches (20 cm) to 40 inches (100 cm) in length, we should expect most tracks to be more than 32 inches (80 cm) in length. The track record, however, gives us a very different perspective. In most areas, the majority of sauropod tracks are much less than 80 percent of adult size, and often the majority are less than 50 percent of adult size.[26] In South Korea the majority of tracks are less than 40 percent of adult size. Does this mean the sauropod bone record is biased toward the preservation of large individuals, thus distorting our calculation of the average body size? Perhaps mortality among juvenile sauropods was low and perhaps those that did die young were eaten, or their bones weathered away before they were buried as fossils.

Clearly, tracks give us potential insights into the distribution and activity of baby sauropods. The Korean sauropod track sample holds the record for the world's smallest brontosaur footprints: 7 inches (18 cm) long. Such measurements indicate animals whose trunks are comparable in size to that of a very large domestic dog, diminutive by sauropod standards. Given that no diminutive species are known, these individuals must have been baby brontosaurs. Their trackways indicate that these juveniles were milling around without any obvious sense of preferred direction. Could it be that the age of these animals accounts for the unusual trackway patterns? Were these animals that were too young to travel or migrate in well-organized groups? And if this was the case, were these baby animals still quite close to their original breeding grounds? Are the random directions and curved trackway orientations a reflection of juvenile behavior? If we draw upon modern analogies, we find that juveniles of many species, including humans, scamper and mill around quite freely, with a less purposeful and sedate sense of direction typical of adults.

WADING IN DEEPER

O'er bog, or steep, through strait . . . with . . . feet pursues his way, and swims or sinks or wades.

John Milton, *Paradise Lost*

Our view of brontosaurs, those most famous of Mesozoic behemoths, has changed radically since the 1930s. They were then regarded as aquatic creatures that wallowed in swamps because they were unable to support their great weight on land. The American paleontologist Roland T. Bird, who is credited with the discovery of the first well-preserved brontosaur tracks, in Texas, found it so hard to accept that brontosaurs were terrestrial that he titled one of his papers "Did Brontosaurus Ever Walk on Land?"[27] As a result he adjusted his interpretation of the environments where the footprints were found to construct watery habitats in which the track makers waded and swam rather than walking normally.

Bird made his initial discovery in 1937 along the Paluxy River, near Glen Rose in Texas, only days after visiting Colorado's Purgatoire site. Bird was looking for theropod tracks on limestone ledges along the riverbank. One day, while eating lunch, Bird took a second look at the bathtub-sized indentations into which he had "thoughtlessly been shoveling the present mud of the Paluxy to get it out of the way." It struck him that these were footprints "of some sort," but they were so large that "the very thought appalled" him.[28] He soon overcame his initial astonishment and set to work to document the footprints, which were to prove the "discovery of the year," according to the vote of the newly founded Society of Vertebrate Paleontology in 1938, and ultimately the discovery of his career. The Texas tracks have been reliably dated at about 100 million to 105 million years ago and represent the activity of dinosaurs on the shores of what was essentially the Cretaceous Gulf of Mexico.

During several years of excavation and study at various sites in Texas, Bird wrote classic popular articles to suggest scenarios of herding, swimming behavior, and predator-prey interaction. Bird's suggestion that parallel brontosaur trackways provide evidence of herd behavior has been universally accepted and validated by additional discoveries. Much more controversial are his interpretations of trackways as evidence for swimming and predator-prey skirmishes.

The swimming-sauropod scenario resulted from the discovery of a trackway consisting mainly of front footprints. Obviously it strains credulity to suggest that the trackway was made by a brontosaur walking on its front feet, like a circus elephant, so some other explanation must be found. Being predisposed to think of brontosaurs as aquatic creatures, Bird suggested that the animal had been buoyed up in 10 or 12 feet (3–4 m) of water with the hind feet floated along behind.[29] There are several problems with this scenario. First, Robert Bakker made a compelling case that brontosaurs were fully terrestrial creatures, much more like elephants than like hippos.[30] Second, the Texas track layers represent a marine environment, so the sauropods would be swimming in the sea, thus suggesting marine adaptations. Moreover, the gradient of the coastline was so gentle that one has to travel tens of miles offshore in order to get into water 12 feet (3–4 m) deep. Though sauropods could probably swim, we must be clear that the track evidence does not indicate such behavior.[31] They did not live on beaches and go to sea like seals.

The reason the scenarios don't work is geological: the tracks of the front limbs are now known to be "ghost prints," prints transmitted through from an overlying layer where the animal originally walked onto what we call underlayers, where footprints may appear quite different in form.[32] Subsequent studies in other regions reveal that trackways in which only the front footprints are found are quite common, as we shall see below.[33] In all such cases the trackway reveals a typical walking configuration, but the back footprints did not sink deep enough to leave clear impressions. The scenario of the aquatic dinosaur is appealing, and still shows up regularly in text books. Ghost prints have given rise to delusional or ghost theories that still haunt even the accessible mainstream literature. We like to read "interesting" interpretations of behavior into footprints if the opportunity arises. The alternative scenario is, after all, quite mundane and pedestrian—they were simply plodding along.

GHOST PRINTS, PHANTOM TRACKS, AND PHANTOM LIMBS

The phantom can feel so real that people who have had their leg removed can easily forget it isn't there.

Rupert Sheldrake, *Seven Experiments*
That Could Change the World

Ghost prints are literally a type of ghostly writing that appears on layers underneath the actual layer on which an animal walks. It is essentially the same phenomenon as the ghost writing that appears on a page beneath the one on which we have written with a ball-point pen. The resulting script can be quite clear, as in the case of carbon paper duplicates. The quality of carbon copies and ghost writing depends on the type of writing implement used and the rigidity of the pad and paper.

Paleontological ghost prints, however, may be harder to date than our carbon duplicate, because the track maker may leave its mark on layers that accumulated years, decades, or centuries before it was born. This would be analogous to writing on a sheet of paper with a ball-point pen, but having underneath the sheet of paper a layer of medieval parchment and perhaps a sheet of Egyptian papyrus below that. In terms of geological time the differences may be slight, but nevertheless the phenomenon of ghost prints is unique to the world of tracks: there is really no paleontological equivalent such as a ghost bone or a ghost skeleton. Geologists, however, do recognize ghost structures, or halos, like relic auras, which are found around or in place of mineral grains that have dissolved, leaving only the faintest trace of their original form.

Ghost prints represent a record of animal existence literally transmitted back through time by causing a disturbance in a layer never actually touched by the limb itself. No wonder ghost footprints confuse trackers. In most cases we cannot retrieve as much detail from a ghost print or ghost trackway as we can from a trackway on the original surface, but some useful information is available. For example, ghost trackways still give us the direction that the animal was moving in, and the length of step and stride.

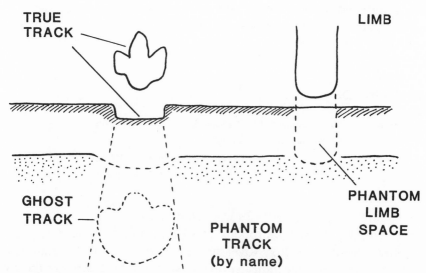

FIGURE 6.4   Ghost tracks impact layers not touched directly by the foot, and so are symbolic of the field of influence beyond the physical limb, much like a phantom limb. Some have even called such ghost tracks phantom tracks.

The German tracker Hartmut Haubold has been interested in the extent to which trackers have been misled by poorly preserved tracks. He has shown that inexperienced and overenthusiastic trackers have given new names to poorly preserved examples of tracks that are already known. Depending on the substrate, the same foot can make different tracks and the same pattern of differences may show up consistently, depending on whether one is looking at tracks made in sand or mud, original surfaces or underlayers. Haubold coined the term "phantom taxa" for inappropriate names given to tracks that should not have been named. Sometimes more complete expressions of these same track types already have perfectly adequate names.[34] The actual phantom taxon may be as different from the true taxon as a ghost print is from a true track, or as we are from our shadows. The phantom taxon might appear as real to the enthusiastic tracker as the phantom limb does to its owner.

So there is a shadowy underworld underfoot. Our footprints leave ghost prints on layers we cannot even see. This ghost writing tells a story similar to that which we can read in the surface tracks, but the message is not always so clear because it is distorted through the filter of substrate. Sometimes we can translate this distortion, but not always. Ghost-print images that bear no obvious resemblance to familiar footprints may be given misleading phantom names. It is perhaps not surprising that ghost prints should sometimes have phantom names, even though this may confuse the uninitiated tracker. Tracks, after all, are the ghosts or shadows of bygone creatures. Sometimes we will find ourselves tracking shadows of shadows.

BIRD'S HERDS

No less than a dozen sauropods had crossed this section. All were progressing in the same direction as a herd.

Roland T. Bird, "A Dinosaur Walks into the Museum"

We should not overlook Bird's important contributions to our current understanding of social sauropods. In 1939 Bird first reported 12 parallel trackways from the Paluxy River site (now Dinosaur Valley State Park) and suggested that they indicated the passage of a herd through the area. He then found another track site, in an area known as Davenport Ranch, where he made a meticulous map of 23 parallel brontosaur trackways, further supporting his case for gregarious brontosaurs. He was puzzled, however, by the lack of tail drag marks, and so proposed that the animals must have been walking in shallow water with their tails floating behind.[35] Many museums still show brontosaur tails floating horizontally just off the floor, reflecting Bird's original intention to place a sheet of glass at this level to simulate water.

At the Davenport Ranch site Bird recorded a mixture of large and small animals. This evidence caused Robert Bakker to propose that the herd was "structured" with large individuals at the periphery and the smallest only at the center.[36] Knowing that African elephants sometimes travel with their young in protected positions in the center of herds, Bakker was encouraged to extend the same scenario to brontosaurs. It is an appealing idea we would "like" to believe, but the track evidence does not confirm that sauropods were protective parents. Nevertheless, in popular magazines such as *National Geographic* and *Natural History* this moving creche scenario has been extended to many other dinosaurs. Such conjecture is misleading, as there is currently no dinosaur footprint evidence to support these hypotheses. I searched hard for such evidence but found only one published photo of a pair of shelduck trackways on either side of the footprints of their clutch of ducklings.[37]

Before dismissing Bakker's structured herd hypothesis, already disputed by John Ostrom, his former professors at Yale, I attempted a careful analysis of the trackway patterns.[38] From such an analysis we can begin to string out two dozen brontosaurs of different sizes along the ancient

FIGURE 6.5   Tracks of a family of shelducks reveal that protective parents shepherded their ducklings across an open beach. Similar behavior has been suggested for dinosaurs, but without corroborating trackway evidence.

shores of Texas, creating a sketchy picture of how the herd might have looked had we been watching it pass more than 100 million years ago. Once the track map had been thoroughly analyzed and color coded it became clear that in some cases small brontosaurs followed the larger ones, but in other cases there was no obvious relationship.[39] The herd also veered a little bit from right to left, but we do not know why. If anything we must conclude that this particular herd was "unstructured." The regularly spaced trackway patterns recorded at the Purgatoire, Portugal, and Paluxy sites could more readily be said to be structured. I concede that we find structure and order almost everywhere, notably in parallel trackways. I can even visualize rhythm entrainment and protective parents, but have to stop short of claiming actual proof based on current trackway evidence.

TRACKERS AND ATTACKERS

Was the flesh-eater merely following the big sauropod's path . . . or was he in actual pursuit of the monster?

Roland T. Bird

One of the great attractions of dinosaurs, especially as portrayed in the media, is that they infuse Mesozoic scenes with drama. What is a dinosaur movie without a *Tyrannosaurus*, or an allosaur attacking some unsuspecting herbivore in the middle of a leafy lunch? Much of the evidence for predator-prey interactions, however, is circumstantial, though it is quite clear that theropods were carnivores. They had raptorial claws and sharp serrated teeth, like steak knives, which have often been found shed around the remains of large dinosaurs on which they fed. Such teeth leave scrape marks on bones, creating a new type of trace fossil for trackers to study.

Given that carnivores frequently fed and scavenged on carcasses, we might expect to find abundant tracks around the remains of dead individuals.[40] This is rarely the case, though, mainly because tracks were obliterated either by a feeding frenzy of trampling or by a later frenzy of paleontological excavation. In the past, bone-o-mania was more prevalent than it is today. Modern excavation techniques are more careful and treat tracks, coprolites (fossil feces), and skin impressions with reverence. Even so, they are often found only as a by-product of bone hunting. Generally, the view that skin, feces, tracks, and the like are rare and insignificant has encouraged a self-fulfilling prophecy: namely, that one doesn't find them if one is not looking for them. Nevertheless, good examples of tracks and trampling, suggestive of predation or scavenging around carcasses, have been reported from sites like the Howe Quarry near Shell, Wyoming. Such predation sites hold the potential to match theropod footprints with teeth that were shed while feeding on prey. Dinosaur skin from this particular site was discarded and destroyed by early excavators. New finds of brontosaur skin proved highly significant, since they revealed frilly ornamentation running down the midline of the back and tail. This discovery is still ignored by the Hollywood movie industry which prefers the traditional smooth-skinned elephantine look for brontosaurs.[41]

Let us look again at a much quoted example of a trackway purported to reveal evidence of a predatory dinosaur attacking its prey, at Dinosaur Valley State Park, near Glen Rose, Texas. Excavated by Bird and his crew in 1939, it revealed parallel trails of a dozen brontosaurs and four carnivores. They removed a large section of rock 29 feet long by 8 feet wide (8.7 by 2.4 m) for the American Museum of Natural History, and several adjacent sections for various other institutions willing to pay the freight charges. Together the slabs weighed more than 80,000 pounds, constituting the largest track excavation ever conducted. The slabs, shipped more than 2,000 miles to New York, must also be the only fossil footprints threatened by torpedoes, for as Bird recalled he fretted over the danger of them offering "fat targets for German submarines" in the Gulf Stream.[42]

Bird's map, published posthumously, in 1985, reveals 3 theropod trackways following the same trend as the 12 parallel brontosaur trackways.[43] Extraordinarily, Bird produced several versions of the site map, some drawn much later from memory! It appears he was something of a cartographic savant, because the different maps match each other quite well. Bird chose to excavate a section of carnivore trackway aligned more or less parallel with a brontosaur trackway. Both trackways indicate animals walking south and turning slightly to the left, in a broad anticlockwise arc. An anomalous feature of the theropod trackway is that it appears to be missing a left footprint at the point where it converges most closely with the brontosaur trackway. This missing footprint has caused more debate than any other missing track in the entire history of dinosaur tracking, and has been the sole foundation of the protracted debate over whether the carnivore was really attacking the brontosaur.

There is no firm consensus as to what the tracks really show. Since the attack scenario is so popular, it is worth going back to the beginning to see how the whole idea evolved. Bird first reported the "three-toed prints" of a two-legged dinosaur that "was a dynamic machine for the pursuit and capture of prey, which he tore to shreds in his powerful jaws and hind feet."[44] Significantly, he concluded that the "smaller flesh eating dinosaurs had come along there first, for I had found a sauropod footprint impressed on one of theirs." If this is true, the whole attack scenario of predator following prey is falsified, since the predator evidently passed through first.

In Bird's second paper, he reported uncovering footprints of a "fourteen-foot flesh eater that was actually stalking" or "apparently following the trail" of a big brontosaur.[45] Three years later, in 1944, Bird still pondered this scenario and asked, "Was the flesh-eater merely following the big sauropod's path . . . or was he in actual pursuit of the monster?"[46] In his last article on tracks, in 1954, Bird inferred that the brontosaurs were "preyed upon by dagger-toothed foes" and worried about their chances "if caught on land" by the huge flesh eaters. In fact, it gave Bird a "wincing tingle to see the deep marks of the murderously sharp talons" of an animal now characterized as "a strapping younger brother" of the king tyrannosaur "capable of attacking and perhaps even killing the biggest of brontosaurs."[47] The predator scenario clearly evolved in Bird's imagination without being fueled by any new footprint evidence. I shall now show the extent to which this is true and how it has caused the popular attack scenario to become entrenched and perennial in paleontological folklore.

Bird's papers are full of whimsical imaginings of Cretaceous scenes. "Looking off in imagination . . . I saw palm trees waving above the fringe of jungle where the carnivores had gone. . . . In my mind's eye I thought I saw the dark bulk of a partly submerged brontosaur . . . then not one but a dozen heads lifted on long necks. . . . Imagination? Who could fail to see these sights here with the aid of the dinosaur tracks?" After reassembling the trackways in the American Museum of Natural History in New York City, he imagined "the brontosaur, great black flanks and belly streaming water, long neck weaving as it tried to escape an open mouthed lunge of the powerful killer."[48]

Nowhere does Bird carefully analyze the trackway evidence. The pivotal evidence, the missing left footprint, only gets mentioned decades later in his posthumously published autobiography. Here Bird actually sketched a theropod attacking the brontosaur's left flank. His recollection of an interview with a reporter is equally revealing. He suggested that "likely . . . the meat-eater was trying to urge the big fellow up onto solid ground . . ." where the brontosaur might be devoured.

"How do you think it might have turned out?" asked the reporter.

"In this case, I don't think." The track evidence "apparently" shows the carnivore "hopping along on . . . the right foot, . . . " leading Bird to ask, "Did he have a rear claw set in the belly or the rump of the plant-eater at his side?"[49] It is not clear whether Bird thought the animals were in water or on land, nor can one really see how the carnivore could have had a left rear claw "set" in its victim's side while making normal tracks with the right foot.

Perhaps it is precisely because so little of Bird's attack scenario hypothesis has been carefully explained that it has become a popular word-of-mouth myth. Any serious tracker who looks at the evidence carefully will probably conclude that the scenario owes much to Bird's imagination. The tracker Jim Farlow, an expert on Texas tracks, made this wry comment on the scenario: "[T]he theropod might have caught up with its intended victim, dug its heels in an effort to slow the big herbivore, . . . and then have been pulled out of its track. The astonished flesh-eater. . . found itself airborne for an instant, and unable to put down its left foot. Keeping its archosaurian wits, the carnivore recovered its balance enough to implant its right foot at the end of an involuntary hop." This, Farlow concludes, is perhaps reading too much into a missing footprint.

Farlow raised many other commonsense objections. For example, the estimated speed of the brontosaur—4 feet (1.2 m) per second, 2.5 miles (4.3 km) per hour—"is rather low for an animal trying to avoid an imminent, and possibly fatal, encounter with a predator." Farlow estimated the predator's weight at about 2,000–3,000 pounds (or about 1.5 tons), whereas the brontosaur probably tipped the scales at anywhere from 5 to 30 times the weight of the carnivore. Farlow's best guess was that it weighed 10 to 15 tons, or was "10 times as heavy as the meat eater." This suggests "the meat eater was a rather ambitious reptile for hunting an animal so much larger than itself."[50] I agree with Farlow that a one-on-one attack scenario is highly improbable, especially as both trackways show normal walking progression. In recent human history we have heard of a "low-speed chase," but this is surely the exception, not the rule.

Bird did not ignore the larger picture entirely. In 1976 he wrote that "a second carnosaur [was] seen as though coming to the aid of the [carnosaur] that had presumably seized a restraining hold on the sauropod." Bird went further to suggest that the sauropods were, "apparently, prompted to flight by the appearance of at least four carnosaurs." As Farlow concluded, "[T]his is a spectacular scenario—a carnosaur pack harassing a sauropod herd—[but] unfortunately, there is little to back it up."[51] So why does the Dinosaur Valley State Park visitors center and the cover of Farlow's guidebook show a picture of a theropod attacking a sauropod's left rear flank?[52] Many publications cling to the attack scenario, and a widely used textbook, *General Science, A Voyage of Discovery*,[53] presents a dinosaur footprint interpretation exercise that is based on this supposed skirmish. Although educational to some degree, the exercise is based on a map that is entirely fabricated – what I consider a pure fantasy. Why does Farlow not simply say, "Bird was wrong!" Perhaps he speaks for many of us when he writes, "I would love for it to be true."[54] We want an appealing scenario, and feel a little cheated if the track evidence does not indicate exciting or unusual behavior. Besides, Bird was such a nice guy, so devoted to paleontology, so harmless, so imaginative and so beloved. How dare we say categorically that he was just plain wrong?

SAUROPOD SERENADE

The Earth has music for those who listen.
William Shakespeare

"Bird was right!" Or so said Dave Thomas to the Society of Vertebrate Paleontologist: "The trackway is consistent with the theropod striking behind the left hip, sliding down the tail and then releasing."[55] I've told Dave I don't believe the attack scenario, but I admire his guts in presenting it yet again—this time with a musical score. Based on an analysis of the map, he suggested that the attacker (carnivore 1) and its prey (sauropod 2) had synchronized steps. In modern chases the predator falls into step with its prey while running, not walking. Thomas went further, suggesting that the dinosaurs were traveling in 4/4 time: "The correlation between two trackways is so precise,. . . " he wrote "so consistent and so long lasting . . . that it can be written as music." Note: Only 12 consecutive steps were studied!

Appropriately, the sauropod is given the more resonant base line and the lighter-weight carnivore, the treble. Thomas stressed that the scene should not be read as two animals moving side by side. Rather, the trackway of the carnivore has to be shifted four steps forward (in space and time) to synchronize with the steps of the sauropod. Thomas perceived that after the first measure, what I call "the chase," the carnivore took three steps, or "two dotted quarter notes," and during "the attack," it took a quarter note to the sauropod's four quarter notes. In the next measure, with the predator and prey still at what I would call close quarters, Thomas inferred that the pre-attack synchronicity of steps is resumed.

In 1995, Thomas, a musician, presented this bold musical theory of Mesozoic predation to a large audience of the world's leading paleontologists and enthusiastically beat time through what he perceived as the chase and attack phases. As he conducted his prelude to an afternoon of predation the audience broke into spontaneous applause followed by cries of "Encore!"

I am happy to concede that there is a rhythmic beat of feet as any animal makes tracks, and that flights of geese and marching battalions display rhythm entrainment, but I am not convinced that the carnivore fell into synchronous steps with the sauropod or that there is compelling ev-

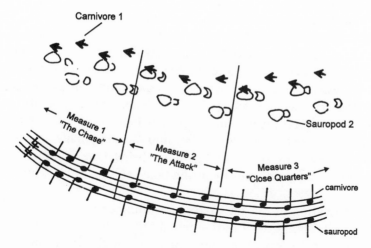

FIGURE 6.6 A story we'd love to believe: Beastly carnivore attacks a giant brontosaur. The trackways of the supposed predator-prey scenario has even been set to music in an attempt to validate the paleontologist Roland Bird's attack hypothesis.

idence of an attack. I have mapped out the scene of a one-on-one hippo battle on an African mudflat and it was an irregular mess, with much evidence of running, slipping, sliding, and changing of direction. However, with the help of James Farlow, Thomas published his ideas in *Scientific American*—or is that *Scientific Musician*?—where we read, "These animals were undoubtedly interacting" and "it now appears perfectly clear that. . . at least one swift carnivore singled out and possibly attacked a huge, lumbering herbivore."[56] *Scientific American* or no, I dissent. How can it be "perfectly clear. . . possibly"? Music hath charms to soothe the savage skeptic.

CREATIVE MIDNIGHT CHISEL WORK: EVOLUTION VERSUS CREATION

He that troubleth his own house shall inherit the wind.
Proverbs 11:29

What do dinosaur tracks have to do with divinity? The answer is simple: they have as much to do with divinity as any of God's creations. They are as much a part of the grand scheme as the stars and the sea. As humans we have a strong inclination to search for meaning. For the Bushmen, the best of all trackers, everything has meaning. This is perhaps because they live so close to the Earth that every fiber of nature's being is vibrantly real.

If we were to speak to them of God, or a great spirit, they would be the first to acknowledge the existence of such a divinity.

To some people the term "God" implies something exclusively religious, mystical, and nonscientific. In a technological world, God has tended to become an elusive phenomenon whose existence cannot be proved scientifically. Those who demand proof of God's existence tend to be those who have made science, rationality, and intellectual endeavor into a religion, and the main avenue through which they find meaning in life. But we can make anything we want to into a religion or a belief. One Olympic athlete surprised immigration officials by listing pole vaulting as his religion on his passport.

The interpretation of certain Cretaceous dinosaur tracks in Texas has become the focal point for a monumental argument about meaning in North American society, one that ripples through classrooms from kindergartens to universities. The debate has become needlessly polarized as science versus religion, or evolution versus creation. In my opinion, the entire debate is largely futile, but it is a real sociopolitical phenomenon in our society and so cannot be ignored. The most famous evolution/creation confrontation was acted out in Tennessee in 1925 after that state enacted legislation to prohibit the teaching of evolution in schools. A young biology teacher named John T. Scopes was prosecuted for allegedly violating this law. The intellectual and antifundamentalist community rallied to his support as the case made the national headlines. Scopes was prosecuted by the fundamentalist lawyer, ex-presidential candidate, and powerful orator William Jennings Bryan, and was defended by the brilliant lawyer Clarence Darrow. Although Darrow was not permitted to call any expert scientific witnesses, he was able to cross-examine Bryan on the literal truth of Genesis and got him to concede that the Earth was much older than most fundamentalist views allow.

Although Scopes was found guilty (he was fined the minimum amount of one hundred dollars), the Scopes trial was seen as a triumph of science over fundamentalism, and "Tennessee was held up for ridicule to the entire nation" for its backward and prejudiced legislation.[57] Bryan died within a few days of the trial, at least in part owing to the stress of public humiliation. The drama of the trial was recreated in the play *Inherit the Wind*, by Jerome Lawrence and Robert E. Lee, and in 1960 was made into a film, in which the role of Darrow was portrayed by the Hollywood star Spencer Tracy. Incidentally, in the play, when the local minister gets into a frenzy and asks God to curse the godless evolutionist and his own daughter, who pleads on Scopes's behalf, his lawyer reminds him that he who troubleth his own house shall inherit the wind. Let us keep this biblical admonition in mind.

When I first encountered the subject of dinosaur tracks and creationism, I was naively unaware that people could argue so vigorously and recognize no common ground. As a paleontologist, I felt that the creationists were being unreasonable when they dogmatically labeled us scientists evolutionists, and went too far when they implied that evolutionists denied the existence of God. Scientific evidence for evolution does not make scientists godless, spiritless beings. I always preferred Teilhard de Chardin's viewpoint, that the world of natural science is the starting point for spiritual investigations. I see plenty of evidence that scientists acknowledge higher divinity, whether it be in Aristotle's dictum that "Beauty is the gift of God" or in the ruminations of cosmologists awe-struck by the brilliance of God's mind.

So what cultural schizophrenia creates two polarized camps, each claiming to know the true meaning of fossil footprints? Perhaps neither side really cares what the other believes until one camp demands that its view be taught in schools. I personally believe we should all be pole vaulters to elevate our bodies, minds, and spirits. Nobody objects to this view because, until now, I have been very quiet about it. When it comes to tracking I am less flexible about what I believe is meaningful. This is why I bring up the subject of creationist views; bringing up these views for discussion does not mean I think the creationist view lacks meaning.

Dinosaur footprints entered the evolution/creation debate in the thirties, when Roland Bird stopped at a remote trading post in Gallup, New Mexico, where he heard reports of footprints made by giant bears and men 12 feet tall. But they proved to be carvings, imported from Glen Rose Texas. Bird headed there and soon verified that in those economically depressed times, "the only money-makin' jobs is cuttin' cedar posts, bootleggin', and quarryin' dinosaur footprints."[58] There was no evidence that those doing the carving were deliberately out to convert anyone to a particular religious belief. Bird paid little attention at first and became entirely accustomed to casual talk of "man tracks." From the outset Bird also knew that most of these "man tracks" had been made by theropods wading in deep mud leaving elongate depressions.

Bird confessed to making the mistake of too often discussing both the real and carved man tracks. Although he clearly stated that no humans had lived during the Age of Dinosaurs, creationists evidently claimed that Bird had said he could not prove the tracks were not human. At first Bird did not object to such claims published in local religious magazines, "if it helped some in their worship of the Almighty."[59] But when a book entitled *Man's Origin's; Man's Destiny* appeared, whose author went "to great lengths to prove that [Bird] did not know what he was talking about," Bird lost patience. The last straw came when the famous "seer," Jeane Dixon, wrote that "Dr. Roland T. Bird . . . reported finding giant petrified human footprints along with those of dinosaurs in a river bed in Texas." Was such blatant distortion accidental, or just poor research by Ms. Dixon, who had obviously not read Bird's papers? When he was again fit to speak, Bird "fired off a scathing letter to this female 'seer' demanding an immediate withdrawal of [his] name from all future reprinting." By return mail he received an immediate apology from her ghost writer.[60]

James Farlow, who was instrumental in publishing some of Bird's letters relating to the controversy, concludes that many in the scientific community regard the creationists as nuisances that if ignored might go away.[61] This approach may not work because once the seed of an idea is planted it will not just "go away." Once Bird mentioned "man tracks," a ghostwriter could script a theory about how scientists are willing to investigate such phenomena. What Bird actually said about the tracks was of secondary importance.

The "man tracks" debate did not go away. In 1986, when dinosaur trackers assembled in Albuquerque, New Mexico, for the First International Symposium on Dinosaur Tracks and Traces, creationists were touting a book entitled *Tracking Those Incredible Dinosaurs . . . and the People Who Knew Them* and a film entitled *Footprints in Stone*.[62] Glen Kuban, a Christian and, at the time, a dedicated amateur paleontologist, decided to make a careful study of the "man tracks." He published his results in the symposium proceedings while two other authors published a refutation of the "man tracks" claims under the memorable title "Dinosaur Tracks, Erosion Marks and Midnight Chisel Work (but No Human Footprints) in the Cretaceous Limestone of the Paluxy River Bed, Texas."[63] The conclusions of these authors, were essentially the same as those of Kuban and Bird, that the tracks had been made by theropods walking in deep mud. Kuban's suggestion that they adopted a crouched position as they walked, as if stalking prey, is perfectly compatible with the behavior characteristic of nerve-sense animals.

Although willing to hear all creationist arguments, Kuban was unable to find any evidence to support their claims. As a result of his work they withdrew their film *Footprints in Stone* from circulation, apparently conceding defeat. Despite the debunking of the "man tracks" hypothesis, the subject was recently revived again in a 1996 NBC production, *The Mysterious Origins of Man*, narrated by Charlton Heston. In this program a number of creationists were presented as scientists, and without their affiliation to creationist institutions being revealed. When one couples this with Heston's reputation as the man who played Moses, deliverer of God's biblical commandments, one wonders how many gullible viewers were persuaded to buy in to the idea of humans coexisting with dinosaurs,

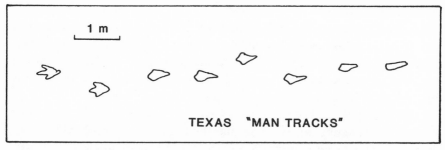

FIGURE 6.7   Theropod dinosaur tracks made in deep mud have been misinterpreted as human footprints or "man tracks." Careful study of these tracks reveals no evidence to support these claims, despite persistent creationist propaganda in favor of this bizarre and highly anomalous hypothesis.

when they are entirely unaware of the true history of the debate. Were such persons to read Leonard Brand's new book, *Faith, Reason and Earth History,* they would find a creationist denouncing the Cretaceous "man tracks" evidence.[64]

Where programs such as *The Mysterious Origins of Man* break down is that they fail to make a distinction between such disparate topics as the age of dinosaurs, the early history of hominids, and the possibility of advanced predynastic civilizations influenced by extraterrestrials. These questions deal with dates on entirely different orders of magnitude—100 million, 100 thousand, and 10,000 years ago, respectively. Such arguments underscore the shaky scientific background on which creationist challenges to geological knowledge are usually based. Moreover, the creationist camp generally fails to marshal arguments for a creative force or spirit that would be taken seriously by the theological, intellectual, and scientific communities. Ironically, it is from within the ranks of academia, rather than the fundamentalist community, that the best arguments for some form of "God" and "creation" emerge.

EVOLUTION AS CREATION:
SCIENCE AND SPIRITUALITY

Not everyone is blessed with the ability to see a higher order of things.
Anne Frank

We can probably all agree that the universe and our earth came into existence a long time ago, and that humans had nothing to do with its "creation." If we can agree on this much, then some common ground exists for many of the disparate views expounded by various religious and scientific philosophies. We can probably also agree that philosophers, theologians, and scientists who have pondered creation and evolution for many millennia and taken the trouble to write extensively on the subject should generally have something worthwhile to say. And what they often say is that there is substantial common ground among the three groups.

I began by pointing out that the universe is probably cyclic, like most other observable phenomena, and that the Big Bang was probably the beginning of the current observable cycle. Since this cycle got under way, some remarkable events have been unfolding. Scientific evidence and ancient creation myths seem to agree on a general trend. From out of the void or nothingness, matter came into existence in the form of galaxies, stars, and planets and the elements that compose them; later, life appeared, leading ultimately to the emergence of humans, whose "special" faculties of awareness and consciousness created the very evolution and creation questions we are seeking to understand. In short, we have seen the evolution of the physiosphere, biosphere, and noosphere in the form of inanimate matter, life, and mind.

Ultimately the big question comes down to this: Why is all this happening, and what, if anything, controls this mysterious unfolding of events? Anyone who claims to have an answer is addressing a question that has intrigued, mystified, and frustrated scientists and philosophers for centuries. We are all entitled to our opinions, but to be credible we should take the question seriously and look at the essence of the ideas in the voluminous literature and opinion on the subject.

A central issue is, what drives this process of creation and evolution? Is it "simply" God, as stated in the Bible and other revered texts? Is the question merely a manifestation of humans' desire to know themselves and their origins—an awareness of something elusive yet so intriguing that it demands explanation? Is there an explanation in terms of statistical probability that we might grasp if we consulted the right mathematician? Are there intelligible "fundamental constants," such as gravitation and the speed of light, and if so, why are they constant? Or is the concept of cause and effect an archaic mode of mechanical thinking, which is not applicable to life and mind? Are the answers to all these questions convergent to some degree? I am inclined to think so and suggest that science and spirituality are different sides of the same coin, and that the quest for the Holy Grail of knowledge and understanding is a common thread. Apart from my intuitive sense that a convergence of science and spirituality is ultimately inevitable and "right," I am comforted that other serious minds have tackled this problem and arrived at similar conclusions.

Let us consider the subject of teleology, the concept that there is design and purpose in nature, and relate it to the psychological mode of thinking—an orientation toward future goals and objectives. We have acknowledged that we humans did not create the universe—so what makes organisms evolve from bacteria to jellyfish to primates, and why do humans argue and write about philosophical, theological, and scientific problems? Because the universe seems capable of organizing itself to produce both consistency and extraordinary complexity and new evolutionary novelty, there is no scientific, philosophical, linguistic, or theological reason not to speak of God, spirit, universal intelligence, or consciousness to describe such awe-inspiring coherence.

Though we comprehend some of this complexity through an understanding of electromagnetism, genetics, psychology, etc., we have to acknowledge that a remarkably intelligent universe has been operational since the beginning of time. Our own emerging intelligence, awareness, and consciousness merely enhance our ability to appreciate this universe and give labels to phenomena. Our consciousness appears to be evolving like a radio receiver capable of tuning into more and more channels. Complex phenomena such as electricity, DNA, or music existed before we understood them, which is precisely why we appreciate them and often sense that evolution appears to be unfolding with a sense of design or purpose toward greater complexity.

These observations relate directly to human evolution. *Homo sapiens* developed a refined aesthetic sense tens of thousands of years ago, expressed in complex tools, cave art, sculpture, and so forth. According to Teilhard de Chardin and others, man's intellectual and aesthetic faculties were essentially as fully developed in the Upper Paleolithic (30,000 years ago) as they are today.[65] The fine points of this conclusion can be debated later, but on a simple level we can conclude that the noosphere, the superorganic consciousness, emerged and was first recognized with the evolution of our species—since we defined the term.[66] We also defined such terms as God and consciousness in ways that earlier species apparently did not.

Along with our ancestors' improved technological and aesthetic repertoire came the development of magical and mythological beliefs, manifested in part in increased awareness of the complexity, magnificence, and power of nature and of inexplicable, or "supernatural," forces. Various cults and belief systems evolved, all of which were appropriate to their historical times and had the common thread of recognizing powers in the universe beyond comprehension. New awareness of time, birth, life, death, and the afterlife all manifested themselves in various ways as *Homo sapiens* crossed the threshold from simple "infant" consciousness to normal self-consciousness. In this transition we fell into time and developed the concepts of past, present, and future that have so shaped recent history.[67]

Since the emergence of complex societies, or "civilization," we have moved rapidly toward an age of rational "enlightenment" in which we have come to develop a scientific understanding of some of those phenomena previously regarded as supernatural. We have not, however, answered any of the ultimate questions about what drives, pulls, or guides the universe, and so we should be careful not to make science the latest substitute religion. Nor should we delude ourselves into thinking that we are no longer superstitious or enchanted by the magic of phenomena beyond our comprehension. Indeed, it is just such enchantment that drives scientists to delve into the beyond.

Even as we gain an improved understanding of natural phenomena, so ever deeper, more complex, and more intriguing questions arise. Here we are reminded of Isaac Newton's opinion that "a limited amount of knowledge leads away from God, but that with increased knowledge one finds the way back."[68] Put another way, a little knowledge of the universe may convince humankind of our ability to grasp its principles, but greater knowledge leaves us incredulous and awestruck by its grandeur complexity and innate intelligence.

The folly of the evolution/creation debate is that it revolves around narrow institutionalized and political perspectives. Clearly, creationists are wrong to regard all scientists or evolutionists as godless, when such giants as Newton and Einstein waxed eloquent about God's grandeur. Clearly, God is not a bearded sage in human form, doling out punishment and reward to humans depending on their adherence to religious or scientific creeds. But this does not mean that God does not exist, simply that such images are too narrow. Good behavior toward neighbors, generosity and good stewardship of the planet may be their own reward, raising the question of why nature, rather than God per se, is intelligent enough to reward being "good" and going with the flow rather than being "bad" and going against the grain. I shall revisit this notion, which we may call "conscious evolution," in the final chapter.

Scientists, or what creationists call evolutionists, could win support for their quest by saying, "Sure let's talk about God!" Let's get down to some serious philosophy and discuss God (or his/her semantic equivalent) in language that is convergent and mutually acceptable. The amazing thing is that the more we learn, the more mysterious, wonderful, and intricate the unfathomable powers of creation and evolution become, and the greater our reverence for the whole process. Just as we have always defined God on the basis of our limited current knowledge, now we must redefine him/her (the power of the universe) to accommodate our new knowledge of quantum physics, Fibonacci numbers, fossil footprints, and whatever new discoveries lie just around the corner. God and the beyond must continually be redefined and reexamined, which, thanks to our inquisitive primate minds, is in fact exactly what is happening all the time.

Here we might benefit from reviewing the words of the thirteenth-century theologian St. Thomas Aquinas, who concluded, "What God . . . is will remain concealed forever."[69] This great divine confessed that the existence of God is not rationally comprehensible and cannot be proved by logical deduction. It is paradoxical that a theologian makes a case that might encourage agnosticism or atheism, as science makes a case for a higher spiritual order of things. The philosopher Hoimar von Ditfurth summarized this paradox in his book *The Origins of Life: Evolution as Creation* when he noted: "We have the Middle ages to thank for the insight that God and the beyond can't be proved [and] we have modern science to thank for the knowledge that the possibility of God's existence and . . . transcendental reality can in no way be refuted."[70]

So what is all the fuss about? We are still learning. We shall look further into the beyond in the final chapter, but for now we shall define "the beyond" as a realm of knowledge and deeper structure of the universe anticipated but not yet understood. A realm, in the words of the philosopher Peter Ouspensky, at best glimpsed or sensed only dimly.

chapter seven

DANCING DINOSAUROLOGISTS

ACT VII: CAST OF CHARACTERS

In this act we find ourselves in the Late Cretaceous (100 million to 65 million years ago), the final epoch of the Age of Dinosaurs, when many famous characters, including *Triceratops* and *Tyrannosaurus rex* put in an appearance. Ubiquitous duck-billed ornithopods such as *Iguanodon* migrated thousands of miles to make their way into the track record and left literally billions upon billions of footprints. Others gain our attention by stampeding through what is now the Australian outback, and still others, with a little encouragement from enthusiastic paleontologists, have literally danced into the limelight. Not to be outdone, pterosaurs impress us with their size and birds show off their newly webbed feet.

The impressive size and power of Mesozoic track makers has not gone unnoticed, even by presidents—one of whom cried, "Long live the tracks" and appropriated millions of dollars to preserve our fossil footprint heritage as part of the modern landscape. We shall even visit the New Mexico backwoods, where the world's only known *T. rex* track is secured behind a chain-link fence. As the Age of Dinosaurs draws to a close, so too does the dance of dinosaurs and dinosaurologists, but not before building to a crescendo of debate. As world-shaking catastrophe strikes, ending the Age of Dinosaurs, we find the spoor which shows the dinosaurs were still around to witness the calamitous event, and we are left to marvel at the phantom images of fairies dressed in white as they dance in the glow of iridium stage lights at the end of Act VII.

MIGRATING ALONG THE DINOSAUR FREEWAY

Ichnologists, who study footprints, are beginning to piece together a dinosaur "freeway" they believe might have run for hundreds of kilometers along the coast of an ancient sea that filled the interior of North America during the Cretaceous Period.

Richard Monastersky, science journalist

During Early Cretaceous, Sir Arthur Conan Doyle's beloved *Iguanodon* and its relatives were among the most ubiquitous of all dinosaurs throughout much of the Northern Hemisphere. These duck-billed ornithopod dinosaurs gave rise to the best-known of their clan, the hadrosaurs (meaning "sturdy reptile"), known for their elaborate nostrils housed in complex crests that are thought to have served as resonating chambers for low-frequency vocal communication, sometimes referred to as "honking" in popular articles.[1] By the middle of the Cretaceous this new generation of hadrosaurs was on the rise.

By this time, about 100 million years ago, or 11.27 P.M. on our 24-hour clock, the Cretaceous Gulf, in the location of the present Gulf of Mexico, had expanded considerably, flooding into the interior of western North America to produce what is known as the Western Interior Seaway; eventually it linked up with the Arctic Ocean. A pterosaur flying from where Chicago now is to where San Francisco now is would have flown over thousands of miles of ocean in midflight. A new type of vegetation, consisting of flowering plants and deciduous trees, was also evolving rapidly, leaving coastal plains verdant, lush, and more colorful than in previous epochs.

Our story in this section is told exclusively by tracks named *Caririchnium leonardii,* denoting a similarity to tracks found the Carir Basin of Brazil by Father Giuseppe Leonardi.[2] In the footprint-bearing strata of the dinosaur freeway, a vast terrain extending from publicly accessible sites at Dinosaur Ridge near Denver to Clayton Lake State Park in Northern New Mexico, not a single fossil bone is known.[3] Other sites on private land such as Mosquero Creek, fill in a rich picture of dinosaur activity along the so-called dinosaur freeway.[4]

Traditionally, duck-billed dinosaurs were regarded as bipedal and were depicted in kangaroolike "tripodal" poses, with their knees splayed and their tails on the ground.[5] Though convenient for balance in museum mounts this pose, referred to as the "Cossack dancer stance," is not supported by abundant new trackway evidence from England, Brazil, Canada and Colorado that reveals small front-foot impressions, implying that quadrupedal locomotion was common.[6]

Along the dinosaur freeway, especially at Mosquero Creek, we find groups of up to 50 or more parallel trackways, representing the largest social orchestration of ornithopods currently known, and surely representing the residual evidence of the passage of a much larger herd. The simple terminology that describes individual footprints can be lost in the reverberation of large herds trampling vast tracts of terrain, so to describe such extensive dinosaur activity trackers have adopted the term "megatracksite," introduced in Chapter 5.

Obviously, the concept of a dinosaur freeway is appealing and a little more user-friendly than the term "megatracksite," but it has led to some confusion. When we first presented the term at a Geological Society of America conference, we spoke continuously of a "regionally extensive" unit of track-bearing strata that indicated the widespread trampling of the Cretaceous coastal plains, the shores of the expanding Cretaceous Gulf.[7] This translated into one headline that trumpeted, "Professor Claims Colorado Was Trampled by Dinosaurs." A writer at *Science News* gave the term another slant, interpreting the dinosaur freeway as a "migration route."[8] This can be neither proved nor disproved. The inference of migration behavior is based on the occurrence of the same dinosaur track types at dozens of sites covering an area on the order of 30,000 square miles (80,000 sq. km).

Could the dinosaurs have been migrating? We cannot prove that a single individual moved throughout the entire area, but we know that the range of the track-making species was very extensive. In the present age, large gregarious herbivores migrate because they cannot sustain themselves for very long on the resources of a single local area. Studies of the distribution of large herbivorous dinosaurs during the Cretaceous further support this argument: the discovery of ornithopods and their tracks in the Spitsbergen region and Alaska shows that some species' ranges extended from north of the Arctic Circle, the 64th parallel, to Montana, well south of the 49th parallel—a latitudinal distribution of 20 degrees. Since it is unlikely that these species lived in the Arctic year-round and endured months of winter darkness, they must have migrated north and south on a seasonal basis.

About 70 million years ago, the Rocky Mountain Front Range thrust up from the Earth, and what once had been the shoreline where the interior seaway met the coastal plain became the great lineament—a linear topographical feature—where the Rocky Mountains meet the High Plains. The dinosaur freeway follows the trend, or direction, of up-thrust Rocky Mountain Front Range. Eagles and other birds of prey use this lineament as a modern migration route, thus flying the same north-south trends followed by their distant dinosaurian relatives—flying above their footsteps,

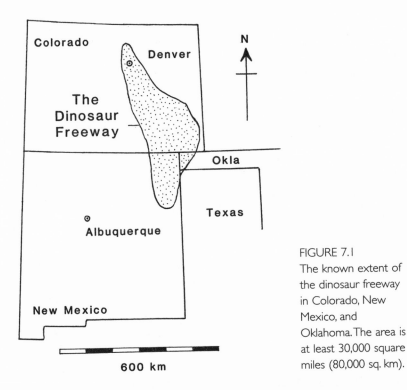

FIGURE 7.1
The known extent of the dinosaur freeway in Colorado, New Mexico, and Oklahoma. The area is at least 30,000 square miles (80,000 sq. km).

one might say. Dinosaur trackers studied the Cretaceous thoroughfare from track evidence before the raptor route was seriously investigated. The existence of well-traveled routes are a perennial feature of the eternal trail. It appears that only the methods of transportation have changed.

A PROMENADE ON DINOSAUR RIDGE

Dinosaur Ridge area is one of the world's most famous fossil localities.

Martin Lockley,
A Field Guide to Dinosaur Ridge

The story of how Dinosaur Ridge and the dinosaur freeway got their names is delightful. The first hints that tracks might be abundant on what came to be known as the dinosaur freeway emerged in the 1930s, when footprints were found along a road known as Alameda Parkway in the foothills of the Rocky Mountain Front Range, near Morrison, Colorado. Geologists and Denver residents were aware of these tracks; they even collected the tracks as patio souvenirs. I saw one particularly nice specimen that I had described from another Colorado location on sale for $2,800. One strike for the commercial bad guys. But the would-be vendors got merit points by returning the track to its rightful place when they were challenged on its origin. The local footprints are so well known that a local author has written a detective story called *Dinosaur Tracks and Murder*.[9]

In 1966 the world-renowned dinosaur experts John Ostrom and Jack McIntosh published a book on the Jurassic dinosaurs of the Morrison area, in which they recommended that the Alameda site be preserved as a National Natural Landmark.[10] Their interest in preserving the site had nothing to do with the Cretaceous footprints but rather with the famous Jurassic bones. Moreover, they recommended that the proposed landmark status of the site be kept a well-guarded secret to prevent theft of the bones. It is hard to say whether this strategy reduced vandalism, for tracks and bones continued disappearing sporadically until the late 1980s.

Thanks to the growing awareness of the finite nature of natural resources, including fossils, and a further incident of track theft, in 1988 a concerted effort was made to preserve the Alameda tracksite. At the time our research group was conducting a study sponsored by the National Science Foundation. It seemed a little ridiculous to have the objects of our study stolen from under our noses, and from a National Natural Landmark no less. So the vandalism provided an incentive for action, and a concerned group of citizens, many of them geologists from the academic and oil-exploration community, created a nonprofit organization, the Friends of Dinosaur Ridge, to preserve the site for science and public education.

Soon afterward the "Friends" formally christened the site Dinosaur Ridge. At the same time that our studies along the Colorado Front Range were revealing the existence of a megatracksite, the Colorado Department of Transportation was building a new freeway (highway C 470) parallel to the ridge. We checked the meaning of the word *alameda* and found that it means "promenade." So the concept of a dinosaur promenade was born, soon to be superseded by the concept of the "dinosaur freeway" and the renaming of C 470 as Cretaceous 470.

MORE FLYING CROCODILES

Here . . . was a track purported to be that of a pterosaur taking off. Unfortunately, the details of the prints failed to match the anatomy of the pterosaur hand, and the trackway was relegated to that of a swimming crocodile.

David Unwin, dinosaur tracker

One of the main problems of paleontologist trackers is identifying and distinguishing among the tracks of various animals that existed at the same time. For example, trackers have long argued over what pterosaur and crocodile tracks look like. Even though modern footprints are available for study, crocodile stalking is a hazardous pastime best left to fictitious heroes like Crocodile Dundee or the ancient Chinese hunters observed by Marco Polo. Pterosaur tracking is also an esoteric pastime, engaged in, only recently, by only a handful of specialists. An instructive lesson in crocodile tracking can be gleaned from the study of footprints at Clayton Lake State Park, one of the most accessible tracksites along the dinosaur freeway. The site was first studied in 1985, by David Gillette, then a paleontologist at the New Mexico Museum of Natural History, and David Thomas, the advocate of the Texan predator-attacks prey scenario which he set to music. They recorded six three-toed footprints positioned in what appeared to be a wide arc; the distance increased between each successive pair of tracks, leading them to postulate that the tracks were made by a pterosaur's wings as it took off.[11]

Such a scenario is highly conjectural, and open to criticism on two fronts. First, were pterosaurs so clumsy as to hit the ground with six consecutive wing beats when trying to take off? Even if they did touch the ground with their wings, as birds sometimes do in snow, would they strike it with such force as to leave three-toed tracks? The intuitive answer would appear to be no. Apart from the obvious danger of breaking a wing, repeated contact with the ground would significantly impede progress. After 100 million years of practice one would assume that the pterosaurs would have learned how to take off less clumsily.

Furthermore, the tracks are very similar to crocodile footprints. In 1986, when Gillette and Thomas formulated their pterosaur hypothesis, the participants of the First International Symposium on Dinosaur Tracks, visited the Clayton Lake site. The local reception committee had arranged a western-style barbecue complete with gun-toting cowboys and cowgirls, so not everyone was fully focused on identifying enigmatic footprints. Most, however, voted for crocodiles, not pterosaurs.[12] But Gillette and Thomas stuck to their guns and published the pterosaur interpretation. James McAllister, a student at the University of Kansas, also published a description of similar tracks from Kansas and attributed them to a swimming dinosaur.[13] Despite a majority vote from the worlds experts for crocodiles, the published record in 1989 labelled the Clayton Lake tracks as pterosautian.

In 1993, the pterosaur expert Chris Bennett, also from the University of Kansas, showed conclusively that the Clayton tracks were a series of several crocodile trackways aligned parallel to one another, even showing sinuous tail traces.[14] So the track makers were clearly not pterosaurian. The moral of the story is that fossil crocodile footprints do exist, even though many purported croc tracks from the Jurassic are pterosaurian, as explained in Chapter 5. Tracks of aquatic animals sometimes show odd configurations, and in the threefold air, land, and water world of pterosaurs, dinosaurs, and crocodiles, we should expect the track record to be different for each group, reflecting the medium to which each was adapted. One might say the medium helps express the message.

DINOSAURS AND DYNAMITE

It is worth looking into reports of unusual indentations to determine if the cause was dinosaurs or dynamite.

Martin Lockley, *Tracking Dinosaurs:*
A New Look at an Ancient World

It is strange but true that in the field of ichnology, dinosaurs and dynamite are sometimes mentioned in the same breath. Both Clayton Lake State Park, and Dinosaur Ridge, a National Natural Landmark, are protected sites. Their value lies in the scientific information they provide about extinct track-making species, and in their potential for educating the public about the Earth and environmental sciences. Fossil footprints have fulfilled this educational potential handsomely by becoming an integral part of the modern landscape, despite their great antiquity. It is also instructive to reflect on the geological history of fossil footprints. Modern tracks exist on the surface of beaches, mudflats, or lake shores for very short periods of time before being destroyed or buried. Once buried they may remain entombed for many millions of years. When eventually exhumed by various agents of erosion, they are again destined for a very short moment in the sun. From a geological perspective, they see the light of day for a mere microsecond before again beginning to erode and be recycled into dust. Our entire sample of fossil footprints—the raw material for this book, for all ichnology—is derived only from those ragged edges of strata exposed at the Earth's surface at this precise moment in geologic time. These edges, like the whitecap crest of a wave rising from a deep ocean, can be seen only in the instant that they break the surface. On our 24-hour time scale they illuminate the paleontological world for no longer than a bolt of lightning.

Fortunately, we are not wholly dependent on the whimsical schedule of the geological erosion to expose track-bearing strata to the light and our inquisitive eyes. We often quarry and excavate rock layers, sometimes using dynamite. Using calculations based on the density of tracks at sites like Dinosaur Ridge, we estimate that somewhere between 2.5 million and 25 million tracks per square mile lie buried beneath the surface. For the dinosaur freeway, which covers 30,000 square miles, we calculate a whopping 80 billion tracks (80,000,000,000) most below the surface. We can use our knowledge of subsurface track layers to predict the location of footprints. At Dinosaur Ridge, the main track-bearing surface, "excavated" inadvertently during road construction in the 1930s, revealed about 200 footprints in an area of about 2,000 square feet (200 sq. m). Excavating another 2,000 square feet of this layer revealed another 200 footprints. Uncovering an area of this size buried by a 3-foot-thick-layer of sandstone requires the removal of approximately 500 tons of rock. This is more than can be expected of enthusiastic weekend volunteers or trail-blazing Boy Scouts. Dynamite is much more persuasive, and well-trained powder monkeys can remove rock layers with the ease and precision of a master chef cutting a slice of cake. This is what we did. In 1992 our expert demolition crew sliced and diced 2,000 square feet and removed 500 tons to expose a couple of hundred new footprints at the Dinosaur Ridge strata without damage to the track layer 3 feet below the surface.

Where precision is not a priority, however dynamite has had some interesting side effects, one of which is to leave big holes in the rock. Such holes are found at Dinosaur Ridge, dating from road construction in the 1930s. Because such blast holes are often regularly spaced, they may end up looking a little like footprints. Anyone familiar with artificially excavated rock outcrops will soon learn to recognize such dynamite damage by distinctive patterns of radiating cracks and to distinguish them from tracks. Nevertheless, trackers sometimes receive reports of footprints that turn out to be blast holes. One of the most controversial examples was found near the Nakasato Dinosaur Center in central Japan, where "indentations" bearing a remarkable resemblance to blast holes had been interpreted—incorrectly, in my opinion—as dinosaur tracks.[15] Like the 1930s blast holes at Dinosaur Ridge, the Nakasato marks are associated with a roadside rock outcrop that was originally exposed with the aid of dynamite. I tease my Japanese colleagues that they have a monument to dynamite, not to dinosaurs.

But such is the power of faith. On the strength of the possibility of dinosaur tracks (and the discovery of a single dinosaur bone nearby) the Nakasato Dinosaur Center was built into a first-rate dinosaur museum. Such are the fruits of supporting scientific investigation and documentation of resources, regardless of whether everyone agrees with all the interpretations. The very possibility that the holes are not tracks but blast holes and that some colleagues dissent drives the search for the real thing. It may begin as another case of believing what we want to believe, but it ends, we hope, with taking another careful look at the evidence.

THE BATTLE OF CARENQUE

"Vivam las pegadas de Carenque!" (Long live the tracks of Carenque!)
Mário Soares, president of Portugal

About 95 million years ago, a large solitary dinosaur marched across a coastal mudflat on the shores of the Atlantic, in an area that is now a part of the Iberian Peninsula. Today the trackway can be seen on the floor of an old limestone quarry in Carenque, a suburb of Lisbon. We do not know much about the dinosaur that made the trackway, because, although the site is visually spectacular, the individual tracks consist only of big round potholes. Nevertheless, the footprints have proved important enough to spark off a battle between the forces of technological progress and advocates of paleontological conservation.

The "Battle of Carenque," began when a freeway construction project reached the vicinity of the quarry and it became clear that a twentieth-century freeway would meet a 95-million-year-old trackway head-on. Much of the freeway had already been constructed on both sides of the quarry, so it appeared impossible to design a detour that would go round the tracks. Regrettably, the trackway would be sacrificed to the advancing thoroughfare of tarmac. Paleontologists and conservationists screamed "Not so fast!" This was a very important piece of Portugal's natural heritage that would be impossible to replace. It was time to call in an unbiased outside observer for an objective opinion. Soon thereafter I found myself with a business-class ticket to Portugal, summoned by the Ministry of Science and Technology.

The preservation of dinosaur tracks is a fascinating challenge. Unlike bones and archaeological remains, which can be dug up and preserved in a museum, it is practically impossible to remove an entire tracksite, the size of a football field, and put it elsewhere. Besides the impracticality, the cost would be prohibitive. If you cannot move the tracks, you are left with a limited number of possibilities: leave them where they are; replicate them; cover them up; or destroy them. Usually the best course of action is to leave them in place, while protecting them from the threat of vandalism. This has been done at many sites around the world by creating interpretive centers or parks. Such strategies have the dual advantage of preserving the site from vandalism while educating the public about the prehistory of the area. In such far-flung locations as the hills of northern

Spain, the outback of Australia, the deserts of Utah, and the outskirts of Denver, the seasoned traveler can visit a variety of tracksites, protected by a combination of signs, fences, shelters, and park rangers.

When I arrived at Carenque on a sunny autumn weekend in 1993, I was immediately struck by the length of the trackway and the excellent view one has from an elevated point high above the quarry floor. Approaching the overlook one has no glimpse nor preview of the Mesozoic scene below. Then, suddenly, it is there in its entirety, standing out starkly against the light rock matrix in which it is impressed. My first question was, how long is this segment of trackway? The answer: 417 feet (127 m), the longest segment of trackway that had actually been measured, anywhere in the world.[16] It was hard to imagine recommending the destruction of the world's longest dinosaur trackway. That we could not immediately identify the track maker was a secondary consideration.

My main conclusion had been reached within a matter of seconds. Two days later I would recommend to the minister of science and technology that the site be preserved and turned into a dinosaur park. The investment in saving the site would bring ample returns as generations of schoolchildren and tourists benefited from trips to the site. The difficult question remained of what to do about the freeway, already under construction. One suggestion was to build a tunnel beneath the trackway—rather like a game of chicken in which a dinosaur meets a freeway head-on and the dinosaur wins. The problem with this was that it would cost the equivalent of 8 million dollars (or several billion Portuguese escudos).

Some people raised the objection that the tunnel would cost too much, and that the money, about three dollars per capita for the entire Portuguese population, would be better spent on social programs. But most were in favor of preserving the nation's natural heritage. One of the strongest supporters was President Mário Soares, who was lobbied persistently by Antonio Galopim de Carvalho, the mischievous but thoroughly lovable head of paleontology at Lisbon's National Museum of Natural History.

On October 2, the president squeezed a field trip to the site into his busy schedule, during which he showed his support by handing the equivalent of ten dollars to Professor Galopim de Carvalho, saying, "Here's my contribution." Despite being pelted by a heavy downpour and being knee-deep in mud, the president and a large entourage of die-hard dinosaur aficionados and new environmentalist converts screamed off to the tracksite in a motorcade of 12 black Mercedes, escorted by a fleet of police motorcycles. Along the way Professor Galopim de Carvalho was virtually on his knees praying for a break in the downpour. Proving that he has friends in high places, the rain ceased, the clouds lifted a little, and the sun shone through.

At the site el presidente stomped unceremoniously through the mud, amid a throng of banner-waving youngsters, leaving his own trail of very messy footprints. He took one look at the tracks and proclaimed, *"Vivam las pegadas de Carenque!"* Cheers went up from all the youngsters, whose banners waved the same message: "Long live the footprints of Carenque." After 95 million years there seemed no good reason to destroy them. When the Portuguese parliament voted on the preservation of the site the vote was unanimously in favor. Dr. Galopim de Carvalho, a prolific writer, wrote a book, the only one ever written entirely on the subject of a single tracksite and the community effort to save it from destruction, called *The Battle of Carenque*. It was a battle he played a large part in winning.[17]

Since this memorable chapter in conservation history Portugal has been on the dinosaur trackers' map. Our expeditionary vehicles, Calypso marks I, II, and III, have docked many times on the Iberian shore to document dozens of sites, in several languages. Christian "Jacques Cousteau II" Meyer and Vanda Santos have dangled off ropes along the picturesque cliffs and sampled some of the best seafood and wine in the province, strictly as separate activities, you understand. Dr. Galopim de Carvalho— a good friend and honorary, executive member of our trackers team—has orchestrated research support for Calypso's eager explorers and the results have flowed freely from the Calypso inkwell onto the pages of *Gaia*, the appropriately titled journal of the National Museum of Natural History in Lisbon. Long may the unraveling of the Portuguese track record continue. *Vivam Galopim—Viva Calypso!*

A DINOSAUR STAMPEDE

[T]he trackways are those of coelurosaurs and ornithopods apparently panicked into a stampede by the approach of a large carnosaur.
Tony Thulborn and Mary Wade, Australian trackers

In the outback of Central Queensland, far from busy freeways that might bring throngs of tourists to a dinosaur attraction, an oddly designed building sits incongruously amid the arid bush, looking a little like a large hang glider. Any visitor stumbling on the site unprepared would no doubt be puzzled as to why such a strange edifice had been erected deep in the outback. The explanation is that the structure encloses and preserves a unique set of dinosaur tracks.

Australian dinosaurs are often regarded as something of an enigma. The reasons for this have to do with the partial or complete isolation of Australia from other continents since the Mesozoic. Clear signs of Australia's biological uniqueness are well known to botanists and zoologists. Even the most basic biology textbooks point to Australia's extraordinary marsupial fauna of kangaroos, wallabies, koala bears, and other exotic species as a classic example of how animals followed different evolutionary paths when isolated from the rest of the world. It is therefore not surprising to find that Australian dinosaur tracks are unlike those from the rest of the world, except in a few cases.[18]

The main Queensland tracksite is at Lark Quarry, one of three excavations at quarries in an old opal mining district that have produced dinosaur tracks. Opal has alternately been regarded as a sign of good and bad luck, but for Australian paleontologists, it must surely be a sign of good luck, for Lark Quarry has become one of the world's top ten tracksites. It revealed to a team led by the University of Queensland professor Tony Thulborn a total of well over 3,000 footprints representing the trackways of at least 150 small bipedal animals. Although not the largest tracksite in the world—it covers little more than 2,000 square feet (200 sq. m)—it reveals the largest concentration of tracks ever counted accurately and put on a single site map.[19] Thulborn and his team must be congratulated for successful completion of the arduous task of removing more than 60 tons of rock overburden and painstakingly cleaning and mapping the surface.

There are two particularly striking features of the site. First, almost all the tracks represent truly diminutive creatures, which according to Thulborn represent small carnivorous dinosaurs and ornithopods whose foot sizes can be compared with those of pigeons and chickens, respectively. (Incidentally, the name Lark Quarry is not connected with the birdlike appearance of the tracks). The second striking feature is that all the track makers were heading in the same direction, and moving at high speed. So what were 150 birdlike dinosaurs doing running in the same direction? Thulborn and his colleague, Mary Wade, postulated that the footprints are evidence of a Cretaceous "dinosaur stampede" caused by the intrusion of a large predator into the area.

Thulborn and Wade based their hypothesis on the presence of two segments of trackway which they interpret as the spoor of a large carnivore. They also postulated that the stampede took place on a small peninsula of land that jutted into a lake. When the small dinosaurs were cut off as the carnivore approached, they got nervous and fled the peninsula, passing the very jaws of death and stepping into the newly formed carnivore footprints. This "panic scenario" attempts to explain the combination of so many small carnivore and herbivore trackways aligned in the same direction which one would not expect to find mingled together in such large numbers under normal circumstances; and also the fact that some of the small birdlike footprints are actually within the larger carnivore prints.

They even explained what appear to be missing footprints in certain trackways as an indication that the animals were moving so fast that some of them became temporarily airborne as they were literally swept off their feet in the panic of flight.

There is no doubt that Thulborn and Wade made a compelling case that the small track makers were all running in the same direction. That the two track types represent herbivores and carnivores is possible but not universally accepted. The assertion that the large track maker was a predator has also been questioned, by Gregory Paul, who suggests in *Predatory Dinosaurs of the World* that the large biped was probably a herbivorous ornithopod.[20] Perhaps this ornithopod walked along the lakeshore sometime before the small dinosaurs all ran in the opposite direction. Such a scenario lacks the drama and musical rhythms of a stampede, but must be considered plausible. The Texas-attack scenario having been called into question, we lack any undisputed track evidence of predatory activity. Even those who, like Jim Farlow and David Thomas, would love such scenarios to be true must admit that they are only possibilities. In the sections that follow we shall inquire further into some of the more sensational claims about dinosaur behavior.

THE KANGAROO FACTOR

There they said there lived an animal as large as a man with a head like a deer and a long tail. It stood on its hind legs like a bird and could hop like a frog. In 1640 Captain Pelsart gave an accurate description of it, only to be greeted with derision at the tall stories told by seamen.

Bernard Heuvelmans, *On the Track of Unknown Animals*

We should not leave Lark Quarry without further discussion of conservation of tracksites. On this subject Australian paleontologists have voiced strong opinions. In an outspoken discussion of the preservation of Lark Quarry, five Australian paleontologists, including Mary Wade asked whether paleontologists have been "sufficiently concerned with the long-term preservation of fossil tracksites." They answered their own question provocatively: "No, because for economic and professional reasons they are unwilling to sacrifice their own research time to the demands of field conservation." They admit that this harsh admonition is a "sweeping statement," and that it is unfair to say that paleontologists are not "concerned." But evidently they are not "sufficiently concerned."[21]

Tracksites are technically owned by whoever owns the land, though in some countries and regions, the national interest may take precedence over local ownership. Certainly paleontologists should be concerned with what happens to sites that will eventually crumble into dust, which is why they collect data and make replicas for posterity. Agents of destruction are relentless, and include earth, air, fire, water, expanding plant roots, trampling by animals and humans, and the most relentless agent of all: old father time. Murphy's Law—if it can happen it will happen—also comes into play: at Lark Quarry a layer of straw designed to protect the track-bearing stratum during construction of the roof was accidentally set on fire during welding, causing damage and discoloration to the track surface. Then there was the "Kangaroo factor." These local inhabitants evidently regarded the roof as a shady shelter erected just for them, and quickly adopted the Cretaceous substrate as their new living-room floor. With typical kangaroo nonchalance they deposited droppings and urine and in some cases even had the gall to pass away, leaving intractable stains on the surface. Solving one problem sometimes leads to another. The paleontologists had hardly anticipated designing a kangaroo funeral parlor or old folks home; they had to erect a fence to turn away those creatures who had come to regard the shelter as a kangaroo gate to the next world.

A KOREAN DUCK POND

Ducks and geese have a long ancestry.
  W. E. Swinton, *Fossil Birds*

One of the great potentialities of fossil footprints is that they reveal details of skin impressions and fleshy anatomy not evident from a study of bones. For example, frogs, ducks, other birds, some turtles, and crocodiles have webbed feet. What of the feet of extinct species like dinosaurs? Roland Bird and others at one time believed that brontosaurs were aquatic. Similar interpretations have been proposed for duck-billed dinosaurs, based on the logic that if it looked like a duck it must have behaved like a duck.[22] The famous duck-billed platypus, for example, is clearly an aquatic creature that exhibits certain ducklike behaviors. But arguments for aquatic duck-bills, based on supposedly paddlelike hands and webbed feet, do not hold up to close scrutiny. It is hard to imagine that a paddle smaller that a Ping-Pong bat could propel a 30- or 40-foot (9–12 m) dinosaur very efficiently. The duck-billed dinosaurs' unwebbed hind feet and the rigid tail are clearly not adaptations for aquatic behavior.

Recent discoveries of the tracks of web-footed birds in southwestern Korea add to the intriguing track record of water birds, and establishes South Korea, a land of Cretaceous lakes, as a key area for the study of water-bird evolution. The footprint evidence consists of many trackways of a small web-footed form named *Uhangrichnus* (from the Uhangri Formation), and a single trackway of a larger species named *Hwangsanipes* (from the Hwangsan Basin).[23] The discovery of more that one track type suggests that web-footed species were probably not rare. According to the conventional view of bird evolution, these tracks could not have been made by ducks—a group that did not evolve until after the Cretaceous.[24] But deciphering the history of bird evolution is a very complex business, and many relationships between ducks, flamingos, herons, and other common waders and water birds are poorly known. Furthermore, none of the present theories takes into account footprint evidence. Could the Korean tracks be evidence of the first duck pond? Probably not, but they are in any event the oldest web-footed bird tracks in the world. As we continue to develop our theme of the conservation of tracksites, we shall see that water-bird trackways have as much right to special paleontological status as the best of the brontosaur tracks, or the oldest spoor of our own hominid clan.

PTEROSAUR PSYNCHRONICITY

[T]he concept of synchronicity indicates a *meaningful coincidence* of two or more events, where something other than the probability of chance is involved.

The Journal of Religious Thought

In October 1996, the press got quite excited over claims we made at the Society of Vertebrate Paleontology meeting in New York of abundant trackway evidence for quadrupedal pterosaurs—the story even made the pages of *Time* magazine.[25] Within a week I found myself changing gears and bound for South Korea to advise on the future of the "duck pond" site. I tried to think of bird tracks, but at 35,000 feet, my mind kept wandering to pterosaurs. Why were there no pterosaur footprints along with all those millions of bird tracks in the Korean duck pond? Why were pterosaur tracks known only from North America and Europe?

October, 28, 1996: On the way to the duck pond site, my colleagues asked what I have been doing in the year since we last collaborated. "Finding lots of pterosaur tracks" I answered. "Surely not!" they replied in astonishment. They had no idea that such footprints existed, much less an idea of what they might look like. I assured them that we had all been equally ignorant until quite recently, and scribbled sketches of pterosaur tracks as we drove along. My colleagues' eyes widened eagerly, and I knew before they declared it that they had seen similar tracks, that we were on

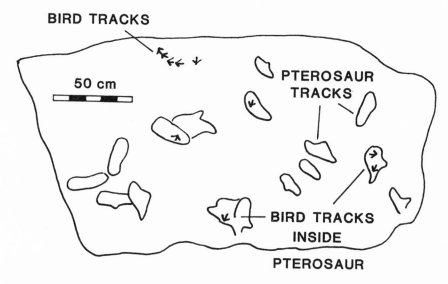

FIGURE 7.2 Giant pterosaur tracks from Korea.

the trail of the first Asian pterosaur footprints. "Where?" I asked. Where else but at the very duck pond site to which we were heading.

On arrival the local dignitary and a throng of journalists were there in full jacket-and-tie regalia, ready to ask if the world's oldest web-footed bird tracks and newly uncovered dinosaur tracks were important enough to warrant designating the site for special protection as an outdoor paleontological park. Of course, of course, it's a very important site! This dinosaur business is sometimes quite extraordinary. A wall of microphones and cameras blocked any view of the newly exposed surfaces I was supposed to interpret. When I was finally on my hands and knees in communion with the Cretaceous mudflat, the reality of what I was looking at began to sink in. For a couple of minutes I was uncertain. These tracks were huge, four times the size of most I had seen, larger than adult human footprints.[26] They had a rather smudgy appearance, but there was no doubt that they were giant pterosaur tracks, which could only have been made by creatures the size of *Quetzacoatalus*, the largest known pterosaur with a 10-12 m (35-40 foot) wingspan—the 747, the mother ship of archosaurian aviation—of the Mesozoic world.

The first pterosaur tracks from Asia, and the world's largest. The journalists were full of questions: So now what do you think of the site? What is a pterosaur? Is it a flying dinosaur? Must we rewrite the Book of Guinness? The problem was that I had no pictures of pterosaurs to show the cameras and eager journalists, and my colleagues, also unfamiliar with pterosaurs, were equally at a loss to conjure up the right images. My sketches were a poor substitute for a decent illustration; then I remembered an illustration used in the *Time* magazine article, which I had checked out from the traveling United Airlines library. The cameras thronged around the two-page spread of a majestic Cretaceous *Quetzacoatalus* winging high above its distant dinosaur relatives. So this is a story about pterosaur tracks in *Time* magazine? Yes! Wonderful! When was it published? I looked at the date. Today—October 28, 1996. How can this be? they asked incredulously. "Because it's *Time*—this is the time for pterosaur tracks!"

THE MYSTERY DINOSAUR

In the Thomas coal mine on the northern periphery of my back yard [in western Colorado] is a thirty three inch track of a mystery dinosaur; at the time of its discovery in 1930, this was the largest footprint in the world. . . . [S]ince his skeleton has not been found let us call him *Xosaurus*, the mystery dinosaur . . . [I] little realized how many paleontological headaches he was going to cause.

Al Look, *In My Back Yard*, 1951

In the wake of the revolutionary sixties, ecology became popular. Paleontologists began to adopt a more holistic view of life on Earth. Animals were seen as dynamic, integral components of complex ecosystems. This new perspective extended to a reevaluation of Mesozoic life and spawned a vibrant new generation of dinosaur books. Once considered defunct failures that had lumbered mindlessly toward inevitable extinction, dinosaurs were quickly recast as agile athletes, smart enough to survive for millions of years. The most vocal champion of dinosaurs was Robert Bakker, whose first paper, "The Superiority of Dinosaurs,"[27] set the tone for much of his subsequent work. Bakker championed the cause of dynamic dinosaurs as the hot-blooded superheroes of the Mesozoic world, and he soon became well known as the central figure in the so called "dinosaur renaissance."[28]

Tracks, which had long been considered mere paleontological curiosities, suddenly assumed new importance as the signatures of living dinosaurs—"the closest thing to dinosaur motion pictures."[29] If dinosaurs really were warm-blooded and athletic, the question on everyone's lips was: "How fast could they run?" This question became the focal point of a historic debate on dinosaur speed that did much to launch what I have called the "Dinosaur footprint renaissance."[30]

There are two ways to estimate the speeds attained by dinosaurs. One is to compare the anatomy of various dinosaurs with modern creatures as diverse as ostriches and elephants, and the second is to study trackways. Anatomical analysis of dinosaurian running potential suggests that small- and medium-sized, chicken- or emulike bipeds were *cursorial* —meaning runners—and thus the dinosaurian equivalents of Olympic athletes.[31] Large bipeds, however, including tyrannosaurs and duck-billed dinosaurs, were labeled as *subcursorial*—the equivalent of joggers that could probably run fairly well if they needed to. By contrast, large quadrupedal dinosaurs such as brontosaurs were plodders and classified as *graviportal*, meaning that they were adapted for carrying considerable weight, but not speedily. When one considers that large brontosaurs may have weighed in at 30 to 50 tons, or five to eight times the weight of a full-grown African elephant, it is hard to imagine them frolicking and skipping their way through Mesozoic life.

Trackways with long strides have confirmed the running ability of small bipeds, but the lack of reliable trackway evidence hampered the earliest estimates of dinosaur speed, and there is still a lively debate as to whether large dinosaurs such as *Tyrannosaurus* and *Triceratops* were capable of a good turn of speed.[32] Enter the *mystery dinosaur,* so named by the paleontologist Barnum Brown in a *New York Times* article as a fundraising stunt.[33] In 1937 Brown and his indefatigable assistant, Roland Bird, had run across large three-toed dinosaur tracks in Cretaceous strata in coal mines in Wyoming, Colorado, and Utah; a cottage industry of mining tracks was supplying a local demand for patio ornaments.[34] Several paleontologists already knew that the tracks were those of large duck-billed dinosaurs, called hadrosaurs. But Brown postulated a gigantic mystery track maker, on the premise that the tracks were larger than anything animals known from actual skeletons could have made. With the word out that top New York paleontologists would pay for big tracks, reports of giant footprints were soon rolling in. A *Guinness Book of Records* mentality developed and reports soon came in of dinosaurs with footprints 4 or 5 feet long (1.2–1.5 m) and colossal steps of 15 or 16 feet (5 m). One man, Al Look, went so far as to rename Brown's mystery dinosaur "*Xosaurus*" and claim evidence of a site where it had stepped on a crocodile![35] Some of these very same step measurements were cited 40 years later as evidence that hadrosaurs could run at high speeds.

In this brief period in the late 1930s Bird and Brown caught tracking fever. In their eagerness to preserve a record of the giant-stepping mystery dinosaur, they excavated a slab containing what Brown interpreted as a

15-foot step from a mine in Colorado. This was one of the most ambitious excavations ever undertaken, requiring the labor of three shifts of miners working round the clock for three weeks. Had the giant steps confirmed a "sprinting duckbill" it would have been an unassailable hadrosaurian world record of 17 miles per hour (27 kph)—a remarkable turn of speed for a five-ton creature.[36] For advocates of high-speed dinosaurs, the estimate was too good to be true. The runaway dinosaur soon ran out of steam, however, when Tony Thulborn showed the 15-foot step to be a stride (i.e., two steps), indicating a pedestrian walking speed of 5 miles per hour (8.5 kph).[37] Ironically, Brown never successfully cracked the mystery of his own creation, and never actually identified the footprints he invested so much effort in procuring. While Thulborn and others have applied the brakes to hadrosaurian dinosaurs run amok, others still insist on considerable bursts of speed and so a vigorous debate continues.

T. REX TRACKS AND DANCING DINOSAUROLOGISTS

The speeds of dinosaurian track-makers appear to have been consistently overestimated.

Tony Thulborn, *The Demise of the Dancing Dinosaurs*

In 1989, Robert Bakker, the self-proclaimed enfant terrible and dinosaur heretic, published a paper entitled "The Return of the Dancing Dinosaur," in which he claimed that dinosaurs "cruised at warm blooded speeds," and that pound for pound they were "stronger, faster, and more maneuverable" than large mammals such as rhinos and elephants.[38] The subject of the speeds attained by dinosaurs has so fascinated those with a bent for the study of locomotion that one biomechanics expert, the British zoologist McNeill Alexander, actually claimed he could outrun a *Tyrannosaurus rex*, and devoted an entire book, *Dynamics of Dinosaurs and other Extinct Giants*, to the subject. Thulborn, the author of the Australian stampede scenario, has also covered the subject extensively.[39]

The question remains: How fast were duckbills, tyrannosaurs, horned dinosaurs, and other extinct giants? The absence of trackways that prove running activity among these behemoths shifts the debate into the theoretical realm: given their size and structure, what are the maximum speeds that they might have attained? Bakker's speculations about "cruising speed" are tinged by his conviction that dinosaurs were "faster" than mammals of equivalent size. Thulborn speaks of "the demise of the dancing dinosaurs," and concludes that Bakker consistently generates overestimates of the speed of dinosaurs.[40] The dance of the dinosaurologists is by no means over, nor is it unimportant, for in the flight-and-fight world of predatory-prey interactions that so enthrall many dinosaur aficionados, it is actually helpful to know whether an animal had the capacity to chase down prey or escape from potential predators.

In 1991, *Scientific American* published the following claim by Alexander: "I am probably fast enough to outrun a pursuing tyrannosaur."[41] It is perhaps insulting to the king of beasts to suggest that potential victims could scamper away with impunity. Underlying such claims, which we cannot test, are popular questions as to whether *Tyrannosaurus rex* was a swift agile predator or a lumbering giant restricted to a life of slow-speed scavenging. At six or seven tons, *T. rex* was presumably no greyhound. *T. rex* was also relatively rare, despite frequent movie appearances. This is what we should expect, for in the fossil record, predatory dinosaurs rarely make up more than 3 to 5 percent of the dinosaur species at any given time.

Tracks attributed to *Tyrannosaurus* have proved both elusive and controversial. The first claim for the discovery of *Tyrannosaurus* tracks was in a quaint paper by William Peterson, of Utah, entitled "Dinosaur Tracks on the Roofs of Coal Mines."[42] But as we now know, these were the tracks of duckbills, not tyrannosaurs. Once in print, however, the notion stuck and the tracks were named *Tyrannosauripus petersoni*.[43] Another problem is that the coal-mine strata are significantly older than those in which *Tyrannosaurus* skeletal remains have been found. So where are the tracks of this king of beasts? A possible tyrannosaur track from a landfill in Colorado has now been buried by millions of broken beer bottles. This did not prevent at least one Japanese film crew from offering to pay a handsome fee to have it excavated again. The next *Tyrannosaurus* track was a Hollywood reconstruction that made its appearance in the 1993 movie *Jurassic Park*. Although hardly authentic, it was a nice touch to see it filled with water on a rainy night and watch the vibration ripples on the water's surface shiver menacingly as the track maker approached. The effect was so good that another single track, cloned from the first, made a second appearance in the sequel, *The Lost World*. This may have been the best bit in the movie.

It was not until 1995 that a convincing example of a *T. rex* track was found in strata of the right age. For the spoor of the world's most famous American carnivore, adored by youngsters from around the world, it seems appropriate that it came to light on a ranch in New Mexico owned by the Boy Scouts of America. True to the track maker's colossal reputation, the footprint measures 34 inches (85 cm) in length—the largest track of a carnivorous dinosaur ever discovered. Although only one footprint was found; 8 feet (2.5 m) on the available rock surface showed no sign of the next step in the sequence, suggesting a stride of at least 16 feet (5 m)

FIGURE 7.3   How to kill a *Tyrannosaurus rex!* Artwork based on original by Jim Whitcraft.

and a speed of 6 to 7 miles per hour (10–12 kph). Given the history of debate about dinosaur locomotion, a single print is more than ample justi-
fication for further discussion on the athletic ability of the king of beasts!
Besides, so far this is all we've got.

The most recent paper on the subject, by the paleontologist Jim Farlow
and his colleagues, seems to be based on the principle "The bigger they
are, the harder they fall."[44] Their hypothesis suggests that if *T. rex* ran too
fast, more than 25 miles per hour (40 kph), it might trip over something
and hit the ground so hard that it would jar itself into oblivion. A stiff
price to pay for exceeding the Cretaceous speed limit. But any large individual can potentially kill him- or herself by falling over, which is why they
avoid the practice. *Discover* magazine seemed to like this idea, and depicted a *T. rex* tripping over a branch.[45] Again this seems a little disre-
spectful to the king of beasts, but who knows? Perhaps it was clumsy and
inept, though I doubt Bob Bakker would agree.

Where did this idea of a tripping tyrannosaur come from? In a strange
departure from its usual sober scientific standards, the prestigious *Journal
of Vertebrate Paleontology* allowed Farlow and his coauthors to show a car-
toonish *T. rex* being deliberately tripped by a mouse-sized Mesozoic
mammal—with a caption that reads "How to kill a *Tyrannosaurus rex*!"
What is the message and symbolism of this David and Goliath scenario,

and what does it mean for science? Mice may bring down the mighty? Miniature mammals caused the demise of the dinosaurs? Surely *T. rex* would turn in his 65-million-year-old grave, if he knew that he had become the object of such ridicule.

In fact, the fossil record has not clarified whether *T. rex* was a predator or a carnivore. A Schadian analysis may shed light on the question. As theropods go, *T. rex* was a very "metabolic" animal, according to the criteria introduced in earlier chapters: it was very large, with a large head, relatively short tail and a short neck, and had its famous short, but very stout, forearms. The back, or posterior, pole of its skull was very wide and it had stout, wide teeth. We know from Schad's work that these are all characteristics of the more metabolic animals. We only have one track so we cannot say whether the trackway was wide or not, like those attributed to megalosaurs, but we do know that the foot was quite fleshy and much wider or bulkier than many skin-and-bone reconstruction would suggest. Large carnivores species in which the metabolic pole is well developed tend not to be at all fussy about what they eat. They are neither exclusively predatory nor exclusively scavengers. I conclude, therefore, that "metabolic" animals like *T. rex* were much more likely to scavenge low-quality food than animals at the nerve-sense end of the spectrum, which prefer fresh live food. I see *Tyrannosaurus* as occupying a position somewhat analogous to the giant Komodo dragon of the lizard world, which unceremoniously devours any available live or dead flesh. Assuming that it did occupy a position toward the extreme metabolic end of the spectrum, it was probably also a drab dark brown or black.

THE WALL: GETTING HIGH IN THE ANDES

[A]ll in all it's just another brick in the wall.
Pink Floyd

Cal Orko limestone quarry, Sucre, Bolivia, August 1, 1998: Anne Schulp points the laser binoculars at Christian "Jacques Cousteau," standing on the quarry floor holding the frayed end of our climbing rope, and clicks the button. The light-speed signal bounces back from Christian's chest, and registers 51.8 meters, meaning there is that much rope between Christian and us—meaning also that 486 feet (148.2 m) have been stolen, cut in the night by banditos with machetes. The frayed end looks like a fluffy shaving brush. On the long, 200-foot-high, near vertical wall above, Stefan "Spider Man" checks for further damage that might be a safety hazard. Another rope is missing. Our total of 3,280 feet (1 km) of rope has been reduced to less than 2,300 feet. For the remainder of the project we will have the extra work of hanging and untying the ropes every day.

Otherwise the project has gone quite well so far. "Jacques Cousteau," or "El Jefe," is in charge and, with admirable Swiss efficiency, has orchestrated the team of a dozen trackers and media specialists from Europe and North America to the rendezvous point in the high Andes of southern Bolivia. There have been problems with getting the climbing gear and casting supplies delivered on time, but they are nothing compared with the problems faced by our colleague Giuseppe Leonardi when trying to work in far-flung parts of South America 20 years ago. Back then he was robbed by bandits at gunpoint, and on one occasion spent weeks trying to get to a Bolivian site only to be turned away by hunger and other life-threatening circumstances. Now, however, we all like Bolivia and the Bolivians and are keen to return to the clear crisp air of the high Andes.

After swinging on ropes to measure and cast dinosaur tracks on various surfaces in Portugal and Switzerland, most team members have had some experience to prepare themselves for "the wall." Giuseppe, now almost 60 years old, who took up climbing in the Italian Dolomites only a few years ago, is in his element. Using the handy walkie-talkie he reminds me of James Bond on a cliff-hanger mission. "Hallo, hallo, this is Giuseppe to base—do you read me? We have here an ornithopod track—please take measurements I will give you now." Our conversation frequently switches to Spanish. It is a great pleasure to work in South America with the man who documented most of the tracksites on this continent. At the base camp Giuseppe switches to French and even a little German with the same facility as Christian, another polyglot. He really didn't speak brilliant Spanish at the start of the project, but his fluency in French, English, and German, along with a smattering of Italian and Portuguese, produces a highly functional and understandable language within a matter of weeks. In any event, much of what we say to the steady stream of journalists and visitors requires the regular repitition of the specific vocabulary of dinosaurs and rocks. Our half dozen languages become interchangeable, and even our plans to use specific words for climbing equipment are largely ignored in the absence of any security hazard. Nicknames and nonsense are more fun anyway.

After a decade or more in and out of the Purgatoire site, the cliffs at Cabo Espichel in Portugal, and the Kodja Pil Ata (holy father of the elephants) site in Turkmenistan, what could we find that was bigger and better? Cal Orko provides the answer in a nutshell. The wall is .75 mile (1.2 km) long, up to 200 feet (75 m) high, and is nearly vertical in places. It reveals two surfaces covered with tracks that from noon until dusk, at least in the southern Bolivian winter, are illuminated so perfectly that the spectacular visual impact of the long and well-defined dinosaur trackways takes one's breath away, even when seen in a snapshot. Indeed it was on the strength of a few stunning photographs that we were lured to the site from far hemispheres.

The track surface represents a lake shore in the final days of the Age of Dinosaurs. The bones tell us of many catfish, turtles, and the occasional crocodile and pterosaur. The trackways are those of the last of the brontosaurs (the titanosaurs) and the last of the armored dreadnoughts (the ankylosaurs), one of which appears to have been running at 6 to 7 miles per hour (11–12 kph). This is the first report ever of a large quadrupedal running dinosaur, so we tell the story and run reenactments of ankylosaur locomotion, at the right speed, for the press and tourist video cameras. Among the numerous theropod trackways, we measure some which can be followed for distances that will add up to new, soon-to-be-ratified world records. It would have been impossible to map this near-vertical wall, longer than a dozen football pitches end to end, had it not been for the recent invention of laser binoculars. These help us map the position of tracks and trackways, which we record on 50 sheets of graph paper that, pieced together, make a fascinating table cloth or king-size bedsheet. But the laser binoculars can only do so much. We still have to climb the wall ourselves to get measurements, tracings, and casts of the important specimens. For several weeks visitors get their money's worth, for in addition to feasting their eyes on the world's most spectacular tracks, they can also watch an international team of spider persons and secret agents rappelling down a 65-million-year-old fossilized mudflat pushed vertical by the upheaval of the Andes.

Preliminary expedition results are splashed over half the world's newspapers. The hotel fax machine sends us amusing reports on what we are doing and the press re-christens us on a weekly basis, as befits secret agent trackers. The Bolivian discoveries prove that tracksites just keep getting better and better, providing more and more useful information. The titanosaurs—last of the brontosaurs—are related, according to recent classifications, to brachiosaurs, wide-bodied brontosaurs with big front limbs. Sure enough, the tracks confirm the predicted wide-gauge, large-manus configuration.

Another site about three hours to the west over challenging terrain reveals 11 parallel trackways of juveniles traveling as a herd. The site is hard to get to and the local villagers, who only see a vehicle once or twice a month, demand payment for having fixed "the road" well enough to allow us in. They threaten to block the road if we don't pony up. Although we have "official permission" and a land cruiser with official insignia, this means little to these back-country folk who speak the Quechwa language of the Incas, not the Castillano of the conquistadors. But they drive a fair bargain, and I am surprised when my casual preliminary offer of 20 bolivianos (four dollars) is happily accepted. At that price even underpaid academics can afford to return.

OF HOOVES AND HORNED DINOSAURS

So what is the locomotor bottom line for a five ton *Triceratops?* The answer may be a maximum speed much higher than that of a charging African elephant or white rhino.

Robert Bakker, *The Dinosaur Heresies*

Two horns of the three-horned *Triceratops*, the best-known of the horned dinosaurs, were first discovered in 1887 in what is now the middle of metropolitan Denver. The specimen was at first mistaken for the horns of an extinct bison, and herein lies a significant message: *Triceratops* and its clan, known as ceratopsians (horned dinosaurs), were in some ways the bison of their day, as I shall soon elaborate. Although they now rank among the most common of late Cretaceous dinosaurs, their tracks have proved very elusive; but by interesting coincidence, some of them were unearthed in Golden, another suburb of Denver, 99 years after the first historic *Triceratops* discovery.

The site, in a maze of clay pits, reveals interleaved clay and sandstone layers tilted upright by the tectonic activity of the Rocky Mountains that loom imposingly above Golden. When the clay is excavated for use by Coors, a giant in the ceramics and brewing industries, the track-bearing sandstone strata are left standing upright in vertical fins like slices of toast. Quarries, clay pits, and other large holes in the ground often make ideal sites for disposal of rubbish. Thus, not long after the world's first-discovered horned dinosaur tracks (possibly *Triceratops*) and a possible *Tyrannosaurus rex* footprint were surfacing for the first time in 68 million years, they were being buried again in a cascade of broken beer bottles. During early expeditions to the site we literally raced against encroaching mountains of glass to document tracks before they went under again.

Why is it that the three-horned *Triceratops* could be mistaken for a bison, and why would we call *Triceratops* the bison of its age? Before we analyze the ceratopsian-bison connection we must briefly air debate concerning the distinctive anatomy of the forelimbs of horned dinosaurs. Among quadrupedal dinosaurs they are unique in having front limbs that evidently cannot be articulated into anatomically convincing positions without their elbows sticking out in what appears to be a very awkward stance. In other words, their hind limbs were erect but their forelimbs

FIGURE 7.4   Vigorous debate surrounds the forelimb position of *Triceratops*. Tracks show that the forelimbs were fairly straight, as on the left. Artwork courtesy of Pat Redman.

were not. So if they couldn't have held their forelimbs erect, the question becomes, how did they carry their forelimbs? This is the topic of the debate. Did they tend to walk erect or did they tend to sprawl? Robert Bakker advocated an erect stance, but did so on the basis of tracks that have since proved to be those of armored dinosaurs, not ceratopsians.[46, 47] Trackways of true ceratopsian tracks should surely shed light on this question.

In our detailed description of the tracks from Golden we concluded that "models of forelimb posture of ceratopsids that postulated sprawling stance are incorrect."[48] Based on footprints we concluded that the horned dinosaurs held their forelimbs erect, like most other dinosaurs. Even basing our evidence on the trackways has not settled the argument. Peter Dodson, an expert on horned dinosaurs, teamed up with James Farlow to produce a reply entitled "Posture of Ceratopsids—A Solution At Last, or Is It?"[49] in which they suggest that the trackways are too wide to indicate an erect posture. This appears to be a case of arguing whether the glass is half empty or half full. Was the *Triceratops* half sprawling or half erect? Although this debate may seem remote from the pressing problems of twentieth-century society, it is regarded as significant among vertebrate paleontologists. Again, it would be nice to know if large quadrupedal dinosaurs could run, like galloping rhinos, as some have suggested, or whether they were pedestrian plodders. Now that we know that Bolivian ankylosaurs could run, the running ceratopsian debate is sure to continue.

We may again consider a Schadian perspective on *Triceratops*, and ceratopsians in general. Clearly *Triceratops* has all the hallmarks of a very metabolic animal. It has an extraordinary head, with a large posterior frill and horns that are highly vascularized, or permeated by blood vessels. In this regard it comes closest to the horn-bearing ungulates of the mammal world such as bison. Remembering that the well-developed posterior physiological, metabolic pole is often matched by an accentuation of the anterior physical pole, we begin to see a pattern. Not only is the head big and the tail very short, but the front limbs are the focus of unusual attention. The front footprints are large, wide, and widely spaced relative to the hind footprints. The unguals (hooves or toenails) are wide and flat, mirroring the shape of the frill, as does the foot shape in general. The head was also held fairly low, and the huge battery of teeth resembles the molars of ungulates, designed for dealing with very tough, presumably low-quality food.

As the millennium closes we have at last bagged trackways of all the major groups of dinosaurs, so we can attempt a complete survey of track-body relationships in the manner advocated. In Chapter 4 we outlined the relationship between biological form and footprint shape in the lizard-hipped saurischians (theropods, prosauropods, and sauropods), demonstrating a progressive trend from small size and narrowness toward greater width. This trend is seen not just between the groups, but within the groups. Now this pattern becomes even clearer. For example the prediction that titanosaur trackways would be wide-gauge with large front footprints is confirmed. Let us attempt a similar analysis for the ornithischians, now that we are familiar with the main groups.

Ornithischians can be broadly divided into two groups: plated and armored forms (stegosaurs and ankylosaurs) and duck-billed and horned dinosaurs (ornithopods and ceratopsians). We find their tracks are all generally as wide as or wider than long, therefore lacking the quality of narrowness seen in the saurischians. Within the group, however, the smaller, mainly ancestral bipedal forms are generally three-toed, with quite narrow trackways, but as we move toward the large end of the spectra, to ankylosaurs and ceratopsians, they become four-toed on the hind feet and five-toed on the front feet, and they develop wider trackways. Steps are generally shorter, which again emphasizes the characteristic breadth, not narrowness. The axes of the footprints point inward in the forward direction, rather than outward (and backward) as in saurischians. The typical characteristics of "metabolic" breadth assigned by Schad to ungulates include vegetarian diet, well-developed molars, and in many cases horns.

The small animals to the extreme nerve-sense side of the spectrum are long-tailed bipeds, which mirror the morphology of theropods. The tracks of small ornithopods are slender, like the tracks of small theropods, and have little padding. They are functionally three-toed, but more widely divergent (greater breadth). When the small ornithopod adopted a crouching position the inner toe and the heel or ankle (metatarsus) sometimes left an impression, or "tail," as sometimes also seen in theropods. So the track says "tail!" Theropods, however, never appear to place their front feet on the ground, unlike even the most bipedal ornithopods. Thus, ornithopods reveal that they belong to a fundamentally quadrupedal group of dinosaurs.

Perhaps even more fascinating than small tracks with tails on small long-tailed animals is the fact that the larger central groups (the stegosaurs, and ornithopods) mirror mammalian ungulates in uncanny ways. The duck-billed dinosaurs have single crests that are centrally located on their heads. By contrast, the "wide" metabolic species have four toes and paired horns just like the cloven-hoofed ruminants. Thus we see that there is a pattern of horn-hoof relationships in ornithischian dinosaurs similar to that outlined for ungulates in Chapter 1. There is a coherent and intelligible flow from posterior to anterior. First the small long-tailed species take on plates and body armor in the central portion of their bodies, either centrally, as in the case of "narrow" stegosaurs, or as side horns, as in the case of "wide" ankylosaurs. Then the emphasis moves to the head, as a narrow central crest or wide horns, in the ornithopods and ceratopsians, respectively. The combination of paired brow horns and nasal horns, sometimes with multiple frill horns in ceratopsians, serves to mark them as the most metabolic group of all dinosaurs, with headgear that attempts to express bison, moose and a bit of rhino all at once. Another example of this

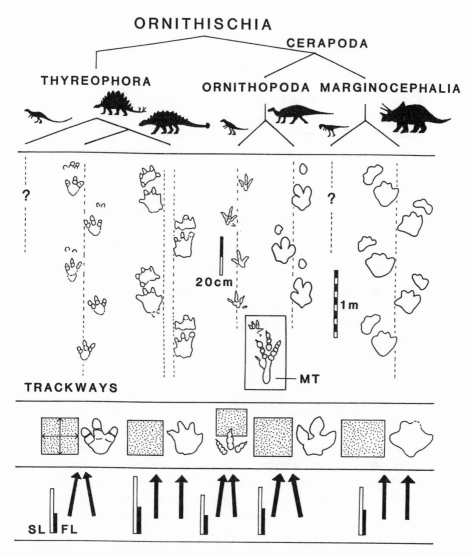

FIGURE 7.5   The spectrum of form in ornithischian tracks. Note how the toe prints tend to turn inward, whereas the toes of the saurischian tracks tend to turn outward. (Compare with Figure 4.5.)

mirroring is that the larger central groups have three toes with the central digit longest, as is the case in the odd-toed, rhinolike species (discussed in the next chapter).

Do the dinosaurs reveal an evolutionary cycle from small ancestors to large descendants later reiterated in the Schadian rodent-carnivore-ungulate sequence? The answer is clearly yes. Whether we look at the theropod-prosauropod-sauropod sequence, the prestegosaur-stegosaur-ankylosaur sequence, or the small ornithopod–large ornithopod–ceratopsian sequence; at sequences within each group; or at the entire order of Dinosauria, we see a similar evolutionary history reiterated again and again. We can now flow with these cycles at any scale and ride these evolutionary waves with an assurance of which way the currents flow. So with our newfound ability to track species down these new pathways of the eternal trail let us use footprints to take one last ride on the morphodynamic dinosaurian highway.

How do we arrange the whole order Dinosauria in a morphodynamic scheme? As ancestral carnivores are to the narrow, nerve-sense side of vegetarian ungulates, the saurischians must lie to the left, and the ornithischians to the right in the scheme. Thus placed, the footprints show a progressive trend from narrow to wide across the entire spectrum of dinosaurs.[50] In both groups the number of digits increases from left to right, and the tendency to move from bipedal to quadrupedal with increased anterior (front-foot) emphasis is clearly seen. Intriguingly we see in the saurischians a shift from parallel foot axes in theropods to an outward or a flat-footed pattern in sauropods, whereas in the ornithischians the pattern progresses from an inward rotation (pigeon-toed) to parallel configuration. The more outward (nerve-sense) group has hind feet that point outward and the inward (metabolic pole) group has hind feet that point inward. Moreover, the claws of all saurischians tend to be narrow and vertically oriented, whereas those of the ornithischians are wide and flat. We can also read a progressive enclosing of the foot in flesh (inward trend) across the spectrum and within groups—but even more fascinating than this, the flesh encloses the foot progressively from back to front, precisely mirroring the whole anatomy shift from posterior to anterior. Here we really begin to see the wisdom of holism and the fundamental coherence of animal form at all levels from toes to horn tips.

The repeatability or reiteration of these patterns makes them very useful scientifically, and allows us to identify characteristics of an animal from such scant or specific evidence as its tracks, or even its toe bones. As these dinosaurian examples show, these systematic patterns are not merely attributable to what statistically minded persons might call variation within a population, for they transcend the boundaries of species and apply to entire groups of vertebrates. Herein lies the reason why such patterns must be considered real and regarded as essential to the tracker's repertoire. They constitute a way of seeing, a new domain of observation, rather than an analytical "method;" the latter too often substitutes for direct perception of our objects of study.

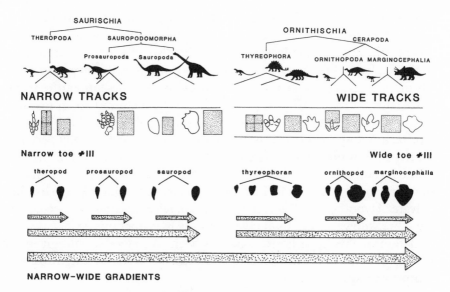

NARROW–WIDE GRADIENTS

FIGURE 7.6 The spectrum of form in dinosaur tracks is a gradient from narrow to wide manifested in the whole group, in all subgroups, and in sub-subgroups. The shape of middle toe bones (digit III) reveals the same gradient.

CELESTIAL MESSAGES ON THE IRIDIUM BAND

Extinction is a hot topic these days.

**Niles Eldredge, paleontologist,
American Museum of Natural History**

Countless layers of rock strata contain coded messages about the universe at the times the layers were deposited. Coal beds condense the ambiance of lush vegetation and humid coastal swamps and fossil sand dunes glare brightly like the desert at noon. But no layer of strata contains a more intriguing message than the thin band of iridium-enriched clay that marks the end of the Age of Dinosaurs. Situated exactly on the boundary between Cretaceous and Tertiary strata, marking the end of the Mesozoic Era, of middle life, and the start of the Cenozoic Era, of recent life, it is also known as the K-T boundary (K indicates "Cretaceous" because C was already used for Carboniferous, and T means Tertiary).

The Cretaceous is thought to have ended with a big bang caused by the impact of a comet or meterorite. This is often called the "K-T event." Our natural curiosity demands that we ask what really caused this catastrophe.

How did it go down? What really killed off the mighty dinosaurs, and many other species? In recent years interest in the K-T event has reached fever pitch, almost entirely as a result of the discovery, in 1980, of iridium in the K-T layer, first in Italy and later worldwide, by the University of California–Berkeley geologist Walter Alvarez. He and his father, the Nobel Prize laureate Louis Alvarez, proposed that the iridium had an extraterrestrial source, evidence, they said, of the impact on Earth of a comet, meteorite, asteroid, or other heavenly body.[51] Iridium is normally very rare on Earth, so it must have come from elsewhere. On the basis of the iridium concentration they calculated that the meteorite would have had to measure about 6 miles (10 km) in diameter. According to the so-called "Alvarez Hypothesis," atmospheric and ecological disaster followed, causing a breakdown of the food chain and many extinctions. (Since the formulation of this hypothesis, there has been considerable interest in the whole idea of meteorite impacts and craters—what we might call the footprints of meteorites!)

At first there was some resistance to the idea of a meteorite impact, especially among certain paleontologists, who asked why some creatures died out and others did not. Some geologists pointed out that iridium can also emanate from volcanic eruptions as well as from outer space. They proposed other possible causes of the catastrophe. For example, at the end of the Cretaceous, lavas known as the Deccan traps erupted over half of India, and some paleontologists felt that they might have been the cause of the catastrophe. The Indian paleontologist Ashok Sahni came up with the catchy idea of opposing D.T. (Deccan traps) and E.T. (extraterrestrial) hypotheses. Recently the meteorite argument has been strengthened by the suggestion that the meteorite footprint has been found: the Chicxulub crater, centered in the sea floor north of the Yucatan Peninsula, which measures about 120 miles (200 km) in diameter.[52] Although some geologists still maintain this is a volcanic crater, the majority are persuaded that it is an impact structure.

Accompanying the extinction of the dinosaurs was a dramatic if temporary disruption of the Earth's flora. In Colorado and New Mexico, the K-T layer is marked by a die-off of flowering plants and a remarkable increase in the abundance of fern spores. Some attribute the floral ecosystem change to fallout from the impact, and suggest that sooty particles indicate the occurrence of wildfires. Microscopic particles in the K-T layer have been subjected to intense scrutiny, and all manner of sooty and mineral debris and pollen grains have been photographed, probed, and measured with sophisticated, high-powered equipment. Each is regarded as a specific paleontological and geological signature of K-T events that helps support or refute E.T. and D.T. scenarios.

A diverting photographic interlude occurred when Farley Fleming and Doug Nichols, two geologists from the United States Geological Survey, were collecting pollen samples at the K-T boundary in Colorado. In order to get good samples they had dug deep into the fresh shale, creating what geologist call a "good exposure." Now they could clearly see the white K-T layer, which they christened "the magic layer," sandwiched between dark coal layers like filling in a slice of rich chocolate cake. The boundary layer looked so photogenic that both geologists began taking pictures.

When Fleming's photographs were developed, however, a most remarkable image appeared. Little ballet dancers in angelic white appeared to be dancing along the iridium band, leaving their dainty footprints on the bright white clay layer, which was illuminated by stage lights. Magic indeed! The explanation was simple on the face of it—a double exposure. Fleming had also captured Nichols taking a picture of the scene—what we might call an exposure of his own double exposure of the rock exposure.

Unbeknownst to the world of geology, homeopaths in England decided to do a standard proving of iridium; "proving" is a procedure where volunteers are given remedies to see what symptoms are produced. The normal proving procedure is to give a highly diluted dose to a group of 20 volunteers in a double-blind test. Two members of the group receive placebos, and neither the patients nor the physicians know what substance they are testing. The physicians record the symptoms reported by the patients. A significant number of those who received iridium reported visions of rainbows, preoccupation with the number seven, enhanced visual acuity, and dreams or visions of angels, birds, and other celestial phenomena. One patient felt he was growing wings.

Readers can make up their own minds about the symbolism associated with this strange convergence between homeopathy and geology. In Greek mythology, Iris was a messenger of the gods and the goddess of the rainbow, a spectrum of seven colors that arises from the interplay of light and dark. (The Spanish word for rainbow is *arco iris*.) Iris is also the name given to the seventh asteroid, discovered in 1847. Among the geological fraternity the K-T asteroid is firmly linked to the white boundary clay and the pure white light depicted in many artistic renderings of the blinding flash produced by the impact. As the colored portion of the eye, the iris constantly adjusts exposure, regulating the flow of energy between the observer and the observed universe. A meteorite crater, the footprint of Iris, has the three-dimensional geometry of an eye—in reverse. What messages the gods may send.

THE LAST DINOSAUR TRACKS

We don't find any dinosaurs within 100,000 years of the iridium band.
Jack Horner and Don Lessem, The Complete T. rex[53]

Clearly the impact of an asteroid traveling at 4 miles (7 km) per second represents a sudden, instantaneous event. Although there is an abrupt change in the flora at certain K-T sites, there is no obvious pile of bones representing evidence of a mass mortality of dinosaurs, pterosaurs and other creatures that went extinct at this time. Some have postulated, therefore, that dinosaurs died out before the K-T event, because no complete or reasonably complete skeletons have been found within 10 feet (3 m) below the boundary. This depth of strata could represent tens, even hundreds, of thousands of years.[54] Others have suggested that dinosaurs may have survived after the K-T impact, because isolated bones and teeth are found above the K-T layer. But such fragmentary remains may have been washed out of older burial sites, and so do not prove dinosaur survival.

Tracks on the other hand provide unequivocal evidence of living dinosaurs. In 1990, I attended a field trip to southern Colorado to view the same magic layer on which the white fairies danced. The other geologists suggested that I, being a tracker, should earn my keep by finding some footprints somewhere near the K-T layer. To their surprise, as well as my own, it took less than a minute for me to find tracks only 15 inches (37 cm) below the iridium layer.[55] [56] No one had scanned these K-T outcrops for tracks. Here was indisputable evidence that at least some dinosaurs had survived until the eleventh hour. The so called "three-meter gap" had all but evaporated completely. In short order we unearthed tracks of hadrosaurs and ceratopsians at four levels within six feet (2 m) below the boundary. These two dinosaur groups represented the dominant herbivores in the latest part of the Late Cretaceous. It looked as if they were alive and well right up to the end.

The K-T layer represents the transition from the Age of Dinosaurs to the Age of Mammals, and the Age of Birds. Dinosaurs left their mark up to the iridium band but not, as far as we know, above it. Dancing immediately on top of the layer we find mammals clad in fairy and angel white. We also find the tracks of birds, winged dinosaurs already breaking with earthbound tradition and exploring the heavens. These creatures and their signatures on the eternal trail form the subject of the final chapters.

chapter eight

OUT OF AFRICA

ACT VIII: CAST OF CHARACTERS

Following the abrupt and devastating crisis 65 million years ago that terminated the Mesozoic, or "midlife," Era and the Age of Dinosaurs, we enter the so called Age of Mammals, the Cenozoic Era (meaning "recent life"), during which time the fauna and flora become progressively more similar to that found in the modern world. Because the Mesozoic was once called the Secondary Era, the Cenozoic is divided into the Tertiary (or third) Period and the Quaternary (or fourth) Period, which is the subject of the final chapters. The subject of this chapter is the Tertiary, which is divided into two subperiods, the Paleogene, from 65 million to 25 million years ago, when archaic mammals predominated; and the Neogene from 25 million to 2 million years ago, when modern mammals came to the fore. These subperiods are further subdivided into epochs: the Paleogene comprises three epochs (Paleocene, Eocene, and Oligocene) and the Neogene comprises two epochs (Miocene and Pliocene. See Fig. 0.1, p. 7).

All modern mammals are our ten millionth cousins, including the ancestors of our cats, dogs, and horses, the latter evolving from minuscule three-toed species no bigger than Scottish terriers. But before we enter the familiar world of modern mammals we must pass through an animal world that was sadly impoverished, with "actors in rehearsal, not yet in dress and make-up, falteringly trying out their roles—awkward understudies trying to revive the play left unfinished by a defunct company. Here is a world where everything is groping towards future opportunities."[1] This haunting and evocative description by the great paleontologist Bjorn Kurten describes the postdinosaurian world left empty in the aftermath of the great extinction.

Had we been transported to this desolate time and space it is likely that birds would have caught our eye and cheered us with song, for they were probably then, as now, more abundant than mammals. It is only because we ourselves are mammals that we have adopted the label "Age of Mammals" to boast our present evolutionary success, and we are even more egocentric in naming the order to which we belong Primates, meaning the premier group on the scale we have erected. But before humans achieved premier travel status on the eternal trail, birds were well established frequent fliers. The first tracks found after the K-T cataclysm, only two feet above the iridium layer, are those of birds—possibly actual survivors of the catastrophe or their immediate offspring.[2]

In addition to these oldest of Tertiary tracks we shall examine the footprints of the strange "flamuck" or "domingo"—a bird reputed to be half duck and half flamingo—found alongside frog prints. We shall also encounter 25-million-year-old camels hitting their stride, and look at why this beast is so different from the other cloven-hoofed creatures that play such a large role in human history and symbolism. We shall also take a look at 3-million-year-old horses that developed interesting gaits by themselves, without the assistance of horse trainers, and ponder the proof that our own hominid ancestors were the first vertebrates to walk upright since the dinosaurs and their avian descendants.

TRACKING THE RISE OF THE HIMALAYAS: MAMMALS MASQUERADING AS DINOSAURS

In the Tertiary tapirs inhabited North America and Europe. . . . Like the earliest horses . . . living tapirs have four toes in the front feet and three behind.

Alfred Sherwood Romer, *The Vertebrate Story*

In some areas very little is known of the age of rocks, and this leads to misinterpretation of the fossil record. In both Peru and China, the lack of fossils had left geologists completely at a loss as to whether they were dealing with strata from the Age of Dinosaurs or from the Age of Mammals. It was only the simultaneous emergence of track specimens from three continents that forced us to look at the global picture and seek a correct understanding of the geological history. In both areas the tracks are found in thick sequences of strata that accumulated rapidly as a result of the erosion of nearby mountain ranges. Such strata can be used to reconstruct the history of mountain building—in this case two of the world's largest mountain ranges, the Andes and the Himalayas. As I shall explain, correct identification of a few ancient tapir-like tracks allowed us to track the rise of the mighty Himalayas.

An example of Peruvian mammals masquerading as dinosaurs shows us just how confusing things can get. The story of correctly dating rock strata in the Andes (and China) began in 1994, when Father Giuseppe Leonardi published a massive atlas on fossil footprints from South America, in which he illustrated various small three-toed tracks from Peru.[3] He and his colleagues concluded that these footprints had been made by duck-billed dinosaurs. Accordingly they also inferred that the rocks must be Cretaceous in age (for that is when duck-billed dinosaurs lived), which required a fairly elaborate revision of the geological history of the area, previously (and correctly) thought to be Tertiary in age.[4] I had been pondering pictures of these footprints when Bradley Ritts, a graduate student from Stanford University sent me pictures of almost identical footprints from China, which were also thought to be Cretaceous in age. It turns out that there was much uncertainty about the age of these tracks in both areas. They could be much younger, perhaps of Tertiary age. The key tracks to solve this puzzle—the ichnological Rosetta Stone—were found in North America, where, in Utah and Colorado, we find the same type of mammalian tracks in rocks that date back about 45 million to 55 million years, the Eocene epoch. The three-toed tracks had actually been made by what paleontologists call odd-toed ungulates (hoofed mammals) related to the horse and tapir.[5]

After searching through obscure literature on early Tertiary, odd-toed mammal tracks, it became clear that in Peru, China, North America, and elsewhere, all such footprints were undoubtedly made by hoofed mammals—not dinosaurs. No rethinking of the history of Andean mountain building was necessary. Moreover, the Chinese mammal tracks, at first thought to be Cretaceous (from the age of dinosaurs) turned out to have been made in sediments shed from the mighty Himalayas as they first began to rise to their present lofty heights.

HORSE TRACKS AT DAWN

The oldest of horses, *Eohippus*, the dawn horse, appears at the beginning of the Eocene epoch; frequent remains of this tiny horse, no larger than a fox terrier, are found in deposits of that date. . .

Alfred Sherwood Romer, *The Vertebrate Story*

Let us look more closely at what type of mammal made these small three-toed tracks. Conventional paleontological wisdom would seem to suggest a tapirlike animal. The tapir is an odd-toed ungulate (perrisodactyl) related to both the horse and the rhinoceros. Tapirs now live in South America, but had ancient relatives in North America in the early Tertiary. An experiment conducted at the Philadelphia Zoo in 1943 neatly demonstrated that tapir tracks closely match fossil footprints, though the tapir was of "uncertain temperament" and objected to having its tracks taken.[6] *Eohippus*, the so called "dawn horse" and a close relative of tapirs, lived during the Eocene epoch ("Eocene" means "dawn of the recent"), and is possibly the most famous of all nonprimate fossil mammals. It is often described as a creature the size of a fox terrier, which is similar in size to the hyrax, a primitive, rabbit-sized subungulate—indeed, *Eohippus* also goes by the name of *Hyracotherium,* meaning "hyraxlike animal."

The evolution of the horse has been studied intensively by many paleontologists, including the famous mammalogist, the late George Gaylord Simpson. It is particularly instructive for students of paleontology to trace the evolution of a small, originally five-toed animal through the three-toed *Eohippus* stage to the large one-toed modern horse *Equus*. These changes can be correlated with the disappearance of forest habitats and the evolution of prairie grasslands, a history that mainly unfolded in North America, though some chapters took place in the Old World. The famous naturalist Thomas Henry Huxley, whom we met in Chapter 2, was quite taken with the discovery of the dawn horse, and in the spirit of fun drew a cartoon of a dawn man ("*Eohomo*") riding his dawn steed.

With all this attention on the dawn horse, what has been said about its footprints? The answer appears to be: virtually nothing—but not because of a lack of evidence. In the late 1950s a beautiful slab of Eocene fossil footprints was found in Strawberry Canyon, Utah. It was donated to the Smithsonian Institution in Washington, where it has resided for more than four decades without being subjected to further study. Close inspection of this magnificent specimen reveals more than 130 footprints, most of them made by an *Eohippus*-sized creature. There are plenty of examples of trackways that provide an indication of step and stride, and some even show skin impressions. All this should excite students of ancestral horses into a serious study of their locomotion. From the beings that made these tracks evolved the ultimate winners of the Kentucky Derby and the steed that bore Paul Revere to his appointment with destiny.

In suggesting that the horse enthusiasts have paid insufficient attention to the track record, I am not just giving them a good-natured prod; I'm making a serious point. Science has apparently been under the delusion that horses were unable to adopt certain gaits without the intervention of horse trainers. I shall soon return to this topic, which revolves around the tracks of a much younger (3 million years old) horse found beside the

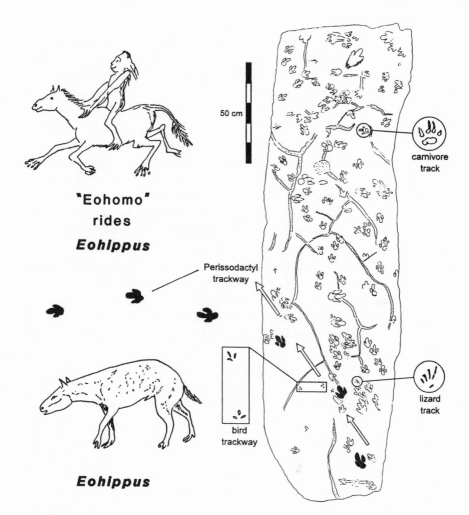

FIGURE 8.1   Tracks of dawn horses, possibly *Eohippus*, and other Eocene animals from Strawberry Canyon, Utah, and Thomas Henry Huxley's humorous sketch of a ficticious rider, *"Eohomo."*

spoor of our own hominid ancestor Lucy. Suffice it to say that track evidence gives credit to the horse for having learned more about locomotion in 50 million years than we seem willing to admit!

OF FROGS AND FLAMINGOS

In Egypt, the frog goddess, Hekt . . . embodied the power of the waters.

Nicholas Sanders, *Animal Spirits*

The first trackway evidence of the dawn horse is found among throngs of feeding birds and the occasional frog along lakeshores of the Eocene age. Anyone seriously interested in starting a fossil collection will not find it too hard to procure a fossil fish or two from an Eocene deposit known as the Green River Formation. The Green River flows through southwestern Wyoming, northwestern Colorado and northeastern Utah, to its confluence with the Colorado River, above the wild white water of Cataract Canyon. The Green River Formation is essentially a fossilized lake basin. In Eocene times this region was quite unlike the present sagebrush country, and was instead the site of huge subtropical lakes inhabited by a diverse flora and fauna that included mammals, birds, crocodiles, boas, turtles, frogs, fish, clams, snails, and hundreds of plant and insect species. As one might expect around lakeshores, by far the most common traces are tracks of waders or shorebirds, and thousands of minuscule, sinuous, winding trails of flatworms and the occasional dawn horse. At one locality there is a record of a single frog hopping its way across the mudflats.

In our investigations of Mesozoic bird tracks we were unable to match the footprints to known bird species because the skeletal remains of the track makers are not known. Can we do any better in the Tertiary age of birds? One of the most abundant and best-known bird fossils from the Green River lake beds is *Presbyornis*, a cross between an ancestral duck and an ancestral flamingo—what one might call a "flamuck" or a "domingo."[7] This bird lived in large colonies, like modern flamingos, and left behind fossilized bones, egg-shell fragments, and fish-bone refuse. The only bird tracks so far studied in any detail are those of a web-footed form, possibly *Presbyornis*. Although the individual footprints are ducklike, they occur in a narrow trackway with a winding trail of dabble marks made as the bird foraged back and forth with its bill.[8] Thus, the bird must also have had, in addition to web feet, long legs and a long neck. In 1970, a picture of the trackway appeared in *National Geographic Magazine*, labeled "*Presbyornis* footprints." This is another example of a precise identification of the track maker.[9]

Further studies of these *Presbyornis* or *Presbyornis*-like tracks resulted in the discovery of more dabble marks and additional information on the shape of the bird's bill.[10] This creature had a feeding pattern somewhat like that exhibited by modern shelducks, which harvest snails and other molluscs from mudflats. The feeding behavior of shorebirds goes back to the Mesozoic. In Cretaceous and Tertiary lake deposits we find together the tracks of waders, countless tiny sinuous trails made by minute flatworms no bigger than threads of cotton, and also the tracks of frogs.[11]

I suspect that frog trackways are more common than the sparse evidence of tracks presently suggests. The green Nile frog was an emblem of abundance, and frogs in general are associated with fecundity because

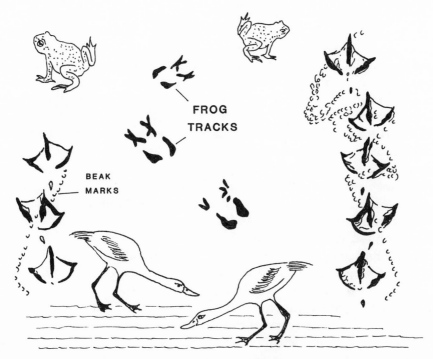

FIGURE 8.2   The tracks of frogs and "flamucks"—birds that appear to be half duck and half flamingo: a lakeshore ecosystem from the Eocene epoch.

they often emerge by the millions with life-giving rains.[12] They might not appear abundant if the lake waters were too saline. They are also small enough to be missed if one is not looking carefully. But frogs are at least as large and heavy as small birds, whose tracks we find abundantly. So for birds, frogs, and flatworms, lakeshore ecology apparently did not change very much from Cretaceous to Tertiary. (Gone are the footprints of dinosaurs and pterosaurs, but in their place we find trackway evidence of throngs of feeding birds, and the occasional dawn horse, frog, and 'flamuck'.)

TRACKING THE SYMBOLISM OF THE CLOVEN HOOF

Cattle, camels, sheep goats, antelopes, deer, giraffes and all other members of the ruminant or cud-chewing family invariably rise hind part first. Other four-footed animals get up front legs first.

George Stimpson, *A Book About a Thousand Things*

This observation by the author George Stimpson focuses our attention on the anterior and posterior polarities of mammal anatomy. Among ungulates the horse is a nerve-sense-dominated animal that holds its head high, rears on its hind legs, and has a narrow, firm, single-toed contact with the ground. Cattle and their kin, by contrast, are placid, highly metabolic ruminants with cloven hoofs and the habit of getting up from the ground hind part first. These cloven-hoofed cud chewers that decline to put their best feet forward are known as even-toed ungulates. Even-toed ungulates are the "modern" form of ungulate. Whereas the "archaic" odd-toed forms were most abundant earlier in ungulate history, especially in the early Tertiary, even-toed forms achieved dominance in the late Tertiary, relatively recently in paleontological terms. These ruminants are also of great significance in human history as ancestors of many of our domestic animals. Cattle and sheep are incorporated into the signs of the zodiac as Aries and Taurus. Bull cults are well known from the famous Neolithic settlement of Catal Huyuk, in Turkey, and from the later Minoan civilization. Of course, rams and shepherds figure prominently in the Bible.

Though astrology was once thought of as unscientific, now it is considered probable that the signs of the zodiac originated from our division of the year into seasons when particular animal activity was manifest, and that this calendar concept was then transferred to the larger precessional cycle.[13] In such a combined astrological/astronomical system of time, the ages of Aries and Taurus each are 2,160 years in duration, from 4490 to 2330 and 2329 to 170 B.C. These very precise astronomical ages derive from a precise twelve-fold subdivision of the 25,920-year precessional cycle of the equinoxes, the time it takes for the night sky to make a complete circle relative to a fixed point on Earth. Each division of this "grand cycle" is analogous to a calendar month. Such cycles are inextricably linked to climate and evolutionary dynamics. Their very recognition in ancient astronomy shows that our ancestors were already using animal icons to divide time into ages,[14] much as paleontologists do on longer time scales when they speak of ages of fish, dinosaurs, mammals, and so on.

Why have these animals taken on such symbolism? Why do mythical creatures like the Minotaur, the Centaur, and the Sphinx represent blends of human form with three major groups of mammals: even-toed ungulates, odd-toed ungulates, and carnivores, respectively. Humans recognize that they possess a mixture of instinctive animal and higher cerebral qualities. Early in the history of civilization both horses and bulls were ridden, and the first observers of humans riding these animals may have been awestruck by what appeared to be a blend of man and beast.

Minoan bull leaping was a ritual sport that reminds us of bull fighting today. It is said that the Minotaur has sexual overtones produced from an unholy coupling between Pasiphae, the wife of King Minos of Crete, and a prize bull. Because Minos had angered the gods, they made Pasiphae fall in love with the bull, and give birth to the hybrid Minotaur. Minos ordered it imprisoned in an underground labyrinth, where it was fed sacrificial youths ordered to be provided by the city of Athens. The hero, Theseus, who volunteered to kill the Minotaur, entered the Labyrinth with a ball of string given him by Ariadne, which he used to find his way back after carrying out the execution. Scholars regard the Minotaur as a symbol of animal passion, the labyrinth as the tortuous path of life, and the string as "the spark of divine instinct which unerringly illuminates the right path."[15] I prefer another interpretation derived from the idea that the bull's horns represent the crescent moon, long connected with the mysteries of childbirth and the power of the feminine link between humans and the Cosmos.[16] King Minos was angry at the power the moon had over his wife and the whole story symbolizes the masculine intellectual struggle to come to terms with the mysteries of feminine and cosmic power. The string is more likely a symbol of the umbilical chord, which connects humans to the dark innerworld of the womb.

The track record of the last 25 million to 30 million years is full of footprints of cloven-hoofed ungulates. As discussed in Chapter 4, hooves reveal various degrees of curvature, which is reflected in the curvature of the track makers' horns. Thus, small nerve-sense-dominated gazelles have straight, pointed tracks that match their straight horns. Deer have gently curved tracks and curved antlers, and bovine ungulates such as cows have more tightly curved or rounded tracks to match their tightly curved horns. Further striking evidence for the integrated relationship between hoof and horn shape comes from sheep that were suspended off the ground: the hooves, freed of the earth, grow into the shape of the animal's horns.

The camel is an odd cloven-foot ruminant different from the others in this group: it is a sense-oriented ruminant that lacks horns and is specially adapted to desert environments. In an elegant study inspired by footprints, the paleontologist David Webb demonstrated how North American camels developed a pacing gait early in their evolutionary history. (They are now extinct.)[17] Pacing locomotion allows for a longer stride, and is much more efficient in covering long distances on flat ground. The camel is not so good at maneuvering at close quarters, and prefers to move in a straight line. It has very specialized "sensitive" hoofs ideally suited to sand but in danger of being lacerated on rocky surfaces. In this it is unlike the other ruminants, which are less choosy about where they walk. Its sensitivity and adaptation to arid environments are shown in its ability to change its body temperature by as much as six degrees, and its ability to

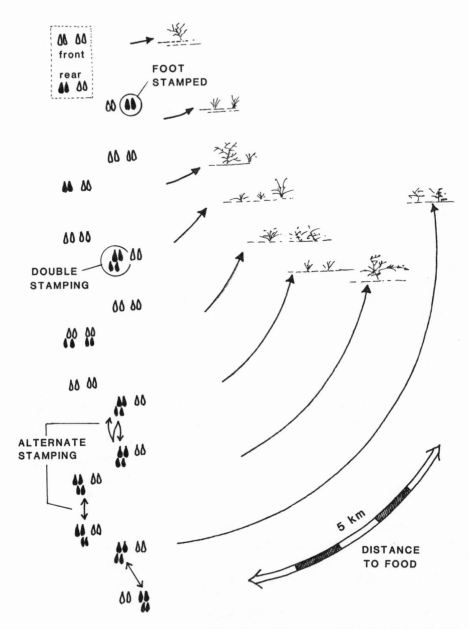

FIGURE 8.3 Foot stamping patterns of gazelles used in communicating different distances to food sources. Black track symbols indicate foot that is stamped.

travel up to 300 miles between water holes, where it may then drink up to a third of its body weight.

The camel and its ruminant relatives are the products of a world turning cooler and more arid. During the Tertiary period climates generally cooled and forests gave way to the grassy prairies, steppes, and savannahs. In the Old and New Worlds (separate since the early Mesozoic) a remarkable convergence is seen between the gazelle on the savannah and the antelope of the prairies, and likewise between the Old World wildebeest and the New World bison of the prairies. (Convergence means similarity of anatomy between species that are not closely related.) These two examples also show the language of "spirit" encoded in their tracks. Gazelle are small nerve-sense-dominated animals that have a small range in comparison with larger ungulates. In 1971 Jean-Claude Armen told the extraordinary story of a feral "gazelle boy," a child adopted by gazelles in the 1960s. As a toddler, the gazelle boy had evidently become separated from a nomadic tribe as they crossed the Spanish Sahara. (Such children were even tied by loose leashes to domestic sheep and goats and so learned ungulate ways at an early age.) Armen followed the gazelle boy and his adoptive herd for several seasons, making the first ever study of a feral child in the wild. Armen tells of the intricate language of foot movement used to communicate information about the distance to food, and see how the gazelle herd foraged in a cycle of regular loops from home base.[18]

By contrast with the gazelle, the "metabolic" wildebeest has a vast range and likes to form part of huge herds, and it also likes to run in circles. I made a map of some wildebeest trackways in East Africa after watching their "wild" behavior. In the spirit of inward metabolic behavior, the trackway turns sharply in on itself, almost tracing the same curves as the animal's curled horns. A camel, on the other hand, would never make such a tight turn by choice. It would be interesting to see if bison and antelope behavior converges to any degree with that of wildebeest and gazelle. Certainly the bison and wildebeest run in huge herds, whereas gazelle and antelope congregate in much smaller groups.

PLIOCENE DRESSAGE: EARLY EQUESTRIANISM

A horse! A horse! my kingdom for a horse!
  Shakespeare, *Richard III*, V.iv

Cats and dogs are among our best friends, but the horse surely ranks even higher in our estimation. Few have been hung for stealing a cat or dog, but horse theft is a capital offense. Even today, theft of the modern horse—the

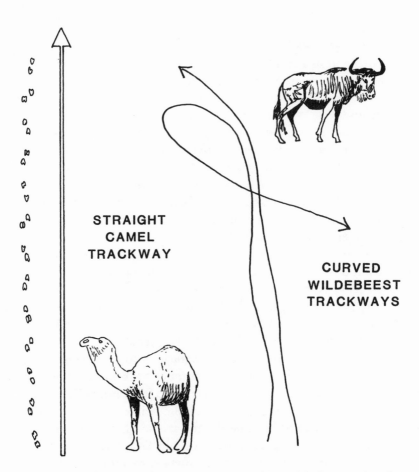

FIGURE 8.4   Differences in trackway pattern between the straight-line camel trackway and the curved trail of the wildebeest.

automobile—is a felony, as heinous a crime from the law enforcement standpoint as the kidnapping of a human. Since the first horses were tamed by steppe-dwelling nomads of Asia, some 4,000 to 5,000 years ago, battles have been lost and won and empires have risen and fallen on the strength of horsepower. Indeed, the very concept "horsepower" shows how much we value the power of this beast that has increased the speed of commerce, war, and communication throughout history. The term "chivalry," from the French *cheval,* horse, conveys a sense of supremacy and courage embodied by knights. The term "horseman" is virtually the only one in which we combine our own name with that of an animal almost without thinking. Cowboy is an amusing variant, which in a way also means horseman! In mythology the horse-man combination of the Centaur was regarded as a powerful amalgam of animal speed and human intelligence, with the ability to understand the will of the gods. Horses were often sacrificed when their riders died, to provide mounts for the journey into the afterlife.

The last horse we tracked was the 45-million-year-old, terrier-sized species known as *Eohippus*. Although tracks of the descendants of this famous "dawn horse" have been reported from time to time, none were the subject of serious attention, until 1984, when a 3-million-year-old set of Pliocene epoch footprints came to light in East Africa at Laetoli, in Tanazania, a site already made famous in 1976 by the discovery by Mary Leakey and her colleagues of the world's oldest hominid trackway. The presence of bones of the ancestral horse *Hipparon* suggested that the track maker was a member of this genus—another example of matching the track maker to its tracks.

Our long history of collaboration with horses has led to a good understanding of their gaits, and this has turned out to be useful when we look at trackways of ancient fossil horses. The biologist Elsie Renders analyzed three of the Laetoli trackways and compared them with those of modern horses. She demonstrated that the track makers were neither walking, trotting, nor pacing, but were moving at a "running walk." Horses are capable of using many different types of gaits, a subject we touched on in Chapter 1, and through a long history of horse training and equestrianism, we have even taught horses some gaits that they didn't adopt naturally—or so we thought. One gait that we thought was invented by us was the "running walk." Thus, Renders expressed some surprise at her discovery, because "it is generally assumed that [the running walk] has been taught by man in those breeds in which it occurs."[19] It seems we were wrong. We are the students, not the teachers, and must humbly accept that we have something to learn from a 3-million-year-old horse.

A fascinating question arises: If some modern breeds adopt this gait and others do not, does this mean that the habit may have existed in some prehistoric horse species and not in others? We are reminded that different species may have different natures or spirits, and these can be read by keen observers and experienced trackers. If different worms and dinosaurs are "programmed" for different behaviors, why not horses also? Renders noted that "the ability of particular horse breeds to select . . . alternative gaits has an anatomical basis, the genetics of which are not understood."[20] Surely it is not all about finding genetic explanations. We speak of horses being spirited to varying degrees. Surely different horses had particular morphologies, behaviors, natures, and spirits before humans began breeding them. What we may be looking at in our 3-million-year-old

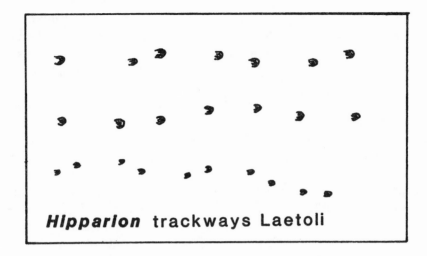

Hipparion trackways Laetoli

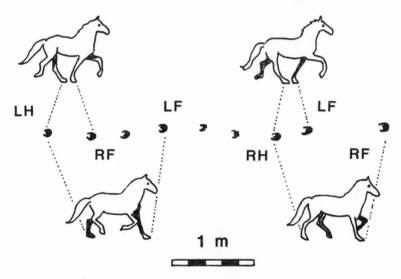

LH   LF   LF
RF   RH   RF

1 m

FIGURE 8.5   Pliocene dressage: trackway pattern and gait sequence for a horse moving in a running walk. This walk evolved in the Pliocene and is instinctive to some modern horse breeds, but not to others. Human trainers can teach this gait, but horses also learned it quite independently long ago. (L, R, F, and H= left, right, front, and hind.)

trackway, is evidence of an intrinsic behavior and spirit that is not simply genetically determined. The case of the Pliocene *Hipparon* clearly demonstrates that horses did their own thing without human intervention. In a self-organizing universe, horses learn to run by themselves, and we learn not to claim all the credit.

LEAKEY TRACKS LUCY

If I were to make a study of the tracks of animals and arrange them in
plates, I should conclude with the tracks of man.

Henry Thoreau

What is a hominid? What human qualities set us apart from other animals, and when did these qualities emerge? These perennial questions are
highly intriguing and largely unanswered. We don't really know. Reasonably complete hominid fossils are rather rare, so our paleontological history is very murky. Such a state of affairs has led to considerable
speculation, and lively, sometimes acrimonious, debate among paleoanthropologists who zealously promote their favorite hypotheses. However,
though we may be uncertain of our roots, in little more than a century we
have pushed back the origin of our own hominid ancestry to somewhere
between 3 million and 4 million years. This time scale would have been
unthinkable a century ago, when time spans of a few thousand years were
considered quite fantastic and unbelievable.

Currently the most famous paleoanthropological fossil is probably
"Lucy," a female member of the genus *Australopithecus,* estimated to date
from 3 million years ago. (The origin of the name, which means "southern
ape," is from Taung, Botswana (200 miles from Johannesburg) in Southern
Africa circa 1920.) Lucy was discovered by Donald Johanson and his
colleagues in Ethiopia in 1974, and according to anthropological folklore
was named after the Beatles song "Lucy in the Sky with Diamonds," which
was playing that day on the camp sound system.[21] Almost as famous, and
slightly older, are the 3.5-million-year-old footprints that Andrew Hill
found in Laetoli, Tanzania, in 1976, which are considered to "rank among
the greatest paleontological discoveries of our century."[22]

Laetoli is one of many sites associated with the great Rift Valley system
of East Africa, and is situated not far from Olduvai Gorge where other famous hominid remains were found by the Leakey family. As mentioned in
Chapter 3, the earth literally gives birth to new life in such fertile and dynamic rift settings, and so they can also be seen as symbols of new life. In
Triassic and Jurassic rifts we saw the emergence of bipedal dinosaurs destined to become among the most successful denizens of the Mesozoic. The
emergence of bipedal hominids in the Cenozoic seems somewhat like a replay of Mesozoic events. But is there any correlation between bipedalism
and success? Certainly the ability to walk upright confers on us a new
anatomy, a new perspective, and hands that are free to hold and carry
things—and create mischief. As we shall see, bipedality is frequently cited

as a highly significant factor in contributing to both hominid success and delinquency.

Thanks to finds made by the Leakeys, Johanson, and their many colleagues, we now generally accept that *Australopithecus* existed some 3 million to 4 million years ago and survived successfully until about 1 million years ago. Conventional wisdom also holds that *Australopithecus* had a small brain in comparison with genus *Homo* (meaning "the same" as us), and did not develop our stone-tool technology or mastery of fire. The earliest representative of "our" genus, *Homo habilis*, or "handy man," is thought to have emerged about 2 million years ago, founding the lineage that gave rise to the species *Homo erectus*, which in turn gave rise to the species *Homo sapiens*, of which two modern and comparatively recent varieties are Neanderthal man and our own Cro Magnon variety.

It is accepted that the ability to walk upright is one of the anatomical hallmarks of being a hominid, and it is widely held that this characteristic is intimately connected with our advanced intelligence. The thinking is that once hominids "came down from the trees" and began to walk upright, their hands became free to carry objects, and eventually to manipulate and manufacture tools. This advantage of being bipedal in turn fed back into sophisticated hand-eye coordination and a ballooning brain. The Tanzanian trackways, however, appear to prove that hominids walked upright long before they used tools, and before their brains ballooned to the size seen in *Homo erectus* and *Homo sapiens*. Thus, though walking upright may be one characteristic of being human, it is evidently not a characteristic that is the direct "cause" of the refined intelligence that evolved to produce sophisticated tool industries, cave art, sculpture, and cooked meat. If these latter are the qualities that we identify with becoming human, then they appear to be attributes of the genus *Homo*, not of *Australopithecus*.

We should not forget, however that, to the best of our knowledge, *Australopithecus* predates *Homo*, and is thought by many to be ancestral to our own genus. If this assumption is correct, then it was at one time the only hominid, carrying the seeds of our future inheritance in its genes, bipedal gait, and incipient hominid consciousness. Perhaps more than anything it is our fondness for Lucy and her australopithecine allies, not to mention her scientific status as a *bona fide* member of the hominid club, that most obviously betrays our instinctive willingness to accept her as one of our own. Even before the discovery of the Laetoli trackways, anthropology had already firmly established the image—perhaps myth—of hominids coming down from the trees, standing on their own two feet, and striding purposefully across the sun-drenched savannah, never to look back to the darkness of the primitive jungle world.[23] The trackways

seem to confirm this march of progress toward a future in which hominids play the role of upstanding citizens, and so "stand" head and
shoulders above other anthropoid stragglers still engaged in evolutionar-
ily archaic monkey business. This new era in primate evolution was con-
veniently recorded 3.5 million years ago when Mount Sideman, located
near Laetoli, triumphantly erupted, laying down a fresh but nonlethal
carpet of volcanic ash on which the new bipedal hominid script was
registered for posterity. Here is the Pliocene equivalent of superstar footprints in the cement of Hollywood Boulevard.

We can easily recognize the oldest hominid signatures, but can we distinguish these 3.5-million-year-old-trackways from those of modern
species of hominid? The answers are, or have been made, rather complex
and ambiguous. Laetoli is a big site that has produced thousands of trackways from several different locations within an area of several square
miles. The site was known since the 1930s, for having yielded skeletal remains, which is why Mary Leakey returned to the area, but prior to 1976,
when the British anthropologist Andrew Hill visited the Leakey camp, no
tracks had been reported. Hill was returning to camp as the sun began to
set when his playful colleagues "began hurling pieces of dry elephant dung
at each other."[24] As Hill ducked to avoid these coprolitic missiles, he found
himself with his nose to the Pliocene rock substrate, looking at animal
footprints enhanced by the slanting rays of the setting sun. This is a case
of one trace fossil being used to find another and just proves that as bipeds
with our heads held high we may easily miss what is beneath our feet.

Once the hominid footprints came to light, the focus of attention
shifted rapidly from bones to ichnology. Out of a total of more than 9,500
tracks discovered at the Laetoli site, almost 90 percent were made by rabbits.[25] Laetoli seems to be the Watership Down of the ichnological world.
In addition, in order of decreasing abundance, we find the tracks of
guinea fowl, hyena, cloven-hoofed bovids, rhinoceros, giraffe, buffalo, elephant, horse (*Hipparion*), various carnivores, monkeys, pigs, and ostriches.

It is important to give credit to those who discovered the site, and correctly describe their contributions and interpretations, for, as we shall see,
the excavation and scientific documentation of these tracks has aroused
lively controversy and an outspoken airing of opinion. Most research has
focused on the three trackways at site G, where at first sight we appear to
see the parallel trackways of a juvenile (G1) and an adult. On closer inspection, however, the larger trail turns out to be a double trackway, revealing that the first track maker (G2) was followed closely by another
individual (G3), that deliberately or inadvertently followed in the footsteps of G2. The G1 footprints on the right side are turned toward the

trackways of the larger individuals, and it has been suggested that the three trackways represent a family group, perhaps with the small individual holding hands with a parent, as shown in some artistic restorations.

The history of research on the site was reviewed by Tim White and Gen Suwa in a controversial paper published in the *American Journal of Physical Anthropology*.[26] They implied that the tracks had been at first misidentified as those of "two superimposed prints of a bovid." But after Leakey allowed limited excavation of the site "it was immediately apparent that the site prints were made by hominids" that had traveled from south to north. Although informative, the White-Suwa paper is strident, provocative, and at times verges on being offensive. The authors describe Leakey's numbering scheme as "patently inadequate and confusing" because she numbered the tracks in the order that they were excavated, not in the normal walking sequence. They also stated that "during the 1979 excavations, Mary Leakey exposed several prints with a hammer and chisel technique," causing at least four prints to be "damaged beyond repair in some critical areas." Such rhetoric has given the field of paleoanthropology a reputation for controversy, and explains why Roger Lewin's book on the subject, *Bones of Contention,* is aptly titled.[27]

Some authors have suggested that the footprints are indistinguishable from those of modern humans, but others have suggested they indicate a primitive strolling or shambling gait. The former might arise from slow progression on a slippery substrate, whereas the latter implies a somewhat primitive, wide gait made by a hominid whose locomotion is not completely modern. Some authors infer a track maker "transitional" between ape and man, thus raising the suggestion that the tracks may not have been made by *Australopithecus*, but instead might represent "a more advanced ancestor of modern man" yet to be discovered in the Pliocene skeletal record.[28] White and Suwa argue in favor of *Australopithecus*, and fit foot skeletons into the footprints to prove their point. This Cinderella-type demonstration appears fairly convincing, at least to some workers. Estimating footprint length at about 14 percent of the height of the track maker, White and Suwa estimate adult height of about four feet four inches to five feet (132–152 cm). This is slightly larger than the estimated size of Lucy, whose foot is only a couple of centimeters shorter than the tracks. As *Australopithecus* is the only hominid known at this time, in the Pliocene such a match offers a parsimonious correlation, in which both means and opportunity are satisfied.

A number of studies have shown that the Laetoli footprints match those of humans that are habitually unshod.[29] In modern humans who wear shoes the toes tend to be jammed together, producing a narrower, elongate foot. In unshod humans the toes fan out and the ball of the foot is wider,

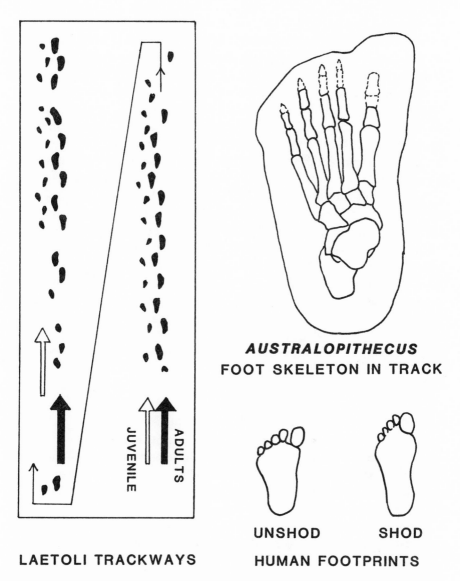

AUSTRALOPITHECUS
FOOT SKELETON IN TRACK

UNSHOD   **SHOD**

LAETOLI TRACKWAYS   **HUMAN FOOTPRINTS**

FIGURE 8.6   Leakey tracks Lucy. Laetoli trackway and outline of a footprint fit with the reconstructed skeleton of an *Australopithecus* foot. Also shown are the outlines of feet of habitually shod and unshod humans.

and an obvious gap develops between the big toe and other toes. In unshod feet there is increased development of the arch, and the indentation made by the ball of the foot immediately behind the big toe is deeper. As we might expect, the Laetoli footprints resemble the feet of habitually unshod humans.

White and Suwa insist, quite vehemently, that although some may regard Lucy's foot as subtly different from that of *Homo sapiens*, this does not mean that the footprints or the bipedal locomotion pattern was significantly different. They also reiterate the perennial refrain that not as much skeletal material is available as paleoanthropologists would like. The Laetoli tracksite, with dozens of footprints, highlights the track-bone discrepancy, for in all of East Africa, it is hard to scrape up a single complete *Australopithecus* foot to fit into the trackway.[30] Again, we are reminded just how common tracks are in comparison with the sparse skeletal record of fossil feet.

Russell Tuttle and his colleagues responded to White and Suwa with a call for less "palaver and politics," but did not quite take the high road themselves. They alleged that "White's brashness had outstripped his welcome in Leakey's home and camp," not least because of the gossip about heavy-handed chisel work, and Mary Leakey's purportedly poor eyesight, which Tuttle and his colleagues suggest was "remarkably keen" at the time of the excavations. They admonish White and his associates, including Donald Johanson, with a challenge to further examine the tracks, rather than reconstructing what they refer to as "spurious feet" of *Australopithecus* and *Homo habilis* from bits of skeleton from here and there. Not content with this admonishment they

> . . . urge that, until White, Suwa, and their colleagues achieve perfection themselves, they stop trying to undermine the reputations and systematic research of colleagues with rumors and pulpy books.
>
> In the past decade, before attempting to resolve questions of hominid evolution, the *modus operandi* of the Leakeys . . . and associates has been to search the Great Rift persistently for more complete fossils . . . while their rivals . . . try to fill the gaps with plaster constructions and novelized palaver during long spells out of Africa. Surely less politics and more poking in the ground are in order for the 21st century.[31]

As a humble paleontologist I was surprised by this level of rhetoric in scientific journals, and can only wonder what role the editors play by not toning this "palaver" down a little. No matter! There are two sides to every story, and controversy, dialogue, and passion are a part of scientific life. I shall try and stay out of the fray as much as possible, but I will comment on one of the observations made by Tuttle and his colleagues, where they suggest that "we need further research . . . before proceeding to conclusions about the locomotor capacities and taxonomic status of the Laetoli track stars."

Here I see a hidden question for trackers, rather than for paleoanthropologists. Could we give the tracks a name? Almost everyone agrees that the tracks were made by hominids. Therefore a name like *Hominipes* would be appropriate—perhaps *H. laetoliensis*. Whether the tracks were made by *Australopithecus, Homo,* or some other unknown hominid, in theory such a name should satisfy. There is no requirement to prove the identity of the track maker before giving the tracks a name, and in future more varieties might be identified. One fact is certain: the tracks have a unique morphology as far as the Pliocene is concerned. But are they significantly different from the tracks of other hominids, such as *Homo erectus* and species living today? If they are not, then all humans, or at least all those that are habitually unshod, make tracks that could be labeled *Hominipes*. I rest my case by saying that I do *not* formally propose this name. It is only a suggestion.[32]

The discovery of the spoor of an ancestor whose tracks are essentially like ours is bound to generate interest and controversy, especially when they are more than 3.5 million years old. In a mere century we have pushed back our hominid ancestry to the middle of the Pliocene, at each stage generating new doubts among those who had previously believed in a shorter history. But such doubt is generated, in large part, as the result of incomplete knowledge. We should perhaps let Mary Leakey have the last word. Commenting on a subtlety in the trackway which hints that one of the hominids hesitated momentarily as if to turn left, she said, "You need not be an expert tracker to discern this motion—the pause, the glance to the left, seems so intensely human. Three million six hundred thousand years ago, a remote ancestor—just as you or I—experienced a moment of doubt."[33]

BLESSED TRAIL OF OUR ANCESTORS

[E]ight elders, morally good, socially wealthy and married, having a great
number of children of both sexes, went to the site of the footprints,
blessing it with their holy prayers.

M. Demas, et al.

This pock-marked, paleoanthropological battleground at Laetoli, this Pliocene Park, this 3.6-million-year-old Watership Down, is what I like to call sacred ground. Standing with feet in contact with Mother Earth is a sacred communion. The symbolic importance of this act is enshrined in the large literature on the Laetoli tracks, and efforts to preserve the site, by reburying the footprints. Although generally successful, after 19 years acacia trees had begun to root in the earth covering the tracks and penetrating the precious surface, threatening to destroy the footprints. In an ambitious and well-organized effort to preserve this sacred ground, the Getty Conservation Institute, in conjunction with the government of Tanzania, mounted a rescue mission in 1995 to re-excavate the tracks, and preserve them properly using the latest conservation techniques.

This is more easily said than done, as there is no optimal approach to preserving fossil footprint sites. Some have questioned the strategy of burying the tracks, others have suggested removing the entire trackway to a museum, and yet others have suggested exposing it for public view beneath a shelter. The suggestion to remove the trackway was abandoned because "removal would be very risky." Besides the tracks have "cultural significance in their context in the savannah landscape of east Africa."[34] Leaving the tracks in place honors the sacred-ground principle. The idea of a shelter was also abandoned, because the site is too remote to be protected. The theft of fossil resources is a common and ubiquitous problem the world over. A shelter in the middle of nowhere would be much like a sign advertising something important, and even if, by some miracle, no one stole the tracks, they would be very vulnerable to erosion by tourists, who could not resist snapshots taken while they walked in their ancestors footsteps.

Having decided on the excavation and reburial strategy, the team first had to tackle a couple of decades of vegetative growth—"150 tress growing in the trackway and adjacent to it were killed by application of the degradable herbicide Roundup."[35] From walkabout to Roundup in 3.6 million years! In a nine-week operation by archaeologists and conservators, the trackway was carefully excavated and cleaned, using a low-powered vacuum cleaner with blowing capacity to prevent abrasion of the surface by repeated brushing. What would the first australopithecine family have thought if they were able to observe *Homo sapiens* spring-cleaning their trails with a vacuum cleaner for nine weeks? This labor of love amounted to more than two days of devotion at each footprint. To clean the feet of another is a sacred act. When one of the world's leading conservation organizations lavishes such tender loving care, it is surely a sign of reverence for the spoor of our ancestors and recognition that this patch of savannah is truly sacred ground.

Once the trackways were exhumed, cleaned, and stabilized, they were again photographed, measured, traced, replicated, and scientifically scrutinized from heel to toe tips. Then they were ceremonially buried, with a combination of local sand and soil and the latest "geotextiles." The first layer of geotextile (a nonwoven, water-permeable polypropylene material) was laid down on sand only one inch (2.5 cm) above the track layer. Another geotextile, Biobarrier, studded with slow-release herbicide nodules, was placed above the first layer, and above this came a capping of Enkamat—a synthetic erosion-control matting. This biotechological combination of natural soils and geotextiles, lovingly built into a mound, will revegetate naturally with grasses, but not with trees. For the first time in history a fossil trackway has been buried below three layers of the latest biotechnology. Fortunately the trackway had not been extensively damaged by tree roots. Though three footprints had been penetrated by roots, the unwanted botanical invasion was surgically removed.

Though the tracks can not be preserved forever, at least not with present technology, the conclusion of the study team is that they were in remarkably good condition after two decades—thoroughly validating the original decision to rebury them after the 1976 research phase. But it never hurts to take a second look, and the study team reported that the track layer was riddled with the tiny burrows of invertebrates, probably termites, made in the Pliocene. In most cases the burrowers had completely avoided the tracks, sometimes moving exactly along the margins, creating a framing effect that highlights the tracks. Why did the burrowers avoid the tracks? Surely termites did not show reverence for the sacred ground on which our ancestors walked? It seems the explanation is much simpler. They avoided the tracks because they found the ground within the footprints already too heavily compacted.

So the Laetoli tracks have been tucked up gently in bed, in new bioblankets, and can rest comfortably for a few more decades. But although quiet, the burial mound—for this is what archaeologists would naturally call one of their own creations— will soon have bedbugs rummaging between the sheets. When opened up the old mound was full of skinks, centipedes, beetles, ants, termites, and even a nest of mice. The new Laetoli burial mound will soon, once again, become a living monument to both mice and men.

The last word goes to the local indigenous people, the Maasai, whom the team wisely brought in to apply traditional methods to help protect the site. The religious leader, or Oloiboni, and community elders organized the sacralization ceremony, in which a sacrificial castrated ram was offered; it represented "love peace and unity . . . [Also sacrificed was] a sheep . . . thought pure and clean due to its characteristic of being obedient, calm and undisturbed. . . . In this way the collective decision was made to make the people of the region revere the site, respect its history and guard it so it remains pure."[36] This act of Maasai tribal unity and collaboration with international environmentalists forever makes the Laetoli tracksite sacred ground.

chapter nine

SPIRIT TRAILS

ACT IX: CAST OF CHARACTERS

Characters in our final chapters perhaps need less introduction than those known only from the depths of geological time. Though some are extinct, most are quite familiar. They include Ice Age (Pleistocene) cave bears, cave lions, and cave artists. We shall also meet mammoth and mastodon, giant wombats, Bigfoot—also known as the Sasquatch or Yeti—various giant flightless birds from New Zealand, and sloths with sandals, about which we shall hear Mark Twain's eloquent interpretations, and we shall also find out what can be learned from tracking extinct animals in prison yards and sewage plants.

We shall also be looking into that gray area of species endangerment where the spirits of departing and recently departed track makers still linger to haunt us and prod us to ask fundamental questions. As we ponder the temporal dimensions of evolutionary rise and fall we will get a glimpse of our own cyclical history, and see how the efforts of hominid ancestors to stand tall have followed cycles of ascent and descent. We shall pose the question, did we arise or descend from the apes? In doing so we may perhaps learn something more of our affinity with all animals, and catch a glimpse of the endless reiteration of evolutionary cycles and spirals which has shaped us, and all of our vertebrate relatives, step by step, increment by increment, from the tips of our upthrust skulls to the soles of our earthbound feet.

MAMMOTHS AND MASTODONS

The tusks that crashed in mighty brawls of mastodons are billiard balls.
The sword of Charlemagne the Just is ferric oxide turned to rust.

The grizzly bear whose potent hug was feared by all is now a rug.
Great Caesar's bust is on the shelf, and I don't feel too good myself.
  Arthur Guiterman "On the Vanity of Earthly Greatness" [1]

When we contemplate the work of 30,000-year-old Paleolithic artists, does it occur to us that we might be looking at a particular animal that is also represented by footprints or bones in the fossil record? When we look at a set of tracks, now fossilized in stone, do we think that these same footprints may have been gazed upon by one of our Paleolithic ancestors when still fresh spoor in soft mud? We are entering a time when such perspectives must be seriously considered. Extinct species ranging from mammoth and cave bear to the giant hominid ape *Gigantopithecus* (possible maker of Bigfoot tracks), Neanderthals, and perhaps *Homo erectus* were all directly observed by members of our own species, and in many cases were recorded in cave paintings and sculpture. We are not the first generation to gaze on dinosaur tracks, though how our ancestors interpreted them is not known. Not so with tracks of extinct animals from the last 50,000 years or so. Our ancestors saw these animals making tracks and probably understood their behavior and ecology better than we ever will.

Mammoth paintings and sculptures found in Europe and Asia provide excellent examples of how modern science has benefited by the record left by Paleolithic artists. In the thirteenth century the bones of the mammoth (from the Tartar word "mamma" meaning "earth") and other extinct mammals of the Pleistocene Ice Age (1.5 million to 12,000 years ago) were regarded as the bones of giants, monsters, or biblical behemoths that lived underground. For example, in 1443, when digging the foundations for a cathedral in Vienna, the thigh bone of a mammoth was unearthed and "hung on one of the gates of the city, which henceforth bore the name 'Giants gate.'"[2]

The abundance of mammoth (genus *Mammuthus*) remains in eastern Europe and Asia spawned a flourishing ivory trade and along with the narwhal is responsible for various versions of the unicorn myth. A true understanding of the age and origin of mammoth skeletons was confounded by the discovery of frozen carcasses in the Siberian tundra over the last several 100 years. Given that they provided edible meat for dogs and other carnivores, it was natural to conclude that they had died quite recently. Although it appeared that mammoth were extinct, there was no way to tell exactly how old they were. Further complications arose when archaeologists found Paleolithic tools in association with mammoth skeletons, suggesting that Paleolithic man had been a contemporary of the ancient mammoth. Others argued that man had not lived so long ago, and that he had simply butchered frozen mammoth, as some Siberian tribes still did into the early twentieth century.

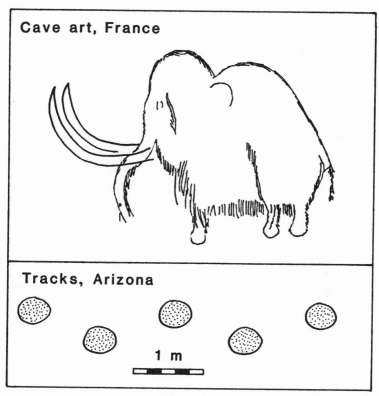

FIGURE 9.1   A mammoth depicted in Font-de-Gaume Cave, Dordogne, France, and tracks of mammoth, or perhaps mastodon, from Elephant Hill, Arizona.

Only when sculptures and cave paintings were found depicting the mammoth with its shaggy coat and characteristic hump was it accepted that man had been a contemporary of the woolly mammoth during the Ice Age. In fact, mammoth anatomy had not been accurately deduced from skeletons or frozen cadavers. The ancient cave painters had "with striking realism . . . rendered the shaggy coats of the huge beasts, the upthrust head, the huge hump on the back, and the characteristic position of the curving tusk."[3]

Cave paintings are a form of ichnological record of the ancient fauna that formed an integral part of the ecosystems shared by Paleolithic humans. Chauvet Cave in France, dating from 31,000 to 32,000 B.P. (before the present), reveals paintings of at least 216 animals, including 47 rhinoceros, 36 lions, 34 mammoths, 26 horses, 12 bears, 10 reindeer, and various other animals.[4] We can treat such records as a sort of census of the local fauna. The numbers are not a precise head count of animals in the area, but they do represent species extant at that time, and may suggest which animals were most abundant and most significant to man.

The fascination of cave art—which we could view as an "alternative," or parallel, ichnology—should not distract us from the actual track record. Footprints of both mammoth and the shorter, smaller-tusked American mastodon (genus *Mammut*), which died out in the Old World at the beginning of the Ice Age, have been reported from deposits in Alaska, Arizona, Nevada, South Dakota, and Michigan.[5] The Arizona site is known as Elephant Hill; the South Dakota site is a mammoth graveyard, where dozens of animals became trapped in a hot springs.[6]

The flying piece of dried elephant dung that allowed Mary Leakey's colleague Andrew Hill to discover the tracks at Laetoli reminds us that elephant tracks on this 3.6-million-year-old savannah belonged to the extinct elephant genus *Deinotherium*, not the present-day genus, *Loxodonta*. Elephants, or proboscidians (trunk bearers), have a long history, and 7- or 8-million-year-old tracks are reported from deposits in Death Valley, California.[7] I have seen tracks of about the same age in Nevada.

In folklore, the elephant is very wise and never forgets. Aristotle described it as the beast that passeth all others in wit and mind. The elephant is possessed of great power and size, but is also a creature of great gentleness and loyalty to its own kind. In India, the Hindu God of wisdom, Ganesha, is represented by a figure with the head of an elephant. The people of Thailand so respect the elephant that they will whisper secrets in its ear and ask for its advice. We owe a debt of gratitude to the mighty mammoth that fed many generations of our ancestors and provided us with a source of ivory, fertilizers, and aphrodisiacs long after it had gone extinct. Let us repay that debt to living elephants by showing our own wisdom in preserving them and their habitat from greed and exploitation.

A SLOTH WITH SANDALS

It may be all very well . . . to talk about the Carson Footprints, and try to saddle them onto the Primeval Man . . . and others who are gone and cannot now defend themselves; it may be all very well, I say, and entertaining, and within the just limits of scientific slander and research.

Mark Twain, "The Carson Fossil Footprints,"
The San Franciscan, **Feb. 16, 1884**

Perhaps the most controversial and historically intriguing Ice Age footprint site in North America is the mammal tracksite in Carson City,

Nevada. Discovered in the early1880s during excavations at the Nevada State Prison, performed by prison inmates, it became notorious when reports claimed that some of the tracks were made by what Mark Twain called "Primeval Man." The site revealed the trackways of mammoth, birds, horse, deer, elk, dogs, and cats made on the shores of an ancient lake. But most important of all, according to early reports, were the imprints of the sandaled foot of a man. Had this interpretation proved correct we would have established unequivocal evidence of humans with footwear living at the same time as Ice Age mammoth, in North America. Mark Twain took an interest in various authentic and bogus discoveries pertaining to human origins, but was even more interested in the public response to such finds, and with characteristic scorn referred to much of the hype and hyperbole as a "ridiculous mania."[8]

But the Carson City tracks were real enough, and the great paleontologist Edward Cope also attributed the footprints to a human,[9] though his great rival, Othniel Marsh, and others attributed the tracks to the giant ground sloth, *Mylodon*.[10] The controversy did not go unnoticed by Twain, who was unable to resist writing a delicious satire in which he concluded that the tracks had been made by drunken members of the first Nevada Territorial Legislature on a rainy night. "It had rained all evening outside . . . and it had rained whiskey all the evening inside. . . . They adjourned. . . . It was then they made the tracks. . . . I shall now cast upon this pale dim void of scientific conjecture, the lurid glare of history. The Primeval Man was absent. . . . The speaker went first. . . . He made the large tracks." Twain got carried away describing how the Speaker picked up straw on his muddy feet that gave the appearance of sandals, which were much admired by other legislatures, and became a source of envy: "there was rancor because of the sublimity of his sandals."[11]

Even though we now know that humans existed in America during the Ice Age and were mammoth hunters, consensus now holds that the Carson City tracks were made by ground sloths—the theories of Mark Twain not withstanding. The discovery of the ground sloth *Mylodon* at the La Brea tar pits in Los Angeles, which are of the same age, supports the identification of the track maker. Despite the 1880s furor, the site was never documented in detail until Jordan Marche of the Los Angeles County Museum published a detailed historic map, in 1986, more than a century after the original excavation.[12] Thanks to the map we can salvage some information. Although sloth tracks are rare elsewhere around the world, they are abundant at this site, outnumbering deer, elk, mammoth, horse, and all other mammal species. Alas, very few of the original tracks are preserved, having been covered or obliterated by prison construction. We can only wonder what Twain would have had to say on the subject of such destruction of evidence by criminals and legislators.

SPOOR OF THE CARNIVORE

Let dogs delight to bark and bite
For God hath made them so;
Let bears and lions growl and fight
For it is their nature too.

Isaac Watts (1674–1748),
"Against Quarelling"

As king of beasts the lion commands special respect, especially among those who have lived close to these mighty carnivores, or heard them roar in the savannah night. Lion imagery is widely associated with nobility and ruling power. Many a rampant lion adorns a noble coat of arms, and we speak of lionhearted kings and warriors. In Egyptian mythology, the lion, often in the form of the Sphinx, guards the entrance to the spirit world. The Chauvet Cave reveals a lion panel and a total of 36 depictions of the cave lion, *Panthera leo spelaea,* that roamed Europe some 30,000 years ago.[13] Ayla, the heroine of Jean Auel's reconstruction of Paleolithic life, *Clan of the Cave Bear* (Crown, 1980), one of the few fictional characters who is a human being of the Paleolithic era, chose the cave lion as a totem after one had gouged an indelible scar (trace fossil) in her thigh. Ayla's close call with this mighty predator vividly recreates the potential danger faced by our Paleolithic ancestors in their daily existence. The lion remains a danger to modern trackers. The ichnologist Theagarten Lingham-Soliar wrote to tell me he was twice ambushed by an angry lion when working at a tracksite in Zimbabwe. The lion was not scared off until shots were fired near its feet.

Archaeology reveals that cave lions, cave bears, hyenas, wolves, and other large carnivores, like humans, made caves their preferred homes. Small wonder that the dark underworld holds such powerful symbolism for humans as the abode of dangerous creatures—monsters from within the earth, spirits that walk through mountains. The Grotto of Aldene in France is famous for containing not only abundant human footprints, but also charcoal crushed underfoot by hyenas some 15,000 years ago and the abundant spoor of cave bear, mainly in the form of scratches on the walls and "bears' nests," which consist of craters hollowed out as hibernation dens.[14] Chauvet Cave also contains actual cave bear footprints, as does Toirano Cave in Italy. The paleontologist Bjorn Kurten, the author of *The Age of Mammals,* is also the author of *The Cave Bear Story: Life and Death of a Vanished Animal* (Columbia University Press, 1976). He claims it was a destiny he could not avoid since the name his parents gave him, Bjorn, happens to be Swedish for "bear."

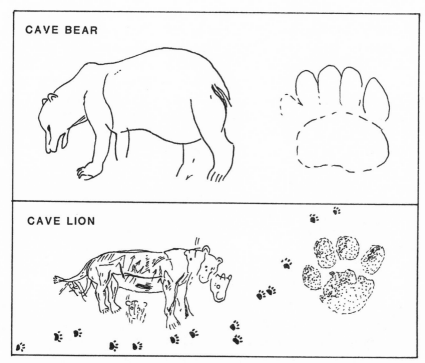

FIGURE 9.2 Spoor of the carnivore: Paleolithic depiction of a cave bear, from Teyjat Cave, Dordogne, and footprint from Chauvet Cave. Drawing and track of cave lion, based on work of W. von Koenigswald and colleagues.

Carnivores are creatures of the night, of caves, and of subterranean dens. The tracks of the extinct 10,000-year-old North American lion *Panthera atrox*—meaning literally "atrocious panther"—were discovered at a cave in Missouri, now known as Cat Track Cave.[15] Other footprints have been reported from caves in Missouri and Tennessee; a cave in Tennessee, is known as Jaguar Cave. A sewage treatment plant in Bottrop, northern Germany, on the banks of the Emscher River, revealed a tracksite with splendid trackways of the European cave lion (*Panthera leo spelaea*); it was the first track record of the species in middle Europe found in the open, outside the animal's cave habitat.[16] Tracks of wolf, bison, horse, and reindeer fill out the ichnological tapestry of a 20,000-to-30,000-year- old setting climatically resembling present-day Lapland.

Though Ice Age mammals grab our attention, we should not forget the long history of carnivores that predates the icy Pleistocene epoch by as much as 10 million to 20 million years. In *The Big Cats and Their Fossil Relatives* (Columbia University Press, 1997), Alan Turner and Mauricio Anton chronicle the history of felines starting with their origins in the

Miocene epoch; they report a site in Spain revealing tracks no larger than those of modern domestic cats. These represent the ancestors of all cats.

The Russian tracker O. C. Vialov proposed the name *Bestiopeda,* meaning "beast foot," for the tracks of all feline and canine carnivores.[17] Although we may look affectionately on our domestic cats and dogs, they are the quintessential beastly carnivores. Our image of beasts and beastliness comes not from our observations of deer and other relatively docile herbivores but from vivid tooth-and-claw images of cave bears and cave lions defending their dens in the prehistoric night.

GIANT WOMBATS DEAD IN THEIR TRACKS

History is the track in the snow left by the creativity wandering in the divine imagination.

R. Abraham, T. McKenna, and R. Sheldrake,
Trialogues at the Edge of the West: Chaos,
Creativity and the Resacralization of the World

Skeptics say that we must find an animal literally dead in its tracks to prove it was the source of a particular set of tracks. In a strict sense this is true, though so far we have seen several dozen compelling matches between tracks and beasts, ranging from monster millipedes to *Tyrannosaurus rex.* Clearly, the identification of track makers becomes progressively easier as we approach the present day, because the corresponding fossils are well known, well preserved, and well dated. Few paleontologists question the assignment of tracks to ground sloths and mammoth, providing they are in the correct strata and the right location. For example, it would be ridiculous to argue that tracks at the Hot Springs mammoth graveyard in South Dakota were not made by mammoth when the deposit is full of their skeletons and little else.[18]

But we can go one better in our search for creatures dead in their tracks. Intriguing nineteenth-century reports of large hippopotamus-sized creatures living in the swamps and wetlands of Australia are consistent in describing hairy creatures with very small ears, a large mouth, the general appearance of an overgrown retriever or bear-sized rabbit, and aquatic, seal-like habits. There is actually compelling evidence of the recent existence of giant wombatlike marsupials that fit this description. Other reports, such as that of a sea leopard killed in the Shoalhaven River in 1870 whose stomach contained a platypus, have contemporary, though nonetheless extraordinary explanations. Moreover, Australian Aboriginal lore speaks of large creatures (*kadikamara*) the size of horses that ate vegetation and spent much time in the water. In colloquial Australian jargon this creature and others of unknown ilk have been referred to as "*bunyips*," or "bogey" creatures.

Paleontological evidence sheds interesting light on these strange sightings, and suggests the existence of relic populations of Pleistocene animals on the verge of extinction. Until comparatively recent times Australia was home to a giant hippo-sized wombat known as *Diprotodon* (referring to its two giant incisor teeth). The modern wombat is a much smaller, marmot-sized creature that has been described as the marsupial equivalent of the woodchuck. Hundreds of virtually complete *Diprotodon* were recovered from a site known as the Callabonna salt pan in the desert interior of South Australia. During the Pleistocene Ice Age, these interior desert areas were lush, green, and fertile. It is thought that the *Diprotodon* graveyards are the result of the recent desertification of these areas. Bernard Huevelmans, in *On the Track of Unknown Animals,* describes how during the last days of *Diprotodon,* vast herds of "these poor beasts, dying of thirst, had apparently ventured onto the hard crust where the lake had recently dried up to no more than a few tempting pools. The crust gave way under their weight and they gradually sank into the soft clay, which preserved their skeletons from the weather. Some had their paws folded beneath them, as if they met their death calmly welcoming the moisture of their muddy grave."[19]

Not everyone agrees with this scenario, however. According to later reports, by the paleontologist Richard Tedford of the American Museum of Natural History in New York, "[T]he great quantity of bones at Callaboona did not result from the catastrophic effect of a protracted drought, but rather was a slow accumulation of individuals or small groups of animals that tried to cross those boggy flats during periods of low water. There is dramatic evidence of this in the churned nature of the clay-sand strata around each skeleton, and especially in the presence of recognizable footprints of *Diprotodon.*"[20] Their feet acted as "cookie cutters" pressing the salt crust into the mud below, so that we find their remains at the end of a series of salty cutout footprints. These now stand as little pedestals pointing the way to where these giant wombats literally died in their tracks.

MOA TRACKS BUT NO MOA TRACK MAKER

The moa was not the brightest bird
and couldn't fly, which sounds absurd
So wingless, this condition led
The moa to grow huge legs instead.

Jon Gadsby and R. Parkinson,
A Book of Beasts

In 1849 sealers camped on Resolution Island off South Island, New Zealand, noticed footprints of a large bird in the snow.[21] They tracked the animal with their dogs and captured it, keeping it alive on their ship for three days before skinning it and roasting it. The skin was saved and was procured by the Englishman Walter Mantell, the son of the famous paleontologist Gideon Mantell, who described *Iguanodon*, the first dinosaur ever discovered in England. Up to that time, the bird, of the genus *Notornis* (its Maori name is *takahe*) had been known only from a subfossil skeleton recently discovered on the North Island. In 1851 another living specimen was captured, by a Maori, on nearby Secretary Island, and again the skin found its way to Mantell. Despite deliberate searches by European scientists, and Maori reports that the bird was "plentiful" in some areas, several decades passed without discovery of another specimen, and the *takahe* was pronounced extinct by some ornithologists. Two other specimens were procured by hunters with dogs in 1879 and 1898, but after that the *takahe* was again pronounced extinct—until 1947, when a small population was found in dense forest at Lake Te Anau, South Island, after a footprint was discovered in the mud. The *takahe* is now much revered and protected by dedicated ornithologists and conservationists.

Well, the moa, as New Zealander's now claim, is truly with us "no moa." The story in a nutshell—told by fossil remains, including bones, skin, footprints, eggs and feathers—is that about a dozen species of moa are known to have existed prior to human colonization of the islands, but the Maori appear to have driven all or most of them to extinction in just a few centuries, between the estimated time of their arrival in New Zealand in about A.D. 1350 and the arrival of Captain Cook in 1769.

The example of the century-long search for the *takahe* serves to remind us of our many perceptual biases. First, rare forest-dwelling creatures such as the *takahe* (and moa) are hard to find, and may too easily be regarded as extinct when in fact they still exist in small, isolated pockets. Second, we see how footprints are not generally accepted as evidence of a creature's existence until after skeletons or living specimens are found. When a skeleton is found, the footprint evidence suddenly becomes acceptable. (We should keep this instructive example in mind when we examine the evidence for Bigfoot in the next section) Third, we learn that local indigenous people may not regard such creatures as rare, as long as they exist. Surely the indigenous Maori concept of rare is not the same as that of ornithologists who spend only a few weeks or months searching unsuccessfully for a particular species. This is to say nothing of the indigenous person's greater ability to "see."

Where we find one example, we may find another. From the same forested region of New Zealand around Lake Te Anau, another zoological mystery entered the annals of science as a result of a paper presented by Dr. Gideon Mantell to the Royal Zoological Society of London in 1850. He reported that despite the belief that there were no indigenous quadrupeds in New Zealand except for the Maori rat, the Maoris spoke of a badger-, otter-, or platypuslike animal named *kaureke* or *waitoreki* that had formerly been abundant.[22] Mantell and his son believed that it still existed or that it was only recently extinct.

In 1861, Sir Julius von Haast, a famous New Zealand geologist, claimed to have seen on many occasions tracks resembling those of the European otter in areas where no non-Maori otter had previously set foot.[23] Other eyewitness reports of otter-, beaver-, or platypuslike creatures and beehive-shaped or beaverlike lodges seem to support the existence of the creature the Maoris spoke of. It was European scientists, not Maoris, who reported that New Zealand was a land without mammals other than bats and introduced rats and dogs.

Evidence for the existence of large flightless birds in New Zealand and elsewhere had a considerable influence on nineteenth-century trackers, leading them to attribute to birds tracks that actually had been made by dinosaurs. When they considered the moa, the Madagascan elephant bird (*Aepornis*), discovered in the 1850s, and the Dodo (*Didus*) of Mauritius, driven to extinction in 1690, paleontologists such as Edward Hitchcock, Gideon Mantell, and their famous contemporaries such as Charles Darwin and Charles Lyell, not surprisingly were strongly persuaded that birdlike dinosaur tracks were attributable to races of giant extinct birds similar to these other more recent ones.

On the subject of the dodo, Stephen Jay Gould has recently lamented the failure of the zoological establishment—at Oxford, no less—to preserve skeletal remains of this great symbol of species extinction within historical memory.[24] If we can lose almost all tangible evidence of an animal that went extinct only 300 years ago, is it surprising that we do not find abundant remains of *Australopithecus* or *Homo erectus*, species that went extinct hundreds of thousands or millions of years ago?

And what will future generations say of the paintings, sketches, and reports of the dodo, when all material remains have been lost? Will they regard our illustrations as manifestations of obscure myth? By 1750 most people living on Mauritius had never heard of the former existence of the dodo, just as some youngsters of today have not heard of the Beatles. How soon we forget. It is only because we are currently interested in zoological classification and the fate of species that we even keep track of such records and reports as the ones discussed here. In a nonzoological century or culture such knowledge could very rapidly be lost.

Some Maori cave paintings depict moas, and similar paintings in Australia may represent the extinct genus *Genyornis*, a type of giant emu.[25]

But how seriously did European scientists take such depictions? Because the moa was apparently extinct when Captain Cook arrived in 1769, some European ornithologists denied that the Maori had ever seen the moa, suggesting that the birds died out before the Maori arrived, around A.D. 1350. Thanks to archaeological evidence of extensive Maori hunting of the moa, we now know that this conclusion is quite wrong. At least one man, a boy of 14 when Captain Cook arrived in 1769, claimed to have taken part in the last moa hunt.

Some recent reports raise the interesting possibility that, like the *Notornis*, some species of moa might still be alive, or very recently extinct. The story of Alice McKenzie of Martin's Bay, South Island, is considered respectable and "the most acceptable to scientists."[26] In 1959, for the Radio New Zealand sound archives, she recorded her memories of coming upon a bird sunning itself on the sand in the winter of 1880, when she was only seven years old. She described the bird as faded bluish gray, with dark, greenish legs with big scales. She claimed to have crept up on the bird and stroked it without its taking much notice, but when she tried to tie a piece of flax around its leg the bird got up and made a harsh grunting noise before biting at her. Her father later came and inspected the footprints, measuring them at 11 inches (27 cm) long.[27] She claimed to have seen the bird or one similar to it again nine years later, in 1889. She saw its footprints in the sand and remarked on the surprised reaction of the cattle she was driving at the time. She also reported that her brother saw the bird and that they encountered its tracks on the beach every winter.

It seems rather unlikely that Alice McKenzie and her brother made up the story, especially the part about the bird's tameness and its tracks appearing regularly in the winter. The possibility that the bird was a *takahe* was apparently ruled out when the *takahe* was rediscovered in 1947, and Alice McKenzie realized that the bird she had seen was entirely different. Despite the skepticism of some observers, several other reports of supposedly extinct moas are consistent with the McKenzie sightings. From the tracker's viewpoint, these reports are similar to the story of the *takahe* in some respects. Here is a creature that is rarely seen, that lives in the bush most of the time, but whose presence is periodically demonstrated by footprints.

There were sufficient moa sightings in the late nineteenth and early twentieth century to encourage the launching of several expeditions. None of these were successful, but this does not prove that the moa is extinct. One Japanese expedition flew around New Zealand's Fjordland broadcasting simulated moa calls, reconstructed on the basis of estimates of measurements of the throat anatomy from skeletons. The fact that this method of attempting to lure the moa was not successful is hardly sur-

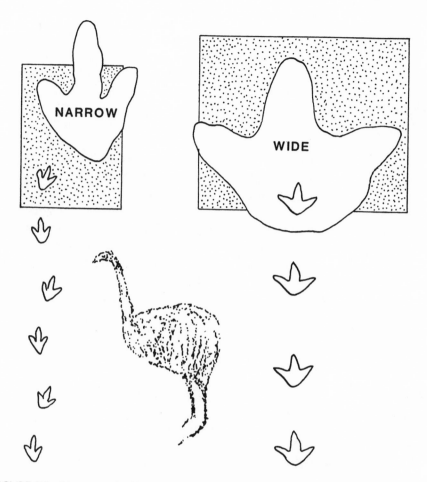

FIGURE 9.3 Narrow and wide moa tracks and a prehistoric Maori painting showing a living track maker.

prising, since any self-respecting moa would presumably not present itself to a noisy helicopter broadcasting hypothetical, and perhaps quite un-moa-like, calls. In the final analysis we can safely conclude that the moa was alive and well until comparatively recent times. Some species probably survived as residual populations at or just before the arrival of Captain Cook, perhaps with remnant populations surviving into the early twentieth century.

Fossilized moa tracks have been known since the 1860s. Archdeacon Williams reported footprints from Poverty Bay, North Island, in 1871.[28] Since that time several other moa tracksites have been reported, all from the North Island. Tracks found at Poverty Bay appear to be those of a relatively small bird, with footprints about 7 inches (18 cm) long. Footprints discovered at Palmerston, North Island, in 1912[29] were 15 inches (15 cm) long. Having examined all the known moa tracks from New Zealand, I conclude that there are two distinct types, presumably representing two distinct species. The larger variety would appear to match the feet of the giant moa, *Dinornis*. We can narrow down the list of 11 possible track-making species to 7 by eliminating the 4 known only from the South Island, because they could not possibly have made any of the North Island tracks. If fossil footprints are ever discovered in the South Island, it will be most interesting to see whether they are different from those from the North Island, as the distribution of moa species would lead us to expect.

Finally we can note that the two track types fall into the categories of small and narrow and large and wide. Given what we now know about morphological form in many groups, this could be more than mere coincidence. One could predict that the larger form would have had relatively long legs, as is the case for *Dinornis*, whereas the smaller species might have had rather short legs, which would be analogous to one of the smaller non-*Dinornis* species.

TRACKING BIGFOOT

What I believe to be the most important piece of evidence in the whole yeti story came to light in November 1951. The central human figures were Eric Shipton and Michael Ward. The evidence was a giant footprint impressed in snow.

John Napier, *Bigfoot: The Yeti and Sasquatch in Myth and Reality*[30]

When assessing evidence for the existence of Bigfoot—the name given to beings half hominid and half beast, with the ability to walk upright—we should bear in mind the fact, discussed previously, that footprint evidence suddenly tends to be viewed as valid only when a skeleton is found. Stories of Bigfoot—or its prints, at least—come from several regions of the world, including central Asia (where the local name is the Yeti, or the Abominable Snowman), and the Pacific northwest of North America, (where it is called Sasquatch).

According to a legend of the Modoc Indians of the Mount Shasta region of Washington State, when the Chief of the Sky Spirits grew tired of the cold sky he carved a hole and pushed down the snow and ice. Down on Earth, he walked about in the snow, which "melted in his footsteps," creating water that ran into rivers.[31] He then created the creatures of the Earth and made the grizzly bear to walk upright and talk like human beings. He was so pleased with his creations that he brought his family down to live on Earth. But an accident befell his daughter and she was blown away by powerful winds and adopted by the grizzlies. She stayed with them for many years, married one of them, and produced offspring that were less hairy than the bears. The grizzlies eventually worried that the Chief of the Sky Spirits would be angry at the loss of his daughter, and so they set out to find him and tell him where he could be reunited with her. He was indeed angry that he had missed seeing his daughter grow and had not witnessed the birth of the new race his grandchildren represented. So he cursed the grizzlies, removed their power of speech, and made them walk on all fours. The hybrid god-grizzly descendants, however, roamed the region and, according to legend, are the ancestors of the native people. For this reason, these peoples will never kill a grizzly bear.

This is just one of several creation stories involving ancestors or other creatures that are half hominid and half beast, with the ability to walk upright. In the Old World, fossil evidence clearly establishes the existence of a 10-foot-tall (3 m) giant ape, named *Gigantopithecus,* discovered by Dutch paleontologist G.H.R. von Koenigswald in 1935, that inhabited the bamboo forests of Southeast Asia for some time during the last million years. The anthropologists Russell Ciochon, John Olsen, and Jamie James, the first American scientists to venture into North Vietnam since the Vietnam war, reported finding evidence that *Homo erectus,* our direct ancestor, was a contemporary of *Gigantopithecus.* The evidence was the discovery of bones of the two species in the same deposit. They speculated that *H. erectus* may well have killed *Gigantopithecus,* and even recreated a typical encounter between *H. erectus* and their ten-foot cousins.[32] A close encounter with a ten-foot *Gigantopithecus,* weighing 1,000 pounds or more, would surely have made a big impression on a five-foot *Homo erectus.* In the long run *Gigantopithecus* evidently came off worse in such encounters, and eventually, it is assumed, became extinct.

An encounter between *Gigantopithecus* and *Homo erectus* of course requires that they walked the Earth at the same time. *Gigantopithecus* is thought to have lived as recently as 200,000 to 100,000 years ago, based on coexistence with *Homo erectus.* But *H. erectus* has recently been reported from deposits only 27,000 years old, so if they both lived at the same time, perhaps *Gigantopithecus* lived much closer to the present time.[33] There are those who believe some *Gigantopithecus* are still alive.[34] Even if they are extinct, images of these giant apes could have lodged in the memories and oral traditions of our immediate ancestors, and perhaps even in the minds of early representatives of our own species.

Such historical perspectives bring us inevitably to a consideration of the Old World Yeti, or Abominable Snowman, and the New World Sasquatch, or Bigfoot. The Yeti has purportedly been sighted on several occasions at high altitudes in the Himalayas and mountains of central Asia, and its footprints in the snow have been reported and photographed by the mountaineer Eric Shipton, whose integrity is described as "unimpeachable."[35] It is interesting coincidence that sightings of the Yeti are consistent with the former range of *Gigantopithecus*—namely, the bamboo and montane forestry of eastern and central Asia—as established by the fossil record.

The skeptic will rightly ask how a giant ape could make a living in snowbound mountains. Writing in *Abominable Snowmen: Legend Come to Life*, Ivan Sanderson, a well qualified and widely traveled professional collector of exotic animals, pointed out that hairy, upright walking, bipedal hominidlike creatures have been reported from many locations worldwide, but particularly from upland (montane) forest regions of North America and eastern and Central Asia.[36] This correlation between animal and ecological setting strengthens the case for the existence of this giant hominid, especially when it is the same range that has been independently inferred for *Gigantopithecus* by Ciochen and his colleagues. Incidentally, these authors judge Sanderson's comprehensive treatment of the subject to be cautious and skeptical, and far from exaggerated or fanciful. Like many who enter a subject with healthy skepticism, Sanderson ended up being persuaded by the evidence, which in this case includes photographs of the famous Shipton trackway discovery, two large "mummified" hands, which have also been photographed, the indirect evidence of hair and feces, and many sworn statements of eye witnesses.[37]

Sanderson examined the position of those who dismiss all reports as hoaxes, showing that almost invariably they have not examined the evidence, been to the areas where sightings were reported, or shown any respect for the intelligence and observational powers of local indigenous people, who consistently reported these creatures from the areas in which they live. For example, Sanderson reports that when British Museum zoologists put the Shipton photograph on display beside the tracks of a bear and a Langur monkey with the irritating caption "You can see for yourself that this Abominable Snowman footprint is that of a bear . . . or a monkey." This attempt to discredit the evidence apparently backfired and insulted the intelligence of the British press and public. Sherpas and other trackers are far more capable of identifying the tracks of native species than office-bound, civil servant, zoologists in London.

Even if there are no surviving Yetis or Sasquatch in Asia and North America today, we can still ask whether they might not have been alive until very recently as endangered species in marginal areas, much like the *takahe* of New Zealand. These purported subhominid giants have been re-

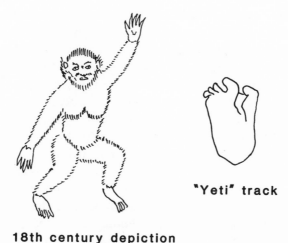

"Yeti" track

18th century depiction

Modern Human track

Neanderthal track

FIGURE 9.4   An artistic rendering of a hairy hominid giant from eighteenth-century China, and purported tracks of the Yeti, with modern human and Neanderthal tracks for comparison.

ported in Asia for hundreds of years and by Native Americans, and Chinese records contain artistic renderings of these creatures. Indigenous people from these areas surely had no motive for sustaining hoaxes for many centuries.

If *Homo erectus* killed off most of *Gigantopithecus*, it begins to make sense that such endangered or recently extinct creatures would be very wary of humans. After all, we are afraid of members of our own species wherever we are unsure whether the natives are friendly. In other words, such survivors would have been smart enough to give us a wide berth. If they do or did exist, they presumably coexisted with our species since we first emerged from the Pleistocene twilight. Here we should admit that we really have only a sketchy idea of our own origins, so it is presumptuous to speak confidently about the evolution of other species whose remains are even rarer than those of our own immediate kin. I hope this point is reinforced by the many examples in this book of how tracks have demonstrated the existence of animals that were or still are unknown from skeletons.

We still know comparatively little of how the present races of humankind became distributed across the globe. We accept that *Australopithecus* coexisted with various representatives of the genus *Homo*, but we admit that the chronology and fossil record of human evolution is sketchy. A 3.5-million-year-old set of footprints at Laetoli is regarded as crucial evidence, but dozens of modern trackways are treated with extreme skepticism when they are found in the backwoods and mountains. Why? This whole book is about footprints of animals that are, in many cases, largely or completely unknown. Would the reader be more inclined to believe in Bigfoot tracks if fossilized examples were found?

There are in fact two reports of possible fossils. In 1995, Robert Pyle, a Yale-educated ecologist, found and recorded a 13-inch-long (33 cm) track that could be from a giant hominid, preserved in volcanic ash near the shore of Deep Lake in Washington.[38] I have not seen the specimen at first hand, but I am not at all convinced by his photograph. The earlier report, in 1879, comes from a Captain Joseph Walker, who found a

> . . . slab of sandstone . . . imprinted [with] the clear form of a gigantic footprint, perfect except for the tip of the great toe. The footprint measured 14 1/2 inches long from the end of the heel to the tip of the toe and was 6 inches wide across the ball of the foot. Captain Walker related how he had found the slab of sandstone formation under about two feet of sand.[39]

This type of track occurrence is what one would expect if the Sasquatch had a subfossil history in the region. The description is reminiscent of the occurrence of moa and *takahe* tracks in New Zealand. The line between something that just went extinct and something that is on the verge of extinction is a fine one, especially when one is dealing with a creature that inhabits large forested areas. Sightings of the North American Bigfoot have been claimed on numerous occasions, and there is even a 200-foot-long segment of 16-mm film, shot in northern California in October of 1967, that shows a large, hairy hominid or ape walking for some distance through open woodlands. This unique footage is known as the Patterson film, and even skeptics find it hard to dismiss as a hoax. Ciochon, Olsen, and James state, "If the Patterson film is a hoax, it is an extremely elaborate one, and what is most surprising, given the immense amount of study devoted to it, is that it has not been definitively exposed."[40]

Grover Krantz, a professor of anthropology at Washington State University, has collected extensive data on Bigfoot sightings, and has a collection of more than 80 plaster casts of footprints purported to originate from the trackways of more than 20 individuals.[41] One of these shows an abnormality in the right footprints, as if the track maker, whom Krantz nicknamed "Cripple Foot," had suffered a foot injury. The dermal ridges (the foot equivalent of fingerprints) seen in some footprints have apparently aroused considerable interest among forensic experts at a number of police departments. According to Krantz, fingerprinters are more convinced of the authenticity of the footprints than most anthropologists and zoologists, but, as Krantz points out, anthropologists have more to lose than fingerprinters if Bigfoot is proved to exist. The views of anti-Bigfooters are often aired in the skeptics' magazine *Skeptical Inquirer*.[42]

Bigfoot is a classic example of a phenomenon that brings believers and skeptics into head-to-head confrontation. Bigfoot either exists or used to exist as *Gigantopithecus* or a relative. Had it never existed, how did it ever get so much press! In 1975 the primatologist Geoffrey Bourne hypothesized—and Krantz agrees with him—that the Bigfoot is probably a *Gigantopithecus*, and that it crossed from Asia to North America by way of the Bering Land Bridge some 10,000 or more years ago. Ciochon, Olsen, and James have described Bourne as "one of the most respected primatologists in the world," but this does not mean that Ciochon and his colleagues believe Bourne and Krantz; in fact they regard theories of surviving *Gigantopithecus* as "threadbare." They describe Krantz as a "fervent Bigfoot believer" who "has stepped outside the bounds of science"[43] by assigning the Sasquatch the species name *Gigantopithecus blacki* on the basis of footprints alone.[44] (*G. blacki*, named after Davidson Black—one of the discoverers in the 1920s of Peking man—is the original species name for *Gigantopithecus*.)

Had Krantz been more familiar with the literature on footprints he might have given the footprints a legitimate scientific name, but he did not follow acceptable procedure (recall that a track name and the name of the proposed track maker should be different). Had Krantz named a track something like *Gigantopithepodus* (meaning "track of a giant ape"), he would be less obviously flying in the face of scientific convention. Such a name would be controversial, but it would at least keep anthropological and ichnological terminology separate. Krantz's name cannot be accepted as valid in our tracker's vocabulary, nor is it helpful in tracking down the elusive Bigfoot.

Before we leave the subject of ancestral hominids, it is worth noting that reports of Neanderthal tracks suggest that they are short and broad, in comparison with those of modern humans.[45] Here we are reminded of the differences between Celtic and Anglo-Saxon footprints, noted in Chapter 1. Studies by Schad[46] suggest that hominid evolution has gone through cycles in which early progressive species are small and high-skulled (narrow), whereas later forms are large and broad-skulled (wide), with more primitive traits. This can be seen in the following comparisons:

Early (Ancestral) Species/Races	Later (Derived) Species/Races
Australopithecus africanus	*Australopithecus robustus*
Homo habilis	*Homo erectus*
Homo sapiens sapiens (Cro Magnon)	*Homo sapiens neanderthalensis* (Neanderthals)
Celts	Anglo-Saxons

Of great fascination here is the pattern of upward extension (narrowing), followed by the reverse, or downward (widening), trend, followed by extinction. Such a pattern can be seen repeatedly in the skulls and bodies of dinosaurs and other groups and can be correlated with foot shape. Such patterns or cycles also remind us of what happens in the aging process. The chimpanzee looks very human in embryonic and juvenile form, but becomes more chimplike as it matures. It is therefore significant that similar patterns reiterated in the skulls and feet of our Cro Magnon and Neanderthal ancestors, and even in modern Celts and Anglo-Saxons. If our own Cro-Magnon species "arose" while the Neanderthals "descended" to extinction, does this mean that the days of the Anglo-Saxons are numbered, as the Celts wait to take over? Not really, for we are talking here of the natural wisdom of coupled cycles. A rise is followed by a fall that is followed by another rise and so on. These skull-footprint relationships remind us that hands and feet are the mirrors of the soul. As above, so below.

A CONVERGENCE OF TRAILS

A morphological pattern, after having found expression at some earlier period, may disappear for a long while and then be taken up and further evolved by organisms of an unrelated stock.

Hermann Poppelbaum, A New Zoology

The phenomenon know as convergence results in the evolution of similar forms in entirely unrelated groups, such as birds, bats, and pterosaurs. So far, the only explanation for convergence has been in Darwinian terms.

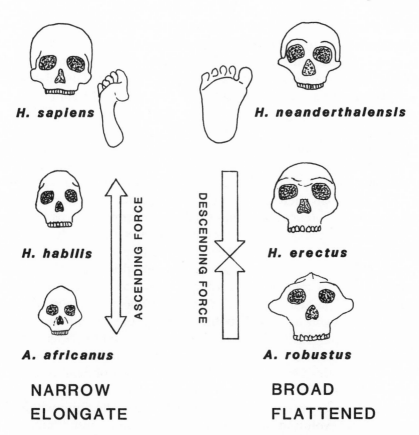

FIGURE 9.5   Hominids show narrow (elongate) and wide (flattened) morphologies in their skulls, and also in their feet. Based on the work of Wolfgang Schad.

The prevailing evolutionary view is that these animals developed similar body plans in response to living in similar environments. But, as we now know, such cause-and-effect explanations are inadequate, and it behooves us to look at the cyclic pattern of ascending and descending forces that characterize the growth cycle of all individuals, species, and larger groups. It should be clear by now that all groups that we have examined, and many more besides, seem to show transitions from small, narrow, environmentally sensitive beginnings to large, wide, environmentally emancipated endings.

We can now look at how these cycles are expressed in the major groups we have examined. This is best shown graphically by a comparison of the tracks of members of various lineages at the culmination of their development. We will find a similar pattern when we compare long-footed

lizardlike species and wide-footed tortoises; ancestral and advanced protomammals; ancestral carnivorous dinosaurs and giant brontosaurs; and
ancestral ornithischians and horned ceratopsians. We can continue the
analogies in comparing songbirds and waders, rodents and ungulates, and
even, perhaps, on a shorter time scale, various groups of hominids. Note
the remarkable convergence between ancestral forms among dinosaurs,
birds, and mammals.

In proposing such a scheme, I am doing more than pointing out similarities between animals that occupy similar habitats. I am suggesting that
the similarities in form reiterate throughout the entire evolutionary history of groups. So there is not just a convergence of various isolated
species, but a coherent reiteration of morphodynamic patterns throughout the evolutionary cycles of entire groups. Between cycles, the old morphologies seem lost, but then they are "taken up" again, as Hermann
Poppelbaum says in *A New Zoology*.[47] One might describe it as evolution
spiraling around a cone, so that each cycle resonates with the forms that
manifest at that point in the cycle. This view is somewhat parallel with Rupert Sheldrake's controversial hypotheses on morphic resonance,[48] which
postulates a type of "field" or growth habit which influences successive
generations to grow into similar shapes. I arrived independently at the
conclusion that this view has merit through the study of fossil vertebrates
and their tracks in the dimension of evolutionary time.

Though Schad's observations of hominids makes the point within the
context of the evolution of our 3-million-year-old hominid family, the
fossil record allows us an overview of hundreds of millions of years. If
similar patterns emerge on these time scales as well, it is hard to deny that
we are glimpsing something like a truly holistic pattern woven into the
very fabric of morphological evolution. To underscore my point I shall illustrate the relationships between the ornithischian dinosaurs and the ungulates, which I alluded to in the previous chapter. Rather than just
compare the end members of the group, I shall also compare the intermediate members and draw the comparisons across the grain of time, i.e.,
from one level on the spiral to the next. Both groups are well known as
large quadrupedal herbivores. Both groups also tend toward the metabolic-limb end of the spectrum rather than the nerve-sense pole. Both
groups are also divided into subgroups, such as the plated-armored and
horned dinosaurs and the odd- and even-toed ungulates. Remember that
the heyday of each of these groups was spread out in time from the Jurassic to the present.

When we examined these groups individually we saw trends toward increased "metabolic" breadth expressed in the increase in number of toes,
widening of the foot, and shifts in morphological emphasis from poste-

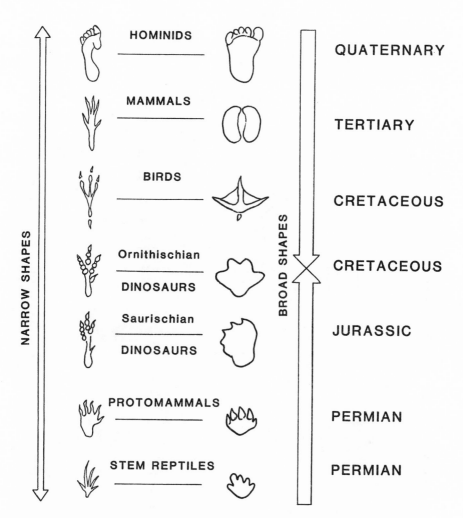

FIGURE 9.6   Narrow/wide patterns in footprints of major vertebrate groups. Note the remarkable convergence in the shape of tracks of small, nerve-sense–dominated dinosaurs, birds, and mice.

rior to anterior. So we should expect to see these patterns parallel one another on the loops of the spiral. What is more intriguing and pleasing from a holistic viewpoint is that when we compare species from one level on the spiral to those on the next, we see other, yet similar, coherent trends emerging. For example, the small ancestral forms all match fairly well for size and are functionally three-toed. The larger three-toed forms all have armor plates or head processes that are centrally located, in each case mirroring the elongated central toe of the hind foot—the archetypal odd-

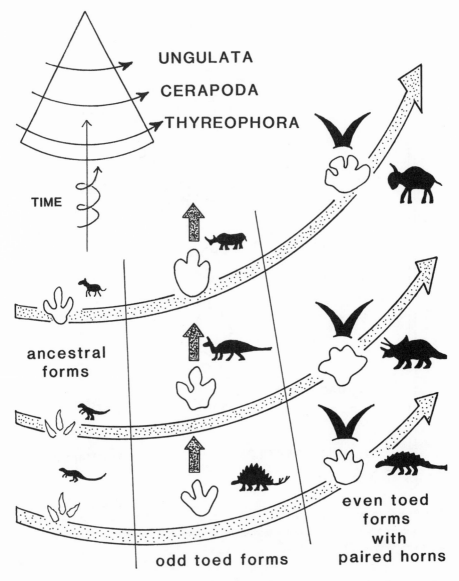

FIGURE 9.7 Spirals of evolution show convergences between two separate lineages of ornithischian dinosaurs and the ungulates.

toed ungulate pattern. More than this, however, the location of these axial processes follows the same forward movement in time "between" groups as within groups. Thus, the stegosaur has spikes on its tail and a row of plates in the center of its back, the duckbill has cranial crests and an inconspicuous axial frill down its back and tail, and the rhino has a "horn" on the end of its nose. This horn is quite superficial in that it is made of hair rather than bone. It is almost as if it is about to fall off the end of its nose having reached as far forward as possible.

The same morphodynamic cycle or "formative movement" is repeated in the four-toed end members of the three groups. Thus, the ankylosaurs have heavy, wide body armor with paired horns that stick out the side of the body, while the ceratopsians have wide frills and paired horns on their heads and a small residual nasal horn, while the even-toed ungulates all have paired horns, but no frills. Again the loss of the nasal horn can be seen as the formative movement working its way through the extreme anterior end of the body to arrive at the condition of almost all even-toed ungulates, which is without centrally located horns. In short, the convergence is not just between individual species; it is more a convergence of complex morphodynamic movements that can be followed in many directions.

There are alternate and more complex ways of depicting these multiple morphogenetic pathways, but the main message is that living systems are complex self-organizing systems, shot through with reiterated patterns that literally have to be seen to be appreciated. By following such patterns we can trace the archetypal odd-toed and even-toed conditions back to the beginning of dinosaur evolution, and perhaps even earlier. Think of a bison next time you look at *Triceratops*, or a horse next time you look at long, equine skull of a duck-billed dinosaur. Think also where we humans came from and how, from time immemorial, our polymorphic character-istics, from reptile brain to mammalian neocortex, have surged and receded in the deepest recesses of being and consciousness.

When we really explore these phenomena of multiple morphodynamic pathways and formative movements, something radical happens in terms of our evolutionary perception. We can no longer clearly see single linear pathways of evolutionary progression. It is as if dozens of morphological pathways swirl and intertwine within and between groups. In truth it is just our linear time consciousness breaking down so that we can see more deeply and holistically into the multifaceted, multitemporal hetero-chronic fabric of nature. The great chain of being becomes a great net-work of interconnectedness. Linear time dissolves, imperceptibly at first, but then more obviously as we realize that the entire past is being ab-sorbed into the present as a function of our expanded evolutionary con-sciousness, which draws all species and eventually the biosphere to our conceptual bosom. In the process we may feel that evolution is speeding up and that we are running out of time. Our perceptions and feelings are valid; this is indeed exactly what is happening.

THE SIGNATURE
OF HUMANITY

ACT X: CAST OF CHARACTERS

We now enter the fascinating era when humans begin to record, or "trace," their very thoughts and consciousness. When our Paleolithic ancestors first painted on the walls of caves, leaving tangible traces of the existence of extinct animals, they acted as agents to record the signature of other species, in the same way that water and sand were the agents that recorded the existence of Cambrian worms. The human psyche, then, is a template or medium through which other creatures, other phenomena, may leave their spoor, or image, in tangible form. Are we now sharing responsibility with the Earth for remembering dinosaurs and mammoths? Are we the backup discs of Earth's own memory? The eternal trail has sprung from the mud and rock layers beneath our feet to the walls of our homes, museums, and temples, and has since scurried, by a million pathways, into the foundations of buildings, the pages of books, and the soils of other planets. We should not forget that the eternal trail is also deeply embedded in the intangible realm of our consciousness, where it forms the shadow side of the manifest expressions that we too often think of as all of reality.

As we look briefly at our ambiguous and murky history we follow the eternal trail through the Paleolithic and Neolithic Stone Ages as it leads in short order to modern forensics, the track record of astronauts on the moon, and the track of Mr. Badger, a character from *Wind in the Willows*, on Mars. We shall ponder what it means to be a tracker or track maker in the twenty-first century, and shall finally ask, "Where do we go from here?" Though we must go where the eternal trail leads, we should be traveling with our eyes open, and in some small way choose how we step off on the next leg of the journey.

From our historical perspective the eternal trail seems to lead forward, following the arc of time's arrow. No animal before us could look forward or backward in time, or so we think. Our concepts of calendar and geological time presumably had no significance for these bygone creatures, yet they responded to the ebb and flow of seasons, the influence of lunar cycles and orbital dynamics, and the surprise of volcanic eruptions and meteorite impacts. What, if anything, is different, now that we can plan for years to take a step, launch ourselves at thousands of miles per hour, and a week later make a carefully planned footprint a quarter of a million miles away? Is this different from living in what we perceive to be the eternal, instinctive now of our dinosaurian ancestors? Are we perhaps entering the fourth temporal dimension where space, time, and the now are being radically redefined, as described by Peter Ouspensky?

It appears to be so, for we can to some extent decide where and when we will next place our feet—or can we? Can we make good on a decision not to trample another species into extinction? We did not invent extinction, or evolution, but we now seem to be active, conscious agents in both processes. What does it mean to be aware of this for the first time? What does it mean to look consciously backward and forward along the eternal trail and see ourselves in motion? I proffer the suggestion that we have traveled for billions of miles and billions of years on the eternal trail only to meet ourselves. We are at last becoming conscious of the evolutionary process at the level of biosphere and cosmos. Our task is not just to know ourselves as individuals, but as a species and as a biosphere. This is where the eternal trail already leads.

MIRROR, MIRROR ON THE WALL

The ability to create symbols—signs that refer to things—is potentially present in any animal that can learn to interpret natural signs, such as a trail of footprints.

Matt Cartmill, "The Gift of the Gab,"
***Discover,* November 1998**

When hominids stood erect they brought their liquid crystal axes into alignment with the Earth's radii, incident light rays, and gravitational field lines. No creatures ever did this before, not even the birds, which, though bipedal, have zigzag necks, backbones, and legs. Is it significant that hominids stand tall and erect from the crown of their heads to the heels of their feet? It is clear that our brains have ballooned tremendously since we adopted this new posture, and it is usually said that we developed innovative culture and technology because our hands became free to fashion tools and other artifacts. The manifestation of our culture is undeniably distinct from that of all other animals.

I think, however, that the ballooning of our brains has as much to do with consciousness as with material morphology and technology. When algae first grew into mounds 3.5 billion years ago, powerful forces were at play transforming the inorganic world into biosphere. These same latent forces are manifest into the prodigious materialization of civilization as we enter the planetization stage of global brain and noosphere. In standing erect we humans face one another and see our own reflection. At some time in hominid history we began to recognize such reflections as mirrors of ourselves, rather than water spirits or other beings. Self-awareness. At some point also tracking ability evolved to fuller self-awareness. These are my tracks, and those are the footprints of other individuals, other species. The realization that the substrate could register, remember, and reflect a portion of one's identity was surely of great significance. No wonder interaction between Earth and its sentient beings was regarded as sacred. The reading of tracks became less instinctive and more of a rational exercise in understanding the identities and behaviors of other creatures. But more than this such reading was accomplished at a distance, when the creatures were not visible. So tracks were the symbolic language of other individuals or species—their signatures, the stories they wrote and stored in Earth's memory banks.

It was not just the ability to transfer an image of an animal from the mind's eye, to the cave wall that defined a new spatial consciousness. This was also the dawn of time awareness. William Irwin Thompson, in *Time Falling Bodies Take to Light*, has made a compelling case that the 13 notches on the horn held by the Venus of Laussel, a Paleolithic carving, represent the first attempt at counting lunar months and rationalizing the feminine reproductive cycle.[1] Perhaps numbers were "invented" at the same time as art, and perhaps the innovators were women. From such thinking it follows that time would have held extraordinary fascination for us as our rational awareness of this dimension unfolded. Such a chapter in our history might perhaps have been marked by the construction of great structures designed for the study of astronomy.

But as above, so below. The trails followed by the moon and planets in their regular orbits were recorded as tracks, or notches, on bone artifacts. There is no doubt that such marks are an integral part of the eternal trail, miniaturized to portable size. The tracking of animals was still important, but now there were new phenomena to track, and these occupied more and more conscious attention. In the transition from Paleolithic to Neolithic 8,000–10,000 years ago, the substrate may have lost some of its significance as far as tracks were concerned, but it gained immense significance for agriculture. The clays of the Tigris, Euphrates, Nile, and other Fertile Crescent floodplains became the seedbeds of the agricultural revolution. As humans developed the art of sowing their fields, they eventually introduced ridge and furrow patterns, and prodded the clay with seed holes spaced at regular intervals, reflecting the measured pace of the farmer's round. The ridge-and-furrow topography resembled the activity of ancient marine worms and the regular rows of the seedbed mirrored the trackways of the agriculturalists who made them. The ichnology of agriculture was taking shape, often framed by deeply etched field boundaries and roads.

To record the bounty of field and farm, the cuneiform tablet was devised. Made from the same clay as the substrate itself, these tablets were laid out like tiny plots of land with rectilinear boundaries and lines of script, like maps or plans of the very fields their information pertained to. In essence, these tablets were symbolic miniatures of the agricultural landscape, just as tracks were symbolic representations of the animal landscape in a previous era. The tablet and the field coevolved as manifestations of an emerging cartographic consciousness.

The term "cuneiform," or "wedge-shaped," refers to the shape of the characters made by the stylus instrument used to inscribe tablets. "Cuneiform" also refers to wedge-shaped bones in the hand and foot. What a marvelous coincidence of circumstance and symbolism. When we wedge our cuneiform feet into the substrate, our tracks reflect our identities and activities in a symbolic shorthand that can be read long after our passage. Early on we adopted the same ritual in writing, deliberately wedging miniature cuneiform footprints into miniature substrates, again leaving trails that can be read long after the script has been written.

In an inspired essay entitled "Clay and Life," William Bryant Logan reminds us of the deeper symbolism and science surrounding this most precious of substances.[2] "Adam" is the Hebrew word for clay, a substance that has always been regarded as malleable and ideally suited for the sculpting of organic forms. Clay, unlike other rock minerals, reaches stability by absorbing water, and establishes equilibrium at normal ambient conditions associated with the biosphere and hydrosphere. It is the end product of the weathering cycle, which converts solid rock into a rich nutrient paste, or "matrix" (meaning womb), from which new life springs so readily. Quoting the biologist Hayman Hartman, Logan makes a case that clay is alive: "There are only two things in the universe that require liquid water for their existence: organic life and clay.[3]

Many theories on the origin of life suggest that clay was the template on which prebiotic chemistry aggregated to organize itself into true organic molecules. If true, this means that life could not have evolved without clay. Was clay the catalyst for the "birth" of organic molecules, or perhaps the very mold from which they were cast? We are reflections of the microscopic lattice of clay. A seed cannot grow without a vessel to hold it, and clay holds water, quite literally, at the level of both its atomic lattice and in its most common utilitarian and symbolic form as the ceramic "vessel." Like a footprint that acts as a germination site for a seed, each layer of lattice in the clay crystal has been described as a series of dimples and holes—"templates with tendencies," as Logan calls them. Vessels waiting to be filled. Footprints waiting for entities to adopt them. "The ground itself is as active as the seed," he reminds us.[4]

So life emerged in intimate partnership with microscopic clay substrates, and clay has never forsaken this role as mirror or counterpart of life. Throughout pre-Neolithic eons it has reflected the passage of countless creatures. Then it became the template on which agriculture was forged, and the substrate on which writing was developed. The partnership of Clay and Life runs deep. A worm's-eye view of the clay substrate would let us witness a vital dance of ever-increasing complexity and novelty. But no dancer is so light that he or she leaves no footprints, and clay with her billion-year memory faithfully records the footprints of all offspring of her vital womb.

FINGERPRINTS OF OUR GENES

Genetic determinism has a very strong hold over the public imagination. . . . Belief in the constancy and fixity of genes substituted for . . . belief in an immortal soul when science replaced religion.

Mae-Wan Ho, *Genetic Engineering:*
Dream or Nightmare?

Let us return briefly to our beloved Bushmen, who somehow possessed a form of holistic consciousness that allowed them to correlate tracks with animals with such consistency that observers claim they were never wrong. Is their world gone forever, preserved only in rare documentaries? I don't believe so. Our tracking ability is more deeply ingrained in our psyches than we know. We rarely get through a day without writing our personalized signatures, or using an identification card that carries our fingerprints. In the world of science, the concept of a distinctive "signature" or "fingerprint" is ubiquitous. We talk about DNA fingerprints in the courtroom and routinely describe chemical and isotopic signatures in scientific journals. The Bushmen would be lost in a DNA laboratory, just as we would be lost in the depths of the Kalahari, but they would no doubt easily identify with having their footprints or fingerprints examined.

The global fingerprint database is far larger than all the footprint specimens, fossil or otherwise, in all the world's museums. The Spanish word for footprints and fingerprints is the same—*huellas*. DNA databases are also growing rapidly, and, among biologists and paleontologists, molecular signatures are considered just as important as anatomical characteristics when it comes to distinguishing between species or individuals. But the molecular approach is part of the picture. The holistic viewpoint is predicated on the notion that almost any signature is a reflection of the whole, just as a fragment of a holographic negative is all that is needed to recreate the whole picture. There are many other ways in which an individual can be identified. Among friends and family we do not need to show our footprints, fingerprints, DNA or voice "print" to be recognized. Our character and even mood can be assessed by a simple glance from someone who knows us. Reading trail signs is simply a matter of experience and practice.

In the world of bureaucracy and law enforcement our fingerprints and DNA signatures are used only to identify us. They are not used to tell anything about our character, certainly nothing that would presently be admissible as evidence in a court of law. Though I note the increased use of handwriting analysis as a guide to character.

"Your honor, it is clear that the fingerprints and handwriting are those of John Doe. Moreover, they indicate that he is left-handed, susceptible to alcoholism, and probably a thief."

Or how about a simpler case?

"The tracks are clearly those of Fred Smith, and the trackway indicates that he was drunk and carrying a heavy object."

What is admissible as evidence in a court of law, or in the mind of a scientist? What are the correlations between tracks and other physical or psychological characteristics? Stanley Coren from the University of British Columbia has written a series of articles demonstrating statistical correlations between fingerprint types and right- and left-handedness.[5] Other studies show a relationship between fingerprint types and a number of hereditary diseases.[6] For years clinical psychiatrists have known of the "simian" line in the palm, similar to a line in an orangutan's palm, which is a sure sign of Down syndrome and related abnormalities. Indeed, entire books in the medical literature are devoted to the relationship between fingerprints, palm prints, and medical disorders.[7]

Despite this literature there is great scientific resistance to the art or science of palmistry, arising in part because most scientists know nothing about the subject, and also no doubt because there are bad palm readers. If there is a scientifically verifiable connection between fingerprints and traits such as handedness and disease propensities, then the whole hand must be an even greater repository of information about an individual. Many experiments could be conducted to show whether a good palm reader can objectively describe a person's characteristics and reveal accurate information that they would have no way of obtaining except directly from the hand.

Good palmists, like good trackers, may be few and far between, but it is worth noting that intelligent, even skeptical, persons such as Mark Twain were willing to testify to the remarkable abilities of the palmist who went by the name of "Cheiro" (Count Louis Hamon). So he cannot have been a mere confidence trickster.[8] The one talented palm reader with whom I am acquainted is certainly not in the business to get rich, nor does he cater to his clients' egos by telling them what he thinks they want to hear. On the contrary, he tells what he sees and invariably surprises clients with penetrating insights that they have to admit "expose" realities in their life and character, which he has no means of deducing, other than from their hands. But as with any art, there are the few who have mastered it and the many who have not.

Many who would dismiss palm reading as unscientific nonsense would readily accept that our genes tell us much about our general health and makeup. But such gene reading is, in my humble opinion, just another type of divination. Looked at the other way round, would you rather have your palm read by someone directly experiencing your whole persona, who could explain why hands are the mirrors of the soul, or by a lab technician at a remote location with a sample of your DNA in a test tube, who might inadvertently switch it with someone else's sample. Why should minuscule genes tell us more about ourselves than our entire hands? If someone were to spend the time to look at gene-fingerprint relationships, they would find connections, and possibly claim that genes control fingerprints as they are said to control everything else.

But in this same sense fingerprints control genes. We know without doubt that genes are affected by feedback from the environment. The feedback into our genes from billions of fingertips fluttering on keyboards is incalculable. Only genetic business interests claim that genes are "in control." In our biological world of simultaneous connections, genes no more control disease or the inclination to violence and murder than do certain lines on your hand, which, incidentally, were recognized long before modern genetics. In a brave new world of genetic regulation we could lock up potential murderers and sociopaths at birth, but we could do the same in a world run by palm readers.

ELECTROSTATIC FOOTPRINTS

Wherever he steps, whatever he touches . . . will serve as a silent witness. . . . This is evidence that does not forget . . . cannot perjure itself. . . . Only human failure to . . . understand it can diminish its value.

W. J. Bodziak, *Footwear Impression Evidence*

We might think that in the built up world in which we live, that there is no longer any call for trackers. How can we track someone over concrete or linoleum? Doesn't our habit of purchasing mass-produced footwear mean that many of us have identical tracks? This is not so, for there are more undersole designs than ever before, and from the moment that we purchase a shoe, we begin to endow it with very specific characteristics.[9] Each person wears his or her shoes differently. A flat-footed person wears down the outside of the heels. Shoe insoles soon develop the characteristic footprint of the bare or stockinged foot. Our shoes become portable individualized footprints that we carry around with us. Add to this the fact that shoes pick up little bits of grit that can show up in footprints and it is easy to see how every footprint is different. In fact, forensic scientists compare the diversity of soles of modern training shoes to so many individualized rubber stamps.[10]

Apart from the obvious potential for tracks made in mud or blood to be transferred to hard surfaces near the scene of a crime, experience has shown that in some cases it is possible to record "latent" two-dimensional footprints that are indistinct and virtually or totally invisible. The technique, called "electrostatic lifting," was devised by a Japanese police officer, Kato Masao.[11] Dust is attracted to high-voltage areas around television sets, and Masao reasoned that a high voltage device could lift latent footprints from surfaces of manmade material. So he designed a machine consisting of a high-voltage source, a metal plate, a hand-held probe, and the film onto which the footprint is lifted. The device works to lift hand- and footprints from carpets, linoleum, doors that have been pushed or kicked in, and other surfaces.

Such devices are now common, and at least two models are available as standard forensic equipment. Other methods of footprint enhancement include chemical treatment of tracks and special photographic techniques. The use of filters and cross-polarized, ultraviolet, and infrared

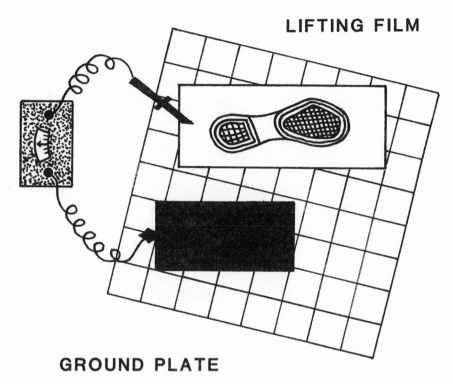

FIGURE 10.1   Electrostatic footprints can be retrieved from hard surfaces such as concrete or linoleum.

light can transform a dull, almost imperceptible footprint into a crisp, colorful image resembling a snapshot of a brand-new sole. Such technology should make the would-be criminal think twice about the risk of detection. The moral of this story, which brought us from Bushmen to the megalopolis of Tokyo, is simply that we eventually learn to track what we need to track in the environments to which we have adapted.

OVER THE MOON: THE TRAIL LEAVES OUR PLANET

This may be an appropriate occasion to take an Olympian overview of ourselves. Life is one of the properties of matter. Its evolution has given rise to consciousness. Someone once said that man is the mind by which the universe contemplates itself. . . . Our reaching out into space is akin to our clutch at a tool, or the poor fish's straining to be an amphibian. In its portent, it is of a different order of magnitude.

David P. Bloch, commentator on the *Apollo 11* mission

On Sunday July 20, 1969, at 3:18 P.M. Houston time, Neil Armstrong and Buzz Aldrin arrived at the Lunar surface. They spent most of Monday ("moon day") in the Sea of Tranquillity. Armstrong actually first set foot in the moon at 3:28 A.M. Monday, Greenwich mean time (or, 9:28 P.M., Sunday, Houston time). As summarized in *Footprints on the Moon,* Armstrong "cautiously . . . stepped with his left foot, a size 9 1/2, in a clumsy awkward step." Then he uttered the immortal words "One small step for a man, one giant leap for mankind."[12]

Not all six Apollo missions arrived on Monday, though more did than the laws of probability might dictate. *Apollo 12,* launched on November 14, 1969, arrived at the moon on Monday, November 17. *Apollo 14* arrived on a Friday, February 5, 1971. *Apollo 15* was launched on Monday, July 26, 1971, and landed on Friday, July 30. *Apollo 16* arrived on Monday, April 24, 1972, and the final mission, *Apollo 17,* left the moon on Monday, December 11, 1972. So the Apollo missions first arrived on Monday and finally left again on Monday.

During the 1239-day, 44.25–lunar month *Apollo* chapter in space exploration, 12 men set foot on the moon and walked a total distance of several miles at sites in or near the Sea of Tranquillity, the Sea of Serenity, the Known Sea (Mare Cognitum), and the Apennine Mountains. Missions 14 and 16 used moon rovers to get around and these produced sets of tire tracks. It is just as well that our species recorded the dates of astronaut track-making activity, for future visitors to the moon will presumably have no idea when the footprints were made if we don't provide them with a record. Twelve members of our species' setting foot on another planet is an historic achievement in space exploration for a primate that was dodging cave lions less than ten millennia ago. Richard Nixon called the *Apollo 11* week the greatest in the history of the world since the Creation. The names of these 12 NASA track makers, in order of arrival, are as follows: Neil Armstrong, Edwin "Buzz" Aldrin, Charles "Pete" Conrad, Alan Bean, Alan Shepard, Edgar Mitchell, David Scott, James Irwin, John Young, Charles Duke, Harrison Schmidt, and Eugene Cernan.

If we knew nothing of human culture and technology, what might we deduce from the tracks of these 12 NASA disciples? They are all very much the same type of footprint, so presumably they were made by the same species. (All were also Caucasian male individuals.) If we assumed that the tracks represented astronauts, who were not barefoot, we could deduce that they had all been issued with identical footwear and so perhaps belonged to the same program. Locomotion was sometimes erratic and playful, showing a tendency for the track makers to hop, skip, or jump. The average weight of the members of the species could be estimated from a knowledge of lunar gravity, the relatively constant condition of the

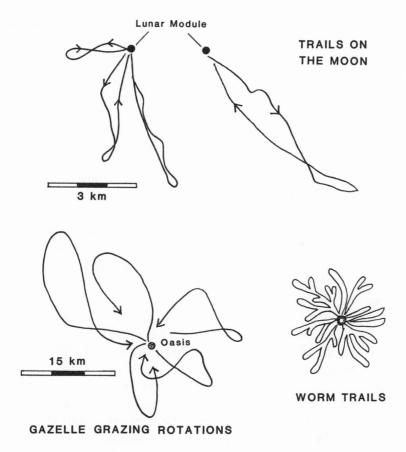

FIGURE 10.2   Over the moon: The eternal trail leaves our planet. Note similarities to trails of ancient worms or foraging gazelle.

substrate, and the depth of the tracks. The amount of dust and meteorite debris that had accumulated on top of the footprints and artifacts since they were made might enable us to estimate their age.

The general behavior patterns deduced from the footprints are all similar in one fundamental respect. All tracks lead out from and return to a single focal point. A smart cosmic tracker would deduce that this central point was a temporary home base or spaceship. This flower-petal pattern of trackway loops radiating from a central point is remarkably similar to the traces left by Cambrian worms foraging out from a central burrow, or the trails of gazelles in their grazing cycles.[13] Like Cambrian worms, the astronauts were restricted in their mobility and were unable to move too far from home base. A couple of the tracksites are associated with a second type of long sinuous, wormlike trail that also loops out and back from the central base. These again resemble the surface traces made by Paleozoic worms. This "species" is rarer, and less adaptable, because we find it dead in its tracks. Depending on our tracking skills, we might deduce that it was a machine, not a life form. We might begin to reconstruct the anatomy of the species that used it, though without prior knowledge of human anatomy, our inferences might prove wrong and very speculative. Again, our conclusions would depend on the extent to which we were able to retrieve other artifacts and information. There are quite a number of mechanical species on the moon, including the unmanned U.S. *Surveyor* craft, Russian *Ranger* craft, and Russian lunar modules. These suggest a range of technical skills, flying abilities, and tendencies to crash. An advanced species would recognize the signs of inexperienced space aviators.

The astronauts were acutely aware of the symbolism of making tracks on a new planet. Buzz Aldrin was fascinated by the way moon dust flew away from one's feet with each footfall, like slow motion fallout from an impact crater. Aldrin, like Paul on the road to Damascus, symbolically represents the disciple converted by profound spiritual revelation. The astronauts found that their style of locomotion changed radically, and they acquired a tendency to hop rather than walk. It was fun and they responded accordingly, with playful, childlike behaviors. Prior to the departure of Apollo 16, Charlie Duke was evidently subconsciously preoccupied with tracks. On a geological trip to Hawaii, he came down with a fever and experienced the following vivid dream, told in Andrew Chaikin's book *A Man on the Moon,* in which he and his partner, John Young, were

> . . . driving their Rover toward North Ray crater, they came up over a ridge and suddenly Duke spotted something that made his heart race. A set of tracks crossed the ground ahead. . . . [They] looked like those from the Rover, but they were definitely different. Duke asked Mission Control, "Can we follow the tracks?"
>
> "Go ahead" was the reply from Earth. The twin trails stretched eastward. . . . They drove onwards for miles . . . until finally, topping another rise, they saw it: a vehicle, looking amazingly like the Rover, stopped on the surface. Aboard were two figures in space suits. After calling Houston to announce their incredible discovery . . . Duke reached the one in the right seat . . . and raised the visor and saw his own face. The one in the left seat was John Young's double. After taking pieces of the space suits and Rover at Mission Control's request [they] drove back to the [lunar module] and blasted off for home. The next thing Duke knew, he was on earth, presenting the samples to the scientists. The test results: The craft was 100,000 year old.[14]

In this case Duke's was the mind by which the universe contemplated itself. If we follow the eternal trail far enough we shall meet ourselves. Duke found the man on the moon and it was himself.

TETRAPODS ON MARS

Someone will say that if we can put a cluster of air bags on Mars, why can't we cure cancer? And we will celebrate the speculation that we are capable of anything. It is probably untrue. We do what we were made to do, no less no more, like everything else in the universe. We are alone and not alone and sometimes in the light.

Roger Rosenblatt,
"Visit to a Smaller Planet," *Time,* **July 4, 1997**

The first life form photographed by the cameras of the *Viking 1* Mars lander was a turtle making tracks in the wind-blown sand of a test site at Great Sand Dunes National Monument in Colorado. It is an extraordinary coincidence that the first fossil footprints ever recorded were purported to be turtle tracks from the Permian sand-dune deposits of Scotland. So our story of tetrapod tracks has come full circle from its first documentation in 1828 to the planting of the robotic feet of *Viking 1* in the Martian soil of Chryse Plantia ("plains of gold"), at 22 degrees north latitude, 48 degrees west longitude on July 20, 1976.[15] During the mission a NASA scientist at Mission Control put on a Viking helmet. *Viking 2* landed ten weeks later, on the September 3, in the Utopia Plantia ("plains of Utopia"), at 48 degrees north, 226 degrees west.

Both Viking landers have three feet; together they have imprinted six footprints on the red planet. The three-toed configuration of the Viking modules reminds us of the riddle of the Sphinx in Sophocles' *Oedipus*. The question posed to Oedipus was "What has four legs in the morning, two legs at midday, and three in the evening?" The answer was "a human" who crawls on all fours as a child, walks upright in maturity and uses a stick in old age. We might equally well declare that the answer to the Sphinx's riddle is tetrapod: first emerging on land as a four-footed creature, evolving into a hominid biped, and creating tripodal vehicles that touch down on other planets.

Three Viking footprints are situated at 22 degrees north, 48 degrees west at the equivalent location on the Earth sphere of the eastern Caribbean, and three prints are at 48 degrees north and 226 degrees west at a site equivalent to the northeastern border of Manchuria. Although these distant, robotic tetrapods, situated more than 200 million miles from Earth, have not taken any steps from their landing sites, they have not been motionless. They have probed the Martian soil and nudged its rocks gently aside with their robotic sampling arms. *Viking 1* left a half dozen mechanical hand prints as it scooped the earth in search of information on soil and rock composition and possible life. *Viking 2* probed the soil in at least ten different places. At one o'clock just to the left of *Viking 2*'s anterior view lay a cluster of rocks the NASA team named for characters in *Wind in the Willows:* Mr. Mole, Mr. Rat, Mr. Toad, and Mr. Badger. All the rocks were candidates for a nudge so that samples of the soil underneath could be obtained. It was Mr. Badger that was finally moved and the soil beneath his feet taken on board and tested.

Eventually, the harsh winds of willowless Mars will remove all trace of the dozen or so footprints made by the two Viking explorers; Mr. Badger's footprint will be obscured, too, and he and his companions Mr. Toad, Mr. Mole, and Mr. Rat will eventually be reduced to dust. By that time the Viking landers will also be weather-beaten skeletons—dead in their tracks. But by that time who knows how many new tracks will have been made on the red planet.

On July 4, 1997, 21 years after the *Viking* mission, the spacecraft *Pathfinder* arrived. As if to celebrate Independence Day with typical American gusto, this ball-shaped cluster of airbags hit the ground running. Arriving at 22 miles per hour (35 kph), the speed of an accomplished sprinter, it hopped gleefully for 50 feet (14 m), then skipped another 23 (7 m), before rolling along the Ares Vallis, an ancient floodplain, for a minute and a half. Already it had set a Martian world record for the hop, skip, and roll, outdistancing its sedentary *Viking* ancestors by many leaps and bounds. Hardly surprising that such a stellar performance was achieved after a run-up of 310 million miles at speeds approaching 17,000 miles per hour (28,000 kph).

The traces made in the Martian dust would look much like the bounce or prod marks made on the ocean floor when spherical shells are rolled along by currents. Once *Pathfinder* had come to rest it unfolded to release its cargo, a six-wheeled minivehicle known as *Sojourner*. Technically, it may have little in common with the large lunar vehicle (although both are referred to as "rovers"), but its traces make it look like an affiliated species.

Future trackers may not be able to tell from *Sojourner*'s wheel traces how fast it traveled, any more than we can tell from Cambrian worm trails or dinosaur tracks exactly how fast these creatures moved. We know that *Sojourner*'s speed is a modest two feet (0.6 m or 60 cm) per minute. But what is time in this context? It is the fastest and slowest vehicle on Mars. One hundred twenty feet per hour is more than half a mile a day, or 200 miles a year. It could theoretically circumnavigate the planet in less than a century—less than three score years and ten, to be precise. Along the way

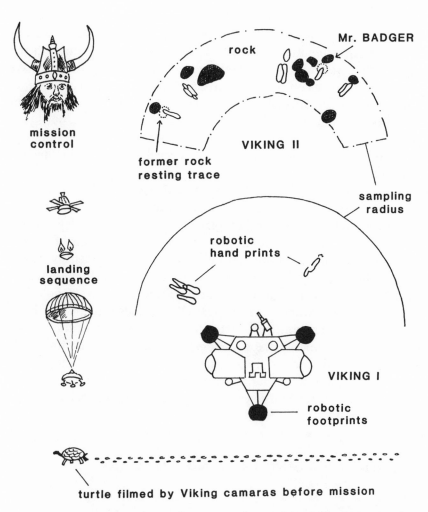

mission
control

landing
sequence

rock

Mr. BADGER

former rock
resting trace

VIKING II

sampling
radius

robotic
hand prints

VIKING I

robotic
footprints

turtle filmed by Viking camaras before mission

FIGURE 10.3   Footprints on Mars made by the mobile arms of *Viking 1* and *Viking 2*. In preparation for these missions technicians wore Viking helmets and filmed turtles making tracks in sand dunes.

it would encounter many unique rocks, each one taking on a new identity through the very act of being observed by extra-Martian primates back on Earth. Some rocks have just been christened by the NASA team—Barnacle Bill is one of these—whereas others, such as Mr. Badger, only a four-year trip away *as Sojourner crawls,* have identities that are coming of age at 21. All are billions of years old, but so are the atoms that make up *Sojourner* and the hominid design team. All that is new is our depth of perception and insight into the eternal trail.

Commentaries accompanying the 1997 landing on Mars were often wistful, lamenting our cosmic loneliness.[16] For example, Roger Rosenblatt, writing in *Time* magazine, asked, "What should one do with the knowledge that we are alone in the universe?" But, I see it otherwise. We identify with the rocks on the red planet, and so give them character and, literally with the utterance of a name, breathe life into them. One might say this way of looking at it is a way to mitigate our loneliness, but I prefer to say that it reflects our growth as a species and our ability to recognize the vitality of all matter. We would be over the moon, quite literally, if we found firm evidence of even the most primitive life on Mars, and indeed recent discoveries suggest that such evidence may now be close at hand in the form of microscopic squiggles in the rock that blur the distinction between the organic and inorganic world.[17] This is to say nothing of controversial claims of structures that some "see" as remains of ancient civilizations. What does it mean that we care so deeply about elusive traces that might indicate the former presence of bacterial-grade life? Do we actually now love rocks enough to give them names? Are rocks people too? What is animate and what is inanimate? What is vital? What, if anything, is not? It seems we are constantly extending our definition of life and vitality in space and time. Perhaps the logical conclusion is that the whole universe is vital and sentient. If rocks are named by the world's top scientists, perhaps we are not so far from recognizing atoms and subatomic particles as vital entities, and so extending the eternal trail back to the Big Bang and beyond.

THIS LAND IS YOUR LAND, THIS LAND IS MY LAND

"This land is your land, this land is my land, from California to the New York highlands. From the redwood forests to the gulf stream waters, this land was made for you and me."

**Woody Guthrie, "This Land Is Your Land,
This Land Is My Land"**

A Bushman walks out one morning from his peaceful night camp to enjoy the sunrise and give thanks for the beauty of his homeland. His tiny feet thread through the brown grass leaving dainty footprints in the cool sand. Suddenly the world explodes. The sun rises with an blinding flash and an ear-splitting explosion from beneath his feet. The bushman's legs vaporize in the searing blast. A land mine shatters the sacred communion of human being and the Earth. This land is no longer your land, this land is mined land. The gods must be crazy, and very angry.

The earth is sacred, it is our home. Half of the world's people removes their shoes as a mark of respect before entering the home of another. The act of taking a step and placing one's foot in contact with the Earth is also sacred. This is why a foot is regarded as a sacred symbol of communion, and why we regard a rabbit's foot as a symbol of good luck. What diseased mentality is at work in the manufacture of land mines, expressly designed to desacralize the Earth and turn it into a predator that indiscriminately ambushes women and children? After 4 billion years of evolution and dependence on Mother Earth, how can we turn her into our enemy? What lunacy is it to plant a rampant cancer beneath her skin—a cancer that afflicts half the world.

The vocabulary of greed reflects this pathology unequivocally. This land is not yours, it is mine. Land mine, land mine, land mine! Land mines are indiscriminate in their effects. They make no distinction between women, children, gazelles, or wildebeest. Some mines, designed only to explode if set off by tanks or large vehicles, could be triggered by elephants. They have no respect for treaties or cease-fires, and they don't know when a war is over. Worse still, many mines are laid in agricultural areas, to deliberately attack and disempower civilian populations, thereby damaging the local economy as well as killing humans and animals. (A litany of the evils of this form of antipersonnel, antihumanity warfare is provided by the Red Cross doctor Gino Strada in a 1996 article in *Scientific American,* "The Horror of Land Mines.") Concerted efforts should be made to defuse vast areas of the Earth afflicted with the chronic cancer of land-mine pollution. Thankfully some efforts are being made in this direction by persons such as the late Princess Diana.

We are quite wrong to look on our Paleolithic ancestors as naive and superstitious for believing that monsters lived in the Earth, and to regard such myths as archaic and irrational. Surely the indescribably cold-blooded horror of land mines is more terrifying, more dangerous, and more real than the occasional fury of a cave lion roaring from its subterranean den, or the myth of giants that dwell in the bedrock where we find huge fossil bones. A master of Chinese medicine and philosophy had the following interesting perspective on the subject of monsters: [18] While we have killed off, or deliberately brought under control, populations of dangerous man-eating predators, we have replaced them with equally deadly machines. Vehicles, many with names such as Jaguar and Cougar, are among the most dangerous twentieth-century predators, accounting for a daily toll on the world's highways. Mines are even more insidious because one can not see them coming. Their names—VS–50, SB–33, M–14, VAR–40, PMN—are impersonal, as befits their antipersonnel character.

The horror and injuries are real. Mines kill or maim more than 15,000 people each year. Thousands of innocent women and children are killed outright, crippled, or reduced to crutches and physical and psychological agony. Some may receive prosthetic limbs, but they still carry phantom pains and dreams of monsters in the Earth. Many survivors still live in proximity to the unpredictable, silent cancer that may erupt at any time. In those places where we should find the footprints of playful children, we find the ugly scars of gaping craters. The convergence between bomb- and mine-cratered battlefields and the proverbial lifeless lunar landscape is chilling. If we cure Mother Earth of the chronic cancer of land-mine pollution, we will cure the human cancer of greed, which is in turn the cause of the disease we dread in ourselves. We should take our children on nature walks to see the tracks of gazelle and wildebeest, and not the ugly crater prints of a VS–50 or a VAR–40. Chief Seattle said: "Earth does not belong to us, we belong to the Earth."

SACRED GROUND

Put off thy shoes from off thy feet, for the place whereon thou standest is holy ground.

Exodus 3:5

There is a sense that we are running out of time. The exponential pace of technological advance has been stressful, and has often outstripped our spiritual growth. It has bred a certain pessimism and jadedness. But I am optimistic that recognizing a problem is the first step to its solution. We need to take the time to ponder what it really means to learn to love Earth and all her species. These facets of being human are embodied in an appreciation of the humanities, arts, pristine nature, philosophy, and culture, as well as in the practice of good science and technological application. Paradoxically, amid our destructive activities we are beginning to love everything from rain forests to rocks on Mars. Such appreciation of the cosmos can be achieved by looking at the big picture of our evolution and our journey on the eternal trail. It is a journey shared by all individuals, of all species, back to bacteria of 3 billion years ago, which are our trillionth cousins.

Study of the eternal trail highlights the significance of ancient tracksites as crossroads in our historical journey through time. The mere act of appreciating their significance means they can be set aside and protected from condominium development, strip mining, or conversion into minefields. A little patch of Earth is treated as sacred. When in 1997 the exhibition "Tracking Dinosaurs" (put together by our research group at the University of Colorado) opened at the National Museum in Cardiff, Wales, a theater group performed a drama in which environmentalists

FIGURE 10.4   *Following the Footsteps of Nature,* a seventeenth-century woodcut from *Atalanta Fugiens* (1687) by Michael Majer, depicts our human journey on the eternal trail and illustrates the notion that tracks are already out there for us to follow and appreciate. Courtesy of the New York Public Library.

fought with a greedy developer to preserve a tracksite—and indeed there is an important Triassic tracksite less than ten miles from the museum.[19] Their message was "We love the tracks." The struggle between those intent on saving the Earth and treating it as sacred and those who blindly pursue material gain by exploitation of our historical and paleontological heritage can develop into an epic battle. It is the pull of the opposing forces of the past (conservation) and the future (perceived progress). How can we strike a balance and decide what is best?

Who decides what is sacred? The answer is that everything is sacred if we only make it so. We can even resacralize the large areas that are infested with land mines simply by making it a priority to do so. Developers and advocates of progress will say that not every site can be preserved, that not every claim by aboriginal people can be honored. There must be a limit, they say; if it is not known to be sacred now, then it can not be designated sacred in the future. This is nonsense: no one can legislate what is and is not sacred. The Laetoli tracksite was not considered sacred until it was discovered in 1976. Now it is one of the most important crossroads on the eternal trail, blessed both by Maasai and Western anthropological traditions

The list of sacred sites at which fossil footprints are known is long and growing. Some—like the site of the Holy Father of the Elephants along the old Silk Road or the "river of lost souls," the Spirit River in Colorado—are already evocatively labeled with appropriate reverence. The sacredness of other locations, such as the mule tracks of Portugal's "Lobster Bay," has already been recognized within a different cultural framework and shrines have been erected to honor the significance of the place and have become part of the physical and spiritual landscape. They are all part of our cosmological heritage.

Sites where the first arthropods and amphibians stepped on shore should also be sacred, as should the surfaces that registered the first steps of reptiles, the arrival of dinosaurs, the meanderings of the first mammals, the proud strutting of the first bird. But why restrict ourselves to these few celebrities. There are the first bipeds, the first tyrannosaurs, the last dinosaurs, the first dogs, cats, camels, mammals, marsupials, the largest and the smallest of every kind, the most exquisitely preserved, the most historically famous, the most enigmatic, the most abundant, and the most rare. Prehistory is literally arising from the bedrock around us in direct proportion to our ability to appreciate it. Everything is sacred, and thank goodness we are beginning to realize it.

As our appreciation of the significance and depth of prehistory matures, so the richness of the eternal trail manifests itself as part of the present landscape. In a tangible sense we enter the dimension of time. In the right locations we may find ourselves in the midst of a plethora of historical and prehistoric monuments—layer upon sacred layer. Two decades ago hardly any fossil footprint sites had been designated for protection and conservation. Now they are springing up everywhere, in Colorado, Utah, New Mexico, Spain, Portugal, Switzerland, France, Australia, South Korea, Japan, and elsewhere. These sacred sites grow in importance and number in direct proportion to our ability to discover and revere them through documentation and appreciation.

It is important to understand that the eternal trail is not a linear sequence of footprints that follows a time line from the origin of life to the present. The eternal trail as I have tried to describe it in this book is the sum total of our appreciation of eternity. It is a multiplicity of interwoven time-transcending pathways. As we humans evolve toward higher or deeper levels of consciousness, so our understanding of the eternal trail becomes richer and deeper. We are creating it as we go along. And we are entering a phase of conscious evolution when we are asking ourselves whether we are moving in the right direction, being good stewards of the planet, loving it enough. The author Norman Friedman, drawing on the ideas of the physicist John Wheeler, put it well in his book *Bridging Science and Spirit:*

> We are not at the mercy of a past we do not control. . . . [T]here are probable pasts, just as there are probable futures. We select only one version of events as our past and ignore the others. John Wheeler comments that it is wrong to think of the past as already existing in all detail. The past is theory. The past has no existence except as recorded in the present.[20]

How true this has been in my academic journey down the eternal trail of fossil footprints. It seems at every juncture there was the possibility to create any number of interpretations of the fossilized tracks that presented themselves, either by working in isolation with my own thoughts and intuition in focus, or more often by entering into the celebration of a rich collaboration with friends and colleagues from all over the world. Much of the eternal trail that I have attempted to summarize in this book has emerged into scientific consciousness in the last decade or so. In previous generations it could not be perceived with clarity, and so did not exist with anything like its present meaning. Nor is its meaning fixed; it will be recreated and reinvented again in the future in accordance with the perceptions of future observers. So scientific paradigms will change shape in new iterations just like the physical iterations of our vertebrate ancestors on each successive spiral. But each new incarnation of paleontological theory will retain something of the shape of its predecessor.

Our evolution as a species is accompanied by an expansion of our awareness of who we are and where we came from to include a deeper and more comprehensive picture of our origins and present makeup. Thus a deeper understanding of paleontology and archaeology inevitably leads to a deeper understanding of biology, cultural history, and psychology. It is not just that high-tech innovations in one field feed back into another, though this happens, but more that our overall intellectual sophistication, our improved intuition and awareness—our evolving mind—rolls back the mists of ignorance on an ever-broader front, ultimately reshaping even our most fundamental philosophies and theories of consciousness. Our inquisitive desire to backtrack along the eternal trail is the same as our drive to unlock the secrets of the universe and our never-ending quest for the Holy Grail.

SAVE THE LAST DANCE FOR ME

Reality is a shimmering presence of infinite planes, a luminous labyrinth of the active now connecting past and future.

Mae-Wan Ho, 1993, *The Rainbow and the Worm*

In his classic book *What Is Life?*, the physicist Erwin Schrödinger tackled the problem of free will and consciousness and argued that, blasphemous as it may at first sound, each individual is in a sense "God Almighty."[21] We are the evolutionary vessels on the way to being all-seeing, all-knowing, and omnipresent. I can think of no better way to illustrate this point than to point out that in less than two hundred years, paleontologists have rewritten what they regard as archaic creation myths so as to bring 3.5 billion years of Earth history into scientific focus. This might seem like a big step toward an all-seeing vantage point, but we should be careful not to mistake our scientific perspectives for an ultimate reality. There are signs that current scientific paradigms are breaking down.

Paleontologists and evolutionary biologists tend to ignore the psychology literature, preferring to believe that the secrets of evolution reside in the material morphology of bones and genes. But psychologists, philosophers, and anthroposophists have been busy probing evolution from other angles, and have arrived at the conclusion that consciousness itself is an evolutionary phenomenon of fundamental significance. Whether it be Richard Bucke's classic book *Cosmic Consciousness,* the prolific writings of Ken Wilber, the gentle words of Anna Lemkow in *The Wholeness Principle*, or the dazzling biological insights of Mae-Wan Ho all of which are rich in scholarship and wisdom, there can be little doubt that the study of bones and genes only tell a part of the evolutionary story, pertaining to the physical realm.[22] Ever since the recognition and definition of the noosphere by such luminaries as Pierre Teilhard de Chardin, or the recasting of human cultural history by William Irwin Thompson, the writing has been on the wall: The study of evolution must ultimately integrate our understanding of the tangible, material biological realm with an understanding of subtler formative forces that are not always seen directly but are expressed in our energetic bodies. Rupert Sheldrake observed that the mechanistic approach to biology pushed God and spirit aside, only to see them enter the field of physics by the backdoor. The biophysicist Mae-Wan Ho has demonstrated just how true this is.

Many a cutting-edge physicist is more intrigued by evolution of consciousness than most biologists and paleontologists engaged in the study of evolution, who tend to think it is a subject with no relation to evolution, something for other subdisciplines to study. Yet when I teach a course on the evolution of consciousness, students leap at the chance to be allowed to think about biological evolution and consciousness as an integral part of reality as a whole.

Perhaps it is trackers' study of phantoms, shadows, and the spirits of creatures that leads them to adopt a different perspective. We are dealing with ghosts that may never be known in material form. Yet their tracks represent creatures that were *real*, that lived, breathed, and expressed their identities in a vital dance over Earth's highly sensitive and receptive skin. Because tracks are far more common than bones, we can state with confidence that in the bedrock of our planet there are dozens, perhaps hundreds, of intangible spirits of extinct creatures for every tangible corpse. Who then can regard the notion of spirits or monsters in the ground as superstitious or unreal? I will show you the spirit of lumbering *Megalosaurus* in Turkmenistan, and the spirit of *Australopithecus* in Tanzania.

Every creature that has walked this Earth and left a record of its footprints is a relative of ours, and through the simple arithmetic of generations and the study of evolution we can precisely establish the closeness of this relationship.[23] This process of gathering all life, and even lovable rocks, into the family tree—the emergence, emancipation, and sacralization of all life—is surely a clear sign of human cosmic consciousness on the rise. The noosphere, like a sleeping giant, is awakening, becoming conscious and self-aware. Humans, especially our most highly exalted luminaries and visionaries—Christ, Mohammed, Buddha, Blake, Teilhard de Chardin, and others—have been the pioneers who forged ahead into the realm of cosmic consciousness. But now the awakening spreads to encompass all life. We are aware not just of our own special human incarnation in the "image of God" but also are becoming aware that all life was created in the divine image. All is sacred.

This realization is the current evolutionary step. Evolution is becoming a conscious process on the level of biosphere. Our growing love for democracy, for one another, and for all life begins to transcend racial and species boundaries. Lingering squabbles and the problems of dangerous weapons and land mines frustrate us and become issues that we increasingly wish to resolve so we can get on with the business of life and evolution. As a counterbalance to excessive materialism the tide of conscious spirituality is on the rise.

This is the new evolutionary paradigm already absorbing its Darwinian predecessor. Progressive cycles of individual, species-wide, even biosphere-wide, maturation from simple to self to universal or cosmic consciousness help explain our cultural evolution. We spiral through magical, mythical, religious, rational, and spiritual growth, only to spiral round again, each time integrating something of the old with the new. As each new pulse of awareness engulfs, floods, and permeates former conceptual edifices, those comfortable with their perches may kick and scream a little, resisting the eager pioneers, who enthuse, "Look what we see over the next horizon." On the eve of the new millennium we have been equipped with unprecedented awareness that allows us to look deep into the interior and exterior realms of the *kosmos,* a power that allows us to become conscious and more fully aware of the evolutionary process.

Evolution constantly dances to a new tune. As the universe and its consciousness expands, the melodies, rhythms, cadence, and timbre change. We respond instinctively to the call to dance to the new music, sometimes without even asking where the melodies come from—who are the composers and players? The tracks, it seems, are already out there, before visionaries like Mozart, or Lennon and McCartney, pluck them from the ether. Once materialized as a trail of sheet music to which our feet instinctively respond, we all recognize that they resonate deeply with our heart strings. This all seems to suggest that a much larger portion of the eternal trail is "out there" beyond the immediate awareness of lesser visionaries, which we all are today, but may not be tomorrow. Like the dimples and dents in clay platelets, which act as footprint molds from which vital molecules are cast, so tracks in the ether exist before we can devise dance steps to fit them.[24] Reality is indeed a shimmering and luminous presence of infinite planes, with no clear demarcation between past present and future. As trackers we may wander whichever way we wish and encounter something new.

To study this eternal trail, this eternal reality, is to do more than catalogue footprints and follow trails to discrete destinations in space and time. It is also to undertake a multidimensional, multidirectional journey down countless pathways to witness countless interactions between entity and substrate. There are dents in the space shuttle from tiny bits of orbiting debris, and there are giant meteorite craters in the bedrock of Earth's crust. There are scars in our flesh from wounds on the battlefield and scars in our psyche from those times when someone or something did not "touch" our heart strings so gently. There are scars on the biosphere from our overeager harvest of Earth's bounty.

The Earth has the ability to remember almost anything, including the scars we have inflicted. We can potentially find the tracks and traces of every subatomic particle that tore through a crystal lattice, every Cambrian worm that crawled on the seafloor, the first amphibians that walked on land, and the first hominids to draw pictures of now-extinct animals on cave walls. The moon and Mars have also recorded the arrival of men and machines. We have ascended very rapidly from the Paleolithic cave womb (yin) of the mother goddess to the masculine (yang) adventure of orgasmic rocket-powered space flight. The trajectories of our male seed like cosmic sperm in search of an egg have arrived at that quintessentially feminine destination. As lunar probes and male feet first penetrated moon's dusty skin the seeds of a new conception were sown. We have flown high but must again come down to Mother Earth.

The touch of our feet on Earth is still a sacred communion, not just symbolically but in the real sense that the interaction is preserved forever in the memory of mind and matter. The champagne toast of lovers is preserved in the fingerprint on crystal and the flight of the fugitive is preserved in electrostatically enhanced sole prints in FBI files. The eternal trail leads everywhere and nowhere; it is both journey and destination. It is yesterday, today, and tomorrow. It is us, our very atomic structure, our mystical consciousness, our vital dance.

NOTES

INTRODUCTION

1. Karl Popper, *The Logic of Scientific Discovery* (Harper & Row, 1959).

2. Robert Jastrow, *God and the Astronomers* (W. W. Norton, 1978), p. 116.

3. Ken Wilber has become essential reading in modern philosophy. Ken Wilber, *Sex, Ecology and Spirituality: The Spirit of Evolution* (Shambhala, 1995).

4. Bruce Holbrook, *The Stone Monkey* (William Morrow, 1881).

5. C. C. Jung, *Synchronicity: An Acausal Connecting Principle* (Princeton University Press, Bollingen Series, 1960).

6. John Horgan, *The End of Science* (Helix Books, 1996).

7. H. von Ditfurth, *The Origins of Life: Evolution as Creation* (Harper & Row, 1982).

8. See, for example, J. M. Masson and S. McCarthy, *When Elephants Weep: The Emotional Lives of Animals* (Delta, 1995).

9. Teilhard de Chardin's evolutionary philosophy was not published until after his death, when it became widely acclaimed. Teilhard de Chardin, *Building the Earth* (Dimension Books, 1965).

10. Wilber, *Sex, Ecology and Spirituality*.

11. Stephen Weinberg, *The First Three Minutes* (Basic Books, 1977).

12. Christian de Duve, *Vital Dust: Life as a Cosmic Imperative* (Basic Books, 1995).

CHAPTER ONE

1. Mae-Wan Ho, *The Rainbow and the Worm*, (World Scientific Publishing, 1993); Wolfgang Schad, *Man and Mammals: Towards a Biology of Form*, (Waldorf Press, 1977).

2. William Irwin Thompson, *Time Falling Bodies Take to Light* (St. Martin's Press, 1981), is a modern anthropological classic.

3. Laurens van der Post, *The Lost World of the Kalahari* (Hogarth Press, 1958).

4. Arthur Conan-Doyle, *The Lost World* (Hodder & Stoughton, 1912).

5. M. Crichton, *The Lost World* (Ballantine Books, 1995). 423p.

6. Elizabeth Marshall Thomas, *The Harmless People* (Alfred A. Knopf, 1958).

7. Ibid.

8. J. Shreeve, "Sexing Fossils: A Boy Named Lucy?" *Science*, 270 (1995):1297–1298.

9. M. G. Lockley, "The Vertebrate Track Record," *Nature* 396 (1998):429–432.

10. M. G. Lockley, "Ichnotopia: A Review of the Paleontological Society Short Course on Trace Fossils," *Ichnos* 2 (1993):337–342.

11. W. A. S. Sarjeant, *in* D. D. Gillette and M. G. Lockley, eds., *Dinosaur Tracks and Traces* (Cambridge University Press, 1989).

12. P. Jackson, "Footloose in Archaeology," *Current Archaeology* 156 (1995):456–457; see also ibid., *Current Archaeology* 144 (1995):466–470.

13. S. Coren, "Are Fingerprints a Genetic Marker for Handedness?" *Behavior Genetics* 24 (1994):141–147.

14. C. F. Johnson and E. Opitz, "Unusual Palm Creases and Unusual Children," *Clinical Pediatrics* 12 (1973):101–112.

15. E. C. Krupp, *Echoes of the Ancient Skies: The Astronomy of Lost Civilizations* (Harper & Row, 1983).

16. Jack Fincher, *Lefties: The Origins and Consequences of Being Left-handed* (Barnes & Noble, 1993).

17. Eadward Muybridge, *Animals in Motion* (Chapman Hall, 1899; reprint, Dover Publications, 1957).

18. Peter Ouspensky, *Tertium Organum* (1920; reprint, Vintage Books, 1970).

19. M. G. Lockley, *Tracking Dinosaurs: A New Look at an Ancient World* (Cambridge University Press, 1991).

20. Edward Hitchcock, 1858, *A Report on the Sandstone of the Connecticut Valley Especially Its Fossil Footmarks* (W. White; reprint, Arno Press, Natural Sciences in America Series, 1994).

21. W. Buckland, 1858, *The Bridgewater Treatises on the power, wisdom and goodness of God, as manifest in the Creation,* 3rd ed., "Geology and Mineralogy considered with reference to natural theology" (Treatise 6).

22. A. Lemkow, *The Wholeness Principle* (Quest Books, 1990) 1995.

23. For a discussion of their ideas, see Ho, *The Rainbow and the Worm,* and Brian Goodwin, *How the Leopard Changed Its Spots* (Touchstone Books, 1995).

24. James Lovelock, *Gaia: A New Look at Life on Earth* (Oxford University Press, 1979).

25. R. Thomas, "Will PIP [polycontrast interface photography] Revolutionise Medicine?" *International Journal of Alternative and Complementary Medicine* (1995):16–17.

26. H. Bortoft, *The Wholeness of Nature* (Lindisfarne Press, 1996).

27. For a more detailed discussion of these ideas, see Schad, *Man and Mammals.*

28. W. Schad, "Heterochronal Patterns of Evolution in the Transitional Stages of Vertebrate Classes," *Acta Biotheoretica* 41 (1994):383–389.

29. A detailed explanation of this principle can be found in B. Rensch, *Evolution Above the Species Level* (Methuen, 1959).

30. Kenneth McNamara, *Shapes of Time* (Johns Hopkins University Press, 1997), is particularly illuminating on the process of evolution in whole organisms.

CHAPTER TWO

1. E. M. Thomas, *The Harmless People* (Alfred A. Knopf, 1958).

2. R. Wilhelm and C. Baynes, trans., *The I Ching* (Princeton University Press, 1967).

3. D. Young, *Origins of the Sacred: The Ecstasies of Love and War* (St. Martin's Press, 1991).

4. William Bryant Logan, *Dirt: The Ecstatic Skin of the Earth* (Riverhead Books, 1995), pp. 123–125.

5. A. Seilacher, *Fossilienkunst: Albumblätter der Erdgeschichte* (Fossil Art: Leaves from the Album of Earth's History) (Goldschenck-Verlag/Werner K. Weidert, Korb, 1995).

6. Della Willis, *The Sand Dollar and the Slide Rule: Drawing Blueprints from Nature* (Addison-Wesley, 1995).

7. William Irwin Thompson, *Coming into Being: Artifacts and Texts in the Evolution of Consciousness* (St. Martin's Press, 1996).

8. Adolf Seilacher, "Vendozoa: Organismic Construction in the Proterozoic Biosphere," *Lethaia* 22 (1989):229–239.

9. Ibid., "Self-Organizing Morphogenetic Mechanisms As Processors of Evolution," *Revista Espanola de Paleontologia* (Spanish Review of Paleontology), special edition, 1991, pp. 5–11.

10. D. Seamon and Zajonc, *Goethe's Way of Science: A Phenomenology of Nature* (State University of New York Press, 1998).

11. W. Schad, in J. Bockemühl, ed., *Toward a Phenomenology of the Etheric World* (Anthroposophic Press, 1977).

12. Ibid.

13. R. Sheldrake, *The Presence of the Past* (Vintage Books, 1988).

14. A. Seilacher, "Fossil Behavior," *Scientific American*, Date TK, pp. 72–80.

15. M. G. Lockley, "Ichnotopia: A Review of the Paleontological Society Short Course on Trace Fossils," *Ichnos* 2 (1993):337–342.

16. Seilacher, *Fossilienkunst*.

17. G. Doczi, *The Power of Limits* (Shambhala, 1981).

18. E. W. Johnson et al., "Non-Marine Arthropod Traces from the Subaerial Ordovician Borrowdale Volcanic Group," *English Lake District Geology Magazine*, no. 131 (1994):395–406.

19. S. Burnshaw, *The Seamless Web* (George Braziller, 1970).

20. W. A. S. Sarjeant, "A History and Bibliography of the Study of Fossil Vertebrate Footprints in the British Isles," *Palaeogeography, Palaeoclimatology, Palaeoecology* 16 (1974):265–378.

21. D. A. Rogers, "Probable Tetrapod Tracks Rediscovered in the Devonian of Scotland," *Journal of the Geological Society of London*, no. 14 (1990)7:746–748.

22. B. Willard, "Chemung Tracks and Trails from Pennsylvania," *Journal of Paleontology* 9 (1935):43–56.

23. K. Caster, "A Restudy of the Tracks of *Paramphibius*," *Journal of Paleontology* 12 (1938):3–60

24. J. W. Warren and N. A. Wakefield, "Trackways of Tetrapod Vertebrates from the Upper Devonian of Victoria," *Australia Nature*, no. 238 (1972):469–470; A. Warren et al., "Earliest Tetrapod Trackway," *Alcheringa* 10 (1986):183–186; I. Stossel, "The Discovery of a New Devonian Tetrapod Trackway in Southwest Ireland," *Journal of the Geological Society London* 152 (1995):407–413.

25. J. Clack, *Palaeogeography, Palaeoclimatology, Palaeoecology* 130 (1997).

26. Ken McNamara, *Shapes of Time* (Johns Hopkins University Press, 1997).

27. M. McMenamin and D. McMenamin, *Hypersea* (Columbia University Press, 1995).

28. D. E. G. Briggs et al, "A Giant Myriapod Trail from the Namurian of Arran, Scotland," *Palaeontology* 22 (1979):273–291.

29. Ibid., "*Arthropleura* Trails from the Westphalian of Eastern Canada," *Palaeontology* 27 (1984):843–855.

30. C. Sagan, *The Dragons of Eden* (Random House, 1977).

31. J. W. Dawson, "Note on Footprints from the Carboniferous of Nova Scotia, in the Collection of the Geological Survey of Canada," *Geological Magazine* 1 (1872):251–253.

32. W.A.S. Sargeant, "A History and Bibliography of the Study of Fossil Vertebrate Footprints in the British Isles," *Palaeogeography, Palaeoclimatology, Palaeoecology* 16 (1974):265–378.

33. M. J. Benton, *Vertebrate Paleontology* (Unwin Hyman, 1990).

34. J.W. Dawson, "Note on Footprints from the Carboniferous of Nova Scotia, in the Collection of the Geological Survey of Canada," *Geological Magazine* 1 (1872):251–253.

35. D. J. Mossman and W. A. S. Sarjeant, "How We Found Canada's Oldest Footprints," *Canadian Geographic Magazine* 100 (1980):50–53.

36. H. T. Martin, "Indications of a Gigantic Amphibian in the Coal Measure of Kansas," *Kansas University Science Bulletin* 13 (1992):103–114.

37. J.H. Calder et al., 1997 Trackways of gregarious tetrapods in a fossil Walchian forest from the Permo Carboniferous of Nova Scotia p. 7 *in* Haubold, H, (ed). Workshop, on Ichnnofacies and ichnotaxonomy of the terrestrial Permian. Martin-Luther-Univ., Halle, Germany 62p.

38. R. Staszko, "The Tournaisian Explosion," unpublished report (Halifax, Nova Scotia).

39. A. S. Romer and L. L. Price, "Review of the Pelycosauria," Geological Society of America Special Papers, no. 28 (1940).

40. Jerry MacDonald, *Earth's First Steps* (Johnson Books, 1995).

41. S. G. Lucas and A. B. Heckert, eds., *Early Permian Footprints and Facies*, New Mexico Museum of Natural History and Science (Albuquerque), Bulletin 6, 1995.

42. Ibid.

43. M. G. Lockley and J. Madsen, "Early Permian Vertebrate Trackways from the Cedar Mesa Sandstone of Eastern Utah: Evidence of Predator Prey Interaction," *Ichnos* 2 (1993):147–153.

44. S. G. Lucas et al., eds., New Mexico Museum of Natural History and Science (Albuquerque), 1998.

45. R. Broom, *The Mammal-like Reptiles of South Africa and the Origin of Mammals* (Witherby, 1932).

46. Konrad Lorenz, *King Solomon's Ring* (Thomas Y. Crowell, 1952).

47. R. M. H. Smith, "Fossils for Africa: An Introduction to the Fossil Wealth of the Newevel Mountains Near Beaufort West," *Sagittarius* 3 (1988).

48. Ibid., "Trace Fossils of the Ancient Karoo," *Sagittarius* 2 (1986): 4–9.

49. Ibid., "Helical Burrow Casts of Therapsid Origin from the Beaufort Group (Permian) of South Africa," *Palaeogeography, Palaeoclimatology, Palaeoecology* 60 (1987):155–170; C. B. Schultz, "A Review of the *Diamonelix* Problem," *Nebraska University Studies in Science and Technology* 2 (1942):1–30.

50. Wolfgang Schad, *Man and Mammals: Towards a Biology of Form* (Waldorf Press, 1977).

51. M. G. Lockley et al., "A Survey of Fossil Footprints Sites At Glen Canyon National Recreation Area (Western USA): A Case Study in Documentation of Trace Fossil Resources At a National Preserve," *Ichnos* 5 (1998):177–211.

52. G. Pemberton et al., "Footsteps Before the Flood: The First Scientific Reports of Vertebrate Footprints," *Ichnos* 4 (1996):321–324.

53. W. Jardine, "Note to Mr Harkness's Paper on 'The Position of the Impressions of Footsteps in the Bunter Sandstone of Dumfries-Shire,'" *Annual Magazine of Natural History*, series 2, 6 (1850):208–209.

54. W. Buckland, *The Bridgewater Treatises on the power, wisdom and goodness of God, as manifest in the Creation*, 3rd ed., "Geology and Mineralogy considered with reference to natural theology" (Treatise 6).

55. T. H. Huxley, "The Crocodilian Remains Found in the Elgin Sandstone, with Remarks on the Ichnites of Cummingstone," Geological Survey Monograph no. 3 (1877):1–52.

56. R. S. Lull, "Fossil Footprints from the Grand Canyon of the Colorado," *American Journal of Science*, series 4, 45 (1918):337–346; C. W. Gilmore, 1926–1928, "Fossil Footprints from the Grand Canyon," *Smithsonian Miscellaneous Collections*, no. 77:1–41; no. 80:1–78; no. 81:1–16.

57. L. R. Brand and T. Tang, "Fossil Vertebrate Footprints in the Conconino Sandstone (Permian) of Northern Arizona: Evidence for Underwater Origin" *Geology* 19 (1991):1201–1204; L. R. Brand, "Variations in Salamander Trackways Resulting from Substrate Differences," *Journal of Paleontology* 70 (1996):1004–1010

58. L. R. Brand, *Faith, Reason and Earth History* (Andrews University Press, 1998).

59. Ibid.

60. M. G. Lockley et al., in Lucas and Heckert, *Early Permian Footprints and Facies*.

61. M. G. Lockley and A. P. Hunt, *Dinosaur Tracks and Other Fossil Footprints of the Western United States* (Columbia University Press, 1995).

62. Two published criticisms of Brand and Tang's controversial hypothesis are M. G. Lockley, *Geology* (1992):666; and D. Loope, *Geology* (1992):667.

63. Brand, *Faith, Reason and Earth History.*

64. Lockley and Hunt, *Dinosaur Tracks and Other Fossil Footprints of the Western United States.*

65. M.G. Lockley et al., "Vertebrate Tracks and the Ichnofacies Concept," in S. K. Donovan, ed. (John Wiley, 1994), pp. 241–268.

66. Schad, *Man and Mammals.*

67. d'Arcy Thompson, *On Growth and Form* (Cambridge University Press, 1961).

68. McNamara, *Shapes of Time.*

69. K. Hopkirk, *Central Asia* (John Murray Books, 1993).

70. M.G. Lockley, et al., in S.G. Lucas and A.B. Heckert, eds., *Early Permian Footprints and Facies.*

71. *Creatures of the Namib Desert,* National Geographic video (1977).

72. P. Leonardi et al., "*Pachypes dolomiticus* n. gen n. sp.: Pareiasaur Footprint from the "Val Gardena Sandstone (Middle Permian) in the Western Dolomites (N. Italy)," Atti Accad. Naz. Lincei, Rend Cl. Sci. fis. Mat. Nat., 8, 57 (1975):221–223.

73. M. A. Conti et al., "Revaluation of *Pachypes Dolomiticus* Leonardi Et Alii. 1975; a Late Permian Pareiosaur Footprint," in Hartmut Haubold, ed. (1997), p. 13.

74. M. Lee, "The Turtle's Long Lost Relatives," *Natural History* 103 (1994):63–65.

75. M. Lockley, "Permian Perambulations Become 'Understandable'" (a report on a tracks workshop), *Ichnos,* in press.

76. Haubold, "Workshop on Ichnnofacies and ichnotakonomy of the terrestial Permian."

77. M. Lockley, "Permian Perambulations Become 'Understandable'" (a report on a tracks workshop), *Ichnos,* in press.

CHAPTER THREE

1. G. Murchie, *The Seven Mysteries of Life* (Houghton Miflin, 1978).

2. H. Wendt, *Before the Deluge* (Victor Gollancz, 1968).

3. W. Buckland, 1858.

4. D. B. Weishampel and W. E. Reif, "The Work of Franz Baron Nopsca (1877–1933): Dinosaurs, Evolution and Theoretical Tectonics," *Jahrbuch der Geologie* 127 (1984):187–203.

5. F. von Nopsca, "Die Fossilien Reptilen," *Fortschritte Geol. Pal.* 2(1923):210.

6. W. Soergel, "Die Fährten der Chirotheria: Eine paläobiologische Studie" (The Tracks of Chirotheria: A Paleobiological Study) (Gustav Fisher, 1925); Weishampel and Reif, "Work of Franz, Baron Nopsca."

7. Frank Peabody, "Reptile and Amphibian Trackways from the Lower Triassic Moenkopi Formation of Arizona and Utah," University of California–Berkeley and Los Angeles, Department of Geological Sciences Bulletin 27 (1948):295–468.

8. G. Tresise, "Sex in the Footprint Bed," *Geology Today* 12 (1996):22–26.

9. C. Ormund, *Complete Book of Outdoor Lore* (Harper & Row, 1964).

10. F. Gettings, *The Book of Palmistry* (Triune Books, 1974).

11. C. F. Johnson and E. Opitz, "Unusual Palm Creases and Unusual Children," *Clinical Pediatrics* 12 (1973):101–112.

12. G. Demathieu and H. Haubold, "Du Problème de l'origine des dinosauriens d'après les données de l'ichnologie du Trias" (On the Problem of the Dinosaurs and the Track Evidence of the Triassic), *Geobios* 11 (1978):409–412.

13. M. J. Benton, *Vertebrate Paleontology* (Unwin Hyman, 1990).

14. R. Ornstein, *The Evolution of Consciousness* (Prentice Hall, 1991).

15. G. Demathieu and H. Haubold, "Reptilfährten aus dem Mittleren Buntsandstein von Hessen (BRD)" (Reptile Trails from the Middle Bunter Sandstone Series in Hesse), *Hallesches Jahrbuch der Geowissenschaft* 7 (1982):97–110.

16. P. Ellenberger, "Une piste avec traces de soies épaisses dans le Trias Inférieur à Moyen de Lodeve (Herault, France): *Cynodontipus polythrix nov. gen., nov. sp.* les Cynodontes en France," (A Track with Traces of Hair in the Lower to Middle Triassic Lodeve [Herault, France]: *Cynodontipus polythrix nov. gen., nov. sp.* the Cynodonts in France), *Geobios* 9 (1976):769–787.

17. Victor Ellenberger, *La Fin tragique des bushmen* (The Tragic End of the Bushmen) (Amiot Dumont, 1953).

18. Peabody, "Reptile and Amphibian Trackways."

19. M. G. Lockley and A. P. Hunt, *Dinosaur Tracks and other fossil footprints of the Western United States.*

20. P. Olsen, *Lamont-Doherty Geological Observatory Newsletter* 19 (1989).

21. P. Ellenberger, "Contribution a la classification des pistes de vertèbres du Trias; les types du Stormberg d'Afrique du Sud" (A Contribution to the Classification of Vertebrate Tracks of the Triassic), Part 1, *Palaevertebrata*, special issue (Montpellier, 1972).

22. P. Olsen and P. Galton, "A Review of the Reptile and Amphibian Assemblages from the Stormberg of Southern Africa, with Special Emphasis on the Footprints and the Age of the Stormberg," *Palaeontographica Africana,* 25 (1984):87–110.

23. E. Hitchcock, "Description of Two New Species of Fossil Footmarks Found in Massachussetts and Connecticut, or of the Animals That Made Them," *American Journal of Science* 4 (1847):46–57.

24. M. G. Lockley et al., "Late Triassic Vertebrate Tracks in the Dinosaur National Monument Area," Utah Geological Survey, Miscellaneous Publications, no. 92–93 (1992):383–391; M. G. Lockley and A. P. Hunt.

25. M. Gell-Mann, *The Quark and the Jaguar: Adventures in the Simple and Complex* (W. H. Freeman, 1994).

26. K. Lorenz, *Behind the Mirror,* trans. R. Taylor (Harcourt Brace Jovanovich, 1978).

27. N. Eldredge and S. Gould, "Punctuated Equilibria: An Alternative to Phyletic Gradualism," Schopf, ed., *Models in Paleobiology* (Freeman, Cooper, 1972), pp. 82–115. This has become recognized as the landmark paper on punctuated equilibrium.

28. P. Ellenberger, "L'Explosion démographique des petits quadrupèdes à l'allure de mammifères dans le Stormberg Supérieur (Trias) d'Afrique du sud aperçu sur leur origine au Permien (France et Karoo)," International Colloquium of the National Center for Scientific Research, Bulletin no. 218 (1975):409–432.

29. E. C. Rainforth and M. G. Lockley, "Tracks of Diminutive Dinosaurs and Hopping Mammals from the Jurassic of North and South America," in M. Morales, ed., *Continental Jurassic* (Museum of Northern Arizona, 1996), pp. 265–269.

30. A. P. Hunt, "The Early Diversification Pattern of Dinosaurs in the Late Triassic," *Modern Geology* 16 (1991):43–60.

31. F. von Huene, "Die Fossilienfährten im Rhät von Ishigualasto im Nordwest-Argentinien" (Fossil Tracks in the Ishigualasto of Northwest Argentina), *Palaeobiologica* 4 (1931):99–112.

32. G. Leonardi, *Annotated Atlas of South America Tetrapod Footprints (Devonian to Holocene)* (Company of Mineral Resources [Brasilia], 1994).

33. A. B. Arcucci et al., "'Theropod' Tracks from the Los Rastos Formation (Middle Triassic), La Rioja Province, Argentina," *Journal of Vertebrate Paleontology* 15 (1995):16A.

34. Georges Demathieu, "Appearance of the First Dinosaur Tracks in the French Middle Triassic, and Their Probable Significance," in D. D. Gillette and M. G. Lockley, eds., *Dinosaur Tracks and Traces* (Cambridge University Press, 1989), pp. 201–207.

35. L. J. Wills and W. A. S. Sarjeant, "Fossil Vertebrate and Invertebrate Tracks from Boreholes Through the Bunter Sandstone Series (Triassic) of Worcestershire," *Mercian Geologist* 3 (1970): 399–414.

36. M. J. King and M. J. Benton, "Dinosaurs in the Early and Middle Triassic? The Footprint Evidence from Britain," *Palaeogeography, Palaeoclimatology, Palaeoecology* 122 (1996):213–225.

37. W. A. S. Sarjeant, "A Re-appraisal of Some Supposed Dinosaur Footprints from the Triassic of the English Midlands," *Mercian Geologist* 14 (1996):22–30.

38. D. Baird, "*Chirotherium Lulli*, a Pseudosuchian Reptile from New Jersey," *Bulletin of the Museum of Comparative Zoology, Harvard* 111 (1954):165–192.

39. Wills and Sarjeant, "Fossil Vertebrate and Invertebrate Tracks. . . "

40. d'Arcy Thompson, *On Growth and Form* (Cambridge University Press, 1961).

CHAPTER FOUR

1. H. E. Huntley, *The Divine Proportion: A Study in Mathematical Beauty* (Dover, 1970).

2. Ibid.

3. G. Doczi, *The Power of Limits* (Shambhala, 1981).

4. E. Hitchcock, "Ornithichnology—Description of the Foot Marks of Birds (Ornithichnites) on New Red Sandstone in Massachusetts," *American Journal of Science* 29 (1836):307–340

5. Ibid.

6. R. T. Steinbock, "Ichnology of the Connecticut Valley: A Vignette of American Science in the Early Nineteenth Century," in D. D. Gillette and M. G. Lockley, eds., *Dinosaur Tracks and Traces* (1989), pp. 27–32.

7. P. Davis, *The Mind of God* (Simon & Schuster, 1992).

8. E. Hitchcock, *Religion of Geology and Its Connected Sciences* (Crosby, Nichols, Lee, 1851).

9. Hitchcock, "Ornithichnology."

10. D.V. Kent et al., "Late Triassic-Earliest Jurassic Geomagnetic Polarity Sequence and Paleolatitudes from Drill Cores in The Newark Rift Basin, Eastern North America," *Journal of Geophysical Research* 100 (1995):14965–14988. P. E. Olsen et al., "High-Resolution Stratigraphy of the Newark Rift Basin (Early Mesozoic, Eastern North America)," Geological Society of America, Bulletin 108 (1996):44–77.

11. C. Zimmer, "Peeling the Big Blue Banana," *Discover*, January 1992, pp. 46–47.

12. In most earth sciences textbooks, the durations of Milankovitch cycles are given as about 26,000 (precession of equinoxes), 41,000 (change in rotational tilt), and 100,000 years (eccentricity). There is a strong correlation between temperature fluctuations, as measured from oxygen isotopes, the 100,000-year cycle, and a longer 400,000-year cycle. See also K. C. Condie and R. E. Sloan, *Origin and Evolution of the Earth* (Prentice Hall, 1998). Estimates of the duration of longer cycles may vary slightly.

13. E. Hitchcock, *A Report on the Sandstone of the Connecticut Valley Especially Its Fossil Footmarks* (W. White, 1858; reprint, Arno Press, Natural Sciences in America Series).

14. Wolfgang Schad, *Man and Mammals: Towards a Biology of Form* (Waldorf Press, 1977).

15. G. Paul, *Predatory Dinosaurs of the World* (Simon & Schuster, 1988).

16. Hitchcock, "Ornithichnology."

17. C. L. Bernheimer, *Rainbow Bridge* (Doubleday, Page, 1924).

18. E Hitchcock, "An Attempt to Name, Classify and Describe the Animals That Made the Fossil Footmarks of New England," *Proceedingss of the Sixth Annual Meeting of the Association of American Geologists and Naturalists* (1845):23–25.

19. P. E. Olsen et al., "Type Material of the Classic Theropod Footprint Genera *Eubrontes, Anchisauripus* and *Grallator* (Early Jurassic, Hartford and Deerfield Basins, Connecticut and Massachusetts, USA)," *Journal of Vertebrate Paleontology* 18 (1998):586–601. See also M. G. Lockley and A. P. Hunt, *Dinosaur Tracks and other fossil footprints of the Western United States.*

20. E. Hitchcock, "Description of Two New Species of Fossil Footmarks Found in Massachussetts and Connecticut, or of the Animals That Made Them," *American Journal of Science* 4 (1847):46–57.

21. Ibid.

22. Hitchcock, "An Attempt to Name, Classify and Describe."

23. R. S. Lull, "Triassic Life of the Connecticut Valley," Connecticut State Geological and Natural History Survey, Bulletin 81 (1953).

24. H. Haubold, *Saurierfährten* (Saurischian Tracks), 2nd ed. (A. Ziemsen Verlag, 1984); R. A. Thulborn, *Dinosaur Tracks* (Chapman Hall, 1990); G. Gierlinski, "Thyreophoran Affinity of *Otozoum* Tracks," *Prezeglad Geologiczny* (Wittenberg Lutherstadt) 43 (1995):123–125.

25. Schad. "Man and Mammals"

26. E. Hitchcock, "An Attempt to Discriminate and Describe the Animals That Made the Fossil Footmarks of the United States, and Especially of New England," *Transactions of the American Academy of Arts and Science* (N.S.) 3 (1848):129–256.

27. Hitchcock, *A Report on the Sandstone of the Connecticut Valley*.

28. T. S. Ahbrandt, "Bioturbation of Eolian Deposits," *Journal of Sedimentary Petrology* 48 (1978):839–848.

29. E. C. Rainforth and M. G. Lockley, "Tracks of Diminutive Dinosaurs and Hopping Mammals from the Jurassic of North and South America," in M. Morales, ed., *Continental Jurassic* (Museum of Northern Arizona, 1996), pp. 265–269.

30. G. Leonardi, *Annotated Atlas of South America Tetrapod Footprints (Devonian to Holocene)* (Company of Mineral Resources [Brasilia], 1994); L. Psihoyos and J. Knoebber, *Hunting Dinosaurs* (Cassell, 1995).

31. R. M. Casamiquela, *Estudios Ichnologicos* (Ichnological Studies) (Colegio Industrial Pix IX [Buenos Aires], 1964).

32. Leonardi, *Annotated Atlas of South America Tetrapod Footprints*.

33. E. C. Rainforth and M. G. Lockley, "Tracking Life in a Lower Jurassic Desert: Vertebrate Tracks and Other Traces from the Navajo Sandstone," in M. Morales, ed., *Continental Jurassic* (Museum of Northern Arizona, 1996), pp. 285–289.

34. Casamiquela, *Estudios Ichnologicos;* Psihoyos and Knoebber, *Hunting Dinosaurs*.

35. Paul Ellenberger, "L'Explosion démographique des petits quadrupèdes à l'allure de mammifères dans le Stormberg Supérieur (Trias) d'Afrique du sud aperçu sur leur origine au Permien (France et Karoo)," International Colloquium of the National Center for Scientific Research, Bulletin no. 218 (1975):409–432.

36. Z. Kielan-Jawaorowska and P. P. Gambaryan, "Postcranial Anatomy and Habits of Asian Multituberculate Mammals," *Fossils and Strata* 36 (1994).

37. M. A. Raath, "First Record of Dinosaur Footprints from Rhodesia," *Arnoldia* 5 (1972):1–5.

38. P. Bernier, "Une lagune tropicale au temps des dinosaures" (A Tropical Lagoon in the Age of Dinosaurs) (Centre National de la Recherche Scientifique, Museum de Lyons, France, 1985); P. Bernier et al., "Trace nouvelle de locomotion de chelonien et figures d'emersion associées dans les calcaires lithographiques de Cerin (Kimmeridgien Supérieur, Ain, France) (A New Trace of the Locomotion of turtles and associated signs of emergence in the lithographic limestones of Cerin (upper Kimmeridgian, Ain, France) *Geobios* 15 (1982):447–467; ibid., "Decouverte de pistes de dinosaures sauteurs dans les calcaires lithographiques de Cerin (Kimmeridgien Superieur, Ain, France)—implications paleoecologiques" (Discovery of the Tracks of Hopping Dinosaurs in the lithographic limestones of Cerin (upper Kimmeridgian, Ain, France) *Geobios*, Special Bulletin 8 (1984):177–185.

39. Thulborn, *Dinosaur Tracks*.

40. J. Fincher, *Lefties: The Origins and Consequences of Being Left-handed* (Barnes & Noble, 1993).

41. S. Ishigaki, "Dinosaur Footprints the Atlas Mountains," *Nature Study* (Japan) 32 (1986):6–9.

42. M. G. Lockley et al., "Limping Dinosaurs? Trackway Evidence for Abnormal Gaits," *Ichnos* 3 (1994):193–202.

43. V. F. Santos et al., "A New Sauropod Tracksite from the Middle Jurassic of Portugal," *Gaia* (Museum of Natural History, Lisbon) 10 (1994):5–14.

44. M. G. Lockley et al., "Dinosaur Tracks from the Carmel Formation, Northeastern Utah: Implications for Middle Jurassic Paleoecology," *Ichnos* 5 (1998):255–267.

CHAPTER FIVE

1. W. Buckland, 1858, *The Bridgewater Treatises on the power, wisdom and goodness of God, as manifest in the Creation*, 3rd., "Geology and Mineralogy considered with reference to natural theology" (Treatise 6).

2. Robert Bakker, "The Real Jurassic Park: Dinosaur Habitats at Como Bluff, Wyoming, in M. Morales, ed., *Continental Jurassic* (Museum of Northern Arizona, 1996), pp. 35–49.

3. M. G. Lockley et al., 1996 "*Megalosauripus, Megalosauropus* and the Concept of Megalosaur Footprints," in Morales, *Continental Jurassic,* pp. 113–118.

4. M. G. Lockley et al., "Late Jurassic Dinosaur Tracksites from Central Asia: A Preliminary Report on the World's Longest Trackways," in Morales, *Continental Jurassic,* pp. p. 271–273.

5. Bakker, "The Real Jurassic Park."

6. M. G. Lockley, A. P. Hunt, and S. G. Lucas, "Vertebrate Track Assemblages from the Jurassic Summerville Formation and Correlative Deposits," in Morales, *Continental Jurassic,* pp. 249–254.

7. See notes 3, 4, and 6 above.

8. M. G. Lockley, "The Moab Megatracksite: A Preliminary Description and Discussion of Millions of Middle Jurassic Tracks in Eastern Utah," in W. R. Averett, ed., *Guidebook for Dinosaur Quarries and Tracksites Tour, Western Colorado and Eastern Utah* (Grand Junction [Colorado] Geological Society, 1991), pp. 59–65.

9. M. G. Lockley, *Tracking Dinosaurs: A New Look at an Ancient World* (Cambridge University Press, 1991).

10. M. G. Lockley et al., "The Distribution of Sauropod Tracks and Trackmakers," *Gaia* (Museum of Natural History, Lisbon) 10 (1994):233–248.

11. P. R. Vail et al., "Seismic Stratigraphy and Global Changes of Sea Level," Memoir of the American Association of Petroleum Geologists, no. 26 (1977):49–212.

12. A. Mayor, "Griffin Bones: Ancient Folklore and Paleontology," *Cryptozoology* 10 (1991):16–41

13. Lockley et al., "Late Jurassic Dinosaur Tracksites from Central Asia. . . "

14. Ibid.

15. Bakker, "The Real Jurassic Park."

16. W. Stokes, "Pterodactyl Tracks from the Morrison Formation," *Journal of Paleontology* 31 (1957):952–954.

17. K. Padian and P. E. Olsen, "The Fossil Trackway Pteraichnus: Not Pterosaurian, but Crocodilian," *Journal of Paleontology* 58 (1984):178–184.

18. M. G. Lockley et al., "The Fossil Trackway *Pteraichnus* Is Pterosaurian, Not Crocodilian: Implications for the Global Distribution of Pterosaurs Tracks," *Ichnos* 4 (1995):7–20.

19. M. G. Lockley, "A New Dinosaur Tracksite in the Morrison Formation, Boundary Butte, Southeastern Utah," *Modern Geology* 23 (1998):317–330.

20. M. G. Lockley, "Track Records," *Natural History* 104 (1995):46–51.

21. E. N. Shor, *The Fossil Feud Between E. D. Cope and O. C. Marsh* (Exposition Press, 1974).

22. T. C. McLuhan, *Touch the Earth: A Self-portrait of Indian Existence* (Abacus, 1971).

23. William Irwin Thompson, *Time Falling Bodies Take to Light* (St. Martin's Press, 1981).

24. M. G. Lockley and A. P. Hunt, "A Probable Stegosaur Track from the Morrison Formation of Utah," *Modern Geology* 23 (1998):331–342.

25. M. G. Lockley et al., *Dinosaur Lake: The Story of the Purgatoire Valley Dinosaur Tracksites Area,* Colorado Geological Survey, Special Publication 40 (1997).

26. John. S. MacClary, "Dinosaur Trails of Purgatory," *Scientific American* 158 (1938):72.

27. R. T. Bird, *Bones for Barnum Brown: Adventures of a Dinosaur Hunter* (Texas Christian University Press, 1985).

28. M. G. Lockley et al., "North America's Largest Dinosaur Tracksite: Implications for Morrison Formation Paleoecology," *Geological Society of America Bulletin* 97(1986):1163–1176.

29. L. Psihoyos and J. Knoebber, *Hunting Dinosaurs* (Cassell, 1995).

30. Lockley, *Tracking Dinosaurs.*

31. F. C. Baker, "Some Interesting Molluscan Monstrosities," *Transactions of the Academy of Sciences of St. Louis* 11 (1901):143–146.

32. D. Norman, *Dinosaur* (Prentice Hall, 1991).

33. Robert Bakker, *The Dinosaur Heresies* (William Morrow, 1986).

34. A. A. Fauvel, "Alligators in China: Their History, Description and Identification," *Journal of the North China Branch of the Royal Asiatic Society* (new series V) 13 (1879):1–36.

35. U. Lanham, *The Bone Hunters* (Columbia University Press, 1973); J. R. Foster and M. G. Lockley, "Probable Crocodilian Tracks and Traces from the Morrison Formation (Upper Jurassic) of Eastern Utah," *Ichnos* 5 (1997): 121–129.

36. M. G. Lockley et al., "'Pegadas de Mula': An Explanation for the Occurrence of Mesozoic Traces that Resemble Mule Tracks," *Ichnos* 3 (1994):125–133.

37. C. A. Meyer et al., "A Comparison of Well-Preserved Sauropod Tracks from the Late Jurassic of Portugal and the Western United States: Evidence and Implications," *Gaia* (Museum of Natural History, Lisbon) 10 (1994):57–64.

38. E. Hitchcock, *A Report on the Sandstone of the Connecticut Valley Especially Its Fossil Footmarks* (W. White, 1858; reprint, Arno Press, Natural Sciences in America Series).1858.

39. M. R. Djalilov and V. P. Novikov, eds., *Fossil Traces of Life in Central Asia* (Donish Publishing Co. [Dushanbe], 1987).

40. A. Romashko, "Tracking Dinosaurs," *Moscow News,* 1983, p. 10.

41. A. A. Woodward, *The Earliest Englishman* (Watts, 1948).

42. J. Winslow and A. Meyer, "The Perpetrator at Piltdown," *Science* (1983):32–43; D. Elliott, *The Curious Incident of the Missing Link, Arthur Conan Doyle and Piltdown Man* (Bootmakers of Toronto, Ocasional Papers, no. 2, 1988).

43. L. Leakey and A. M. Goodall, *Unveiling Man's Origins* (1969); Stephen Jay Gould, "The Piltdown Conspiracy," *Natural History* 89 (1980):8–28.

44. H. Gee, "Box of Bones 'Clinches' Identity of Piltdown Palaeontology Hoaxer," *Nature* 381 (1996):261–262.

45. Pierre Teilhard de Chardin, *The Phenomenon of Man* (Harper & Row, 1959); ibid., *The Future of Man* (Harper & Row, 1959); ibid., *Building the Earth* (Dimension Books, 1965).

46. See Stephen Jay Gould's Introduction to Peter Medewar, *The Strange Case of the Spotted Mice and Other Classic Essays on Science* (Oxford University Press, 1996).

47. S. Zhen et al., "A Review of Dinosaur Footprints in China," in D. D. Gillette and M. G. Lockley, eds., *Dinosaurs Past and Present* (Cambridge University Press, 1988), pp. 187–198.

48. Pierre Teilhard de Chardin and C. C. Young, "On Some Traces of Vertebrate Life in the Jurassic and Triassic Beds of Shansi and Shensi," *Bulletin of the Geological Survey of China* 3 (1929):131–136.

49. H. de Terra, *Memories of Teilhard de Chardin* (Scientific Book Club/W. Collins, 1964).

50. Gould, Introduction to Medewar, *The Strange Case of the Spotted Mice.*

51. V. Vernadsky, "The Biosphere and the Noosphere," *American Scientist* 33 (1945):1–12

52. J. D. Barrow and F. J. Tipler, *The Anthropic Cosmological Principle* (Oxford University Press, 1986).

CHAPTER SIX

1. A. Conan Doyle, *The Lost World* (Hodder & Stoughton, 1912).

2. R. D. Batory and W. A. S. Sarjeant, "Sussex *Iguanodon* Footprints and the Writing of *The Lost World*," in D. D. Gillette and M. G. Lockley, eds., *Dinosaur Tracks and Traces* (Cambridge University Press, 1989).

3. W. A. S. Sarjeant et al., "The Footprint of *Iguanodon*: A History and Taxonomic Study," *Ichnos* 6 (1998): 183–202.

4. Ibid; V. F. Santos et al., "The Longest Dinosaur Trackway in the World? Interpretations of Cretaceous Footprints from Carenque, near Lisboa, Portugal," *Gaia* (Museum of Natural History, Lisbon) 5 (1993):18–27.

5. G. J. Retallack, "Dinosaurs and Dirt," *Dinofest International* (Philadelphia Academy of Science, 1997): 345–359.

6. M. G. Lockley, *Tracking Dinosaurs: A New Look at an Ancient World* (Cambridge University Press, 1991).

7. J. L. Wright, Ph.D. "Fossil Terrestial Trackways: Function, Paphony, and Palaoecological Significance." Department of Geology, Bristol University, 1996.

8. J. B. Delair, "Notes on Purbeck Fossils, with Descriptions of Two Hitherto Unknown Forms from Dorset," *Proceedings of the Dorset Natural History and Archeological Society* 84 (1963):92–100; J. D. Wright et al., "Pterosaur Tracks from the Purbeck Limestone Formation of Dorset, England," *Proceedings of the Geologists Association* 108 (1997):39–48.

9. C. M. Sternberg, "Dinosaur Tracks from the Peace River, British Columbia," *Annual Report (1930) of the National Museum of Canada*, pp. 59–85.

10. A. F. de Lapparent, "Footprints of Dinosaurs in the Lower Cretaceous of Vestspitsbergen-Svalbard," *Arbok* (Norwegian Polar Institute) 1960 (1962):14–21.

11. K. Carpenter, "Skeletal Reconstruction and Life Restoration of *Sauropelta* (Ankylosauria: Nodusauridae) from the Cretaceous of North America," *Canadian Journal of Earth Sciences* 21 (1984):1491–1498.

12. P. J. Currie and W. A. S. Sarjeant, "Lower Cretaceous Dinosaur Footprints from the Peace River Canyon, British Columbia, Canada," *Palaeogeography, Palaeoclimatology, Palaeoecology* 28 (1979):103–115.

13. L. Psihoyos and J. Knoebber, *Hunting Dinosaurs* (Cassell, 1995).

14. P. J. Currie, "Dinosaur Footprints of Western Canada," in Gillette and Lockley, *Dinosaur Tracks and Traces,* (1989): 293–300.

15. R. McCrea and P. Currie, "A Preliminary Report on Dinosaur Tracksites in the Lower Cretaceous (Albian) Gates Formation near Grande Cache, Alberta," *New Mexico Museum of Natural History and Science Bulletin* 14 (1998):155–162.

16. P. J. Currie, "Bird footprints from the Gething Formation (Aptian, Lower Cretaceous) of Northeastern British Columbia, Canada," *Journal of Vertebrate Paleontology* 1 (1981):257–264

17. M. G. Mehl, "Additions to the Vertebrate Record of the Dakota Sandstone," *American Journal of Science* 21 (1931):441–452; B. K. Kim, "A Study of Several Sole Marks in the Haman Formation," *Journal of the Geological Society of Korea* 5 (1969):243–258..

18. M. G. Lockley, et al. "The Track Record of Mesozoic Birds: Evidence and Implications," *Philosophical Transactions of the Royal Society of London* 336 (1992):113–134; C. F. Vidarte Fuentes, "Primeras Huellas de aves en el Weald de Soria (Espana) Nuevo icnogenero et nueva icnoespecie *Archeornithipus meijidei*," *Estudios Geologicos* 52 (1996):63–75.

19. M.G. Lockley, et al. "The Track Record of Mesozoic Birds: Evidence and Implications," *Philosophical Transactions of the Royal Society of London* 336 (1992): 113–134.

20. S. K. Lim et al., "Preliminary Report on Sauropod Tracksites from the Cretaceous of Korea," *Gaia* (Museum of Natural History, Lisbon) 10 (1994):109–117.

21. S-Y. Yoon and J-S. Soh, "Traces of Time Past: Footprints of Dinosaurs and Primitive Birds," *Seoul* (January 1991): 6–11.

22. M. G. Lockley et al., "The Distribution of Sauropod Tracks and Trackmakers," *Gaia* (Museum of Natural History, Lisbon) 10 (1994):233–248.

23. D. B. Weishampel et al., *The Dinosauria* (University of California Press, 1990).

24. V. Morrell, "Announcing the Birth of a Heresy," *Discover* (August 1987): 26–50.

25. D. Weishampel, et al. *The Dinosauria.*

26. M. G. Lockley, "Dinosaur Ontogeny and Population Structure: Interpretations and Speculations Based on Footprints," in K. Carpenter et al., eds., *Dinosaur Eggs and Babies* (Cambridge University Press, 1994), pp. 347–365.

27. R. T. Bird, "Did Brontosaurus Ever Walk on Land? *Natural History* 53 (1944):61–67.

28. J. O. Farlow and M. G. Lockley, "Roland T. Bird, Dinosaur Tracker: An Appreciation," in Gillette and Lockley, eds., *Dinosaur Tracks and Traces,* pp. 33–36.

29. Bird, "Did Brontosaurus Ever Walk on Land?"

30. R. T. Bakker, "The Ecology of Brontosaurs," *Nature* 229 (1971):172–174.

31. M. G. Lockley and A. Rice, "Did *Brontosaurus* Ever Swim Out to Sea?" *Ichnos* 1 (1990):81–90.

32. J. G. Pittman, "Dinosaur Tracks and Track Beds in the Middle Part of the Glen Rose Formation, Western Gulf Basin, USA," *Geological Society of America Field Trip Guide* 8 (1990):47–53.

33. M. G. Lockley et al., "On the Common Occurrence of Manus-Dominated Sauropod Trackways in Mesozoic Carbonates," *Gaia* (Museum of Natural History, Lisbon) 10 (1994):119–124.

34. H. Haubold, "Ichnotaxonomie und Klassifikation von Tetrapodenfährten aus dem Permisch—Halleschen" (Ichnotaxonomy and Classification of Tetrapod Tracks from the Permian in Halle), *Jahrbuch der Geowissenschaft* (Yearbook of Geology)(Halle, Germany), Series B, 18 (1996):28–86; ibid., 1997.

35. Bird, "Did Brontosaurus Ever Walk on Land?"

36. R. T. Bakker, "The Superiority of Dinosaurs," *Discover* (March 1968): 11–22. See M. G. Lockley, *Tracking Dinosaurs: A New Look at an Ancient World* (Cambridge University Press, 1991), for a dissenting opinion.

37. E. A. R Ennion and N.Tinbergen, *Tracks* (Oxford University Press, 1967).

38. J. Ostrom, "Social and Unsocial Behavior in Dinosaurs," *Field Museum of Natural History* (Chicago) *Bulletin* 55 (1985):10–21; M. G. Lockley, "Dinosaur Trackways," in S. J. Czerkas and E. C. Olsen, eds., *Dinosaurs Past and Present* (Los Angeles County Museum, 1987), pp. 80–95.

39. M. G. Lockley, "Track Records," *Natural History* 104 (1995):46–51.

40. M. G. Lockley et al., "Theropod Tracks from the Morrison Formation, Howe Quarry, Wyoming," *Modern Geology* 23 (1998):309–316.

41. S. Czerkas, "Discovery of Dermal Spines Reveals a New Look for Sauropod Dinosaurs," *Geology* 20 (1992):1068–1070.

42. Farlow and Lockley, "Roland T. Bird, Dinosaur Tracker: An Appreciation."

43. R. T. Bird, *Bones for Barnum Brown: Adventures of a Dinosaur Hunter* (Texas Christian University Press, 1985).

44. Ibid., "Thunder in His Footsteps," *Natural History* 43 (1930):254–261

45. Ibid., "A Dinosaur Walks into the Museum," *Natural History* 47 (1941):74–81.

46. Ibid., "Did Brontosaurus Ever Walk on Land?"

47. Ibid., "We Captured a Live Brontosaur," *National Geographic Magazine* 105 (1954):707–722.

48. Bird, *Bones for Barnum Brown.*

49. Ibid.

50. J. O. Farlow, *A Guide to Lower Cretaceous Dinosaur Footprints and Tracksites of the Paluxy River Valley, Somervell County, Texas, South Central Section* (Geological Society of America, 1987).

51. Bird, *Bones for Barnum Brown.*

52. J. O. Farlow, *The Dinosaurs of Dinosaur Valley State Park* (Texas Parks and Wildlife Press, 1993).

53. D. G. Hurd et al., *General Science: A Voyage of Discovery* (Prentice Hall, 1989).

54. Bird, *Bones for Barnum Brown.*

55. D. A. Thomas, "Bird Was Right: The Most Peculiar Trackway," *Journal of Vertebrate Paleontology* 15 (1995):56A.

56. D. Thomas and J. O. Farlow, "Tracking a Dinosaur Attack," *Scientific American*, no. 277 (1997):74–79.

57. T. M. Berra, *Evolution and the Myth of Creation* (Stanford University Press, 1990).

58. Bird, *Bones for Barnum Brown.*

59. Ibid.

60. Ibid.

61. Ibid.

62. Gillette and Lockley, *Dinosaur Tracks and Traces;* J. D. Morris, *Tracking Those Incredible Dinosaurs and the People Who Knew Them* (Creation Life publishers, 1980); S. Taylor, *Footprints in Stone,* Films for Christ Association, 1973.

63. G. Kuban, "Elongate Dinosaur Tracks," in Gillette and Lockley, *Dinosaur Tracks and Traces,* pp. 57–72; D. H. Milne and S. D. Schafersman, "Dinosaur Tracks, Erosion Marks and Midnight Chisel Work (but No Human Footprints) in Cretaceous Limestones of the Paluxy River Bed, Texas," *Journal of Geoscience Education* 32 (1983):111–123.

64. L. R. Brand, *Faith, Reason and Earth History* (Andrews University Press, 1998).

65. Teilhard de Chardin, *Building the Earth* (Dimension Books, 1965).

66. V. Vernadsky, "The Biosphere and the Noosphere," *American Scientist* 33 (1945):1–12.

67. William Irwin Thompson, *Time Falling Bodies Take to Light* (St. Martin's Press, 1981).

68. Isaac Newton quoted in G. Staguhn, *God's Laughter: Physics, Religion and the Cosmos* (Kodansha International, 1992).

69. Ibid.

70. H. von Ditfurth, 1982, *The Origins of Life: Evolution as Creation*, Harper & Row, 1982).

CHAPTER SEVEN

1. D. B. Weishampel et al., *The Dinosauria* (University of California Press, 1990).

2. G. Leonardi, "Le impromte fossili di dinosauri," *Sulle orme dei dinosauri* (Errizo Editrice, Venice, 1980).

3. M. G. Lockley and L. Marquardt, *A Field Guide to Dinosaur Ridge* (Friends of Dinosaur Ridge/ University of Colorado–Denver, 1995); M. G. Lockley and A. P. Hunt, *Fossil Footprints of the Dinosaur Ridge Area* (Friends of Dinosaur Ridge/University of Colorado at Denver, 1994).

4. M. G. Lockley et al., "The Dinosaur Freeway: A Preliminary Report on the Cretaceous Megatracksite, Dakota Group, Rocky Mountain Front Range and Highplains; Colorado, Oklahoma and New Mexico," in R. Flores, ed., *Mesozoic of the Western Interior* (1992), pp. 39–54.

5. M. Wade, "The Stance of Dinosaurs and the Cossack Dancer Syndrome," in D. D. Gillette and M. G. Lockley, eds., *Dinosaur Tracks and Traces* (Cambridge University Press, 1989), pp. 73–82.

6. D. Norman, "On the Anatomy of *Iguanodon atherfieldensis* (Ornithischia: Ornithopoda)," *Bulletin de l'Institut Royal des Sciences Naturelles de Belgique* (*Bulletin of the Royal Institute of Sciences of Belgium*) 56 (1986):281–372; Leonardi, *Le impromte fossili di dinosauri*; P. J. Currie, "Dinosaur Footprints of Western Canada," in Gillette and Lockley, *Dinosaur Tracks and Traces*, pp. 293–300; M. G. Lockley, "Dinosaur Footprints from the Dakota Group of Eastern Colorado," *Mountain Geologist* 24 (1987):107–122.

7. M. G. Lockley and J. G. Pittman, "The Megatracksite Phenomenon: Implications for Paleoecology, Evolution and Stratigraphy," *Journal of Vertebrate Paleontology* 9 (1989):30A.

8. Lockley et al., "The Dinosaur Freeway."

9. J. Dellinger, *Dinosaur Tracks and Murder* (North West Publications, 1995).

10. J. Ostrom and J. McIntosh, *Marsh's Dinosaurs* (Yale University Press, 1966).

11. D. D. Gillette and D. Thomas, "Problematic Tracks of late Albian (Early Cretaceous) Age, Clayton Lake State Park," in Gillette and Lockley, *Dinosaur Tracks and Traces*, pp. 337–342.

12. D. Unwin, Tracking the dinosaurs. *Geology Today* (1986): 168–169.

13. J. A. McAllister, "Dakota Formation Tracks from Kansas: Implications for the Recognition of Subaqueous Traces," in Gillette and Lockley, *Dinosaur Tracks and Traces*.

14. S. C. Bennett "Reinterpretation of Problematic Tracks at Clayton Lake State Park: Not One Pterosaur but Several Crocodiles," *Ichnos* 2 (1992):37–42.

15. M. G. Lockley et al., "Dinosaur Tracks and Radial Cracks: Unusual Footprint Features," *Bulletin of the Natural Science Museum* (Series C) 15 (1989):151–160.

16. V. F. Santos et al., "The Longest Dinosaur Trackway in the World? Interpretations of Cretaceous Footprints from Carenque, near Lisboa, Portugal," *Gaia* (Museum of Natural History, Lisbon) 5 (1993):18–27.

17. A. M. Galopim de Carvalho, *Dinosáurios e a Batalha de Carenque* (Dinosaurs and the Battle of Carenque) (Editorial Noticias [Lisbon]), 1994).

18. R. A. Thulborn, "Ornithopod Dinosaur Tracks from the Lower Jurassic of Queensland," *Alcheringia* 18 (1994):247–258.

19. R. A. Thulborn and M. Wade, "Dinosaur Trackways in the Winton Formation (Mid-Cretaceous) of Queensland," *Memoirs of the Queensland Museum* 21 (1984):413–517.

20. G. Paul, *Predatory Dinosaurs of the World* (Simon and Schuster, 1988)

21. N. Agnew et al., "Strategies and Techniques for the Preservation of Fossil Tracksites," in Gillette and Lockley, *Dinosaur Tracks and Traces,* (1989): 397–407.

22. R. S. Lull, *Hadrosaursian Dinosaurs of North America* (Geological Society of America, Special Papers, no. 40, 1942).

23. S-Y. Yang et al., "Flamingo and Duck-like Bird Tracks from the Late Cretaceous and Early Tertiary: Evidence and Implications," *Ichnos* 4 (1994):21–34.

24. M. G. Lockley, "The Track Record of Mesozoic Birds: Evidence and Implications," *Philosophical Transactions of the Royal Society of London* 336 (1992):113–134.

25. M. Nash, "The Age of Pterosaurs," *Time* (October 28, 1996): 84–85.

26. M. G. Lockley, et al., "First Reports of Pterosaur Tracks from Asia, Chullanam Provice, Korea," Paleontological Society of Korea, special publication no. 2 (1997):17–32.

27. R. T. Bakker, "The Superiority of Dinosaurs," *Discover* (March 1968): 11–22.

28. Ibid., "The Dinosaur Renaissance," *Scientific American* 232 (1975):58–72.

29. G. Paul, *Predatory Dinosaurs of the World* (Simon & Schuster, 1988).

30. M. G. Lockley, "The Dinosaur Footprint Renaissance," *Modern Geology* 16 (1991):139–160.

31. W. P. Coombs, "Theoretical Aspects of Cursorial Adaptations in Dinosaurs," *Quarterly Review of Biology* 53 (1978):393–418.

32. R. McN. Alexander, "Estimates of Speeds of Dinosaur," *Nature* 261 (1976):129–130.

33. Barnum Brown, "The Mystery Dinosaur," *Natural History* 41 (1938):190–202, 235.

34. W. Peterson, "Dinosaur Tracks in the Roofs of Coal Mines," *Natural History* 24 (1924):388–391.

35. A. Look. In My Back Yard. Denver Univ. Press (1951): 316.

36. Lockley, *Tracking Dinosaurs;* D. A. Russell and P. Beland, "Running Dinosaurs," *Nature* 264 (1976):486.

37. R. A. Thulborn, "Estimated Speed of a Giant Bipedal Dinosaur'" *Nature* 292 (1981):273–274; M. G. Lockley et al., "Hadrosaur Locomotion and Herding

Behavior: Evidence from Footprints from the Mesa Verde Formation, Grand Mesa Coalfield, Colorado," *Mountain Geologist* 20 (1983):5–13.

38. R.T. Bakker, "The Return of the Dancing Dinosaur," in S. Czerkas and E.C. Olsen, eds., Dinosaurs Past and Present (Washington University Press, 1987), vol. 1, pp. 38–69.

39. A. Thulborn, *Dinosaur Tracks* (Chapman Hall, 1990). Lockley, *Tracking Dinosaurs.*

40. R. A. Thulborn, "The Demise of the Dancing Dinosaur" *The Beagle: Records of the Northern Territory Museum of Arts and Sciences* 9 (1992):29–34.

41. R. McN. Alexander, *Scientific American* 264 (1991).

42. Peterson, "Dinosaur Tracks in the Roofs of Coal Mines."

43. M. G. Lockley and A. P. Hunt, "A Track of the Giant Theropod Dinosaur *Tyrannosaurus* from Close to the Cretaceous/Tertiary Boundary, Northern New Mexico," *Ichnos* 3 (1994):213–218.

44. J. O. Farlow et al., 1995 "Body Mass, Bone Strength Indicator, and Cursorial Potential of *Tyranosaurus rex*," *Journal of Vertebrate Paleontology* 15 (1995):713–735.

45. *Discover*, April 1996.

46. C. M. Sternberg, "Dinosaur Tracks from the Peace River, British Columbia," *Annual Report (1930) of the National Museum of Canada* (1932), pp. 59–85.

47. Robert Bakker, *The Dinosaur Heresies* (William Morrow, 1986); K. Carpenter, "Skeletal Reconstruction and Life Restoration of *Sauropelta* (Ankylosauria: Nodusauridae) from the Cretaceous of North America," *Canadian Journal of Earth Sciences* 21 (1984):1491–1498

48. M. G. Lockley and A. P. Hunt, "Ceratopsid Tracks and Associated Ichnofauna from the Laramie Formation (Upper Cretaceous: Maastrichtian) of Colorado," *Journal of Vertebrate Paleontology*, 15(1995):592–614.

49. P. Dodson and J. O. Farlow, "Posture of Ceratopsids—A Solution at Last, or Is It?" Dinofest International Symposium, Arizona State University, p. 45.

50. M. G. Lockley, "Dinosaur Trackways," in D. E. G. Briggs and P. Crowther, eds., *Paleobiology: A Synthesis* (in press). D. Nani, Symbolic dynamics in the evolution of vertebrates: some general considerations concerning the dynamics of form in vertebrates in the light of comparative anatomy and teratology. Magazine of morphology (Bologna 1, 1998): 75–105.

51. L. W. Alvarez et al., "Extraterrestrial Cause for the Cretaceous-Tertiary Extinctions," *Science* 208 (1980):1095–1108.

52. A. Hildebrand et al., "Chicxulub Crater: A Possible Cretaceous Tertiary Boundary Impact Crater on the Yucatan Penninsula," *Geology* 19 (1991): 867–869.

53. J. Horner and D. Lessem, *The complete* T. rex (Simon & Schuster, 1993).

54. Ibid.

55. Lockley, *Tracking Dinosaurs.*

56. M. Lockley and A. Hunt, "Dinosaur Tracks and other fossil footprints of the Western United States."

CHAPTER EIGHT

1. Bjorn Kurten, *The Age of Mammals* (Columbia University Press, 1971).

2. M. G. Lockley, and A. P. Hunt, "Dinosaur Tracks and other fossil footprints of the Western United States."

3. G. Leonardi, *Annotated Atlas of South America Tetrapod Footprints (Devonian to Holocene)* (Company of Mineral Resources [Brasilia], 1994).

4. C. Noblet et al., "Nouvelle découverte d'empreintes laissées par des dinosaures dans la Formation des Couches Rouges (bassin de Cuzco-Sicuani, Sud du Perou): conséquences stratigraphiques et tectoniques," *Couche Rouge Académie des Sciences* 320, (series 2) 320 (1995):785–791.

5. M. G. Lockley, B. Ritts, and G. Leonardi, "Mammal Track Assemblages from the Early Tertiary of China, Peru, and North America," *Palaois* (in press).

6. R. G. Chafee, "Mammal Footprints from the White River Oligocene," *Naturae Naturae* (Philadelphia Academy of Natural Sciences) 116 (1943):1–13.

7. A. Feduccia, "Presbyornis and the Evolution of Ducks and Flamingos," *American Scientist* 66 (1978):1298–1304.

8. R. Erickson, "Fossil Bird Tracks from Utah," *Observer* (Science Museum of Minnesota) 5 (1967):140–146. S-Y. Yang et al., "Flamingo and Duck-like Bird Tracks from the Late Cretaceous and Early Tertiary: Evidence and Implications," *Ichnos* 4 (1994):21–34.

9. M. P. Kahl, "East Africa's Majestic Flamingos," *National Geographic* 137 (1970):179–294.

10. Yang S-Y et al. "Flammingo and duck-like bird tracks from the Late Cretaceous and Tertiary; Evidence and Implications." *Ichnos* 4 (1994): 21–34.

11. S. F. Robison, *Bird and Frog Tracks from the Late Cretaceous Blackhawk Formation in East Central Utah*, Utah Geological Association Publications 19 (1991):325–334.

12. N. J. Sanders, *Animal Spirits* (Little, Brown, 1995).

13. The measurement is taken by marking where star constellations are situated relative to sunrise at the equinoxes. It takes 25,920 years for the background constellations to make a complete cycle from their position at an equinox to the same position again.

14. William Irwin Thompson, *Time Falling Bodies Take to Light* (St. Martin's Press, 1981).

15. N. J. Sanders, *Animal Spirits* (Little, Brown, 1995).

16. Thompson, *Time Falling Bodies Take to Light*.

17. D. Webb, "Locomotor Evolution in Camels," *Forma et Functio* 5 (1972):99–112.

18. J-C. Armen, *Gazelle-boy* (The Bodley Head, 1974).

19. E. Renders, "The Gait of *Hipparion* Spp. from Fossil Footprints in Laetoli, Tanzania," *Nature* 308 (1984):179–181.

20. Ibid.

21. Don Johanson and Mait Edey, *Lucy: The Beginnings of Mankind* (Simon & Schuster, 1981).

22. R. T. Tuttle, "The Pitted Pattern of Laetoli Feet," *Natural History* (1990):60–65; M. D. Leakey and R. L. Hay, "Pliocene Footprints in the Laetoli Beds at Laetoli, Northern Tanzania," *Nature* 278 (1979):317–323; R. L. Hay and M. D. Leakey, "The Fossil Footprints of Laetoli," *Scientific American* 246 (1982):50–57.

23. Misia Landau has become famous in anthropological circles for pointing out the heroic, storytelling flavor of this version of hominid history. See R. Lewin, *Bones of Contention* (University Chicago Press, 1987).

24. R. Lewin, *In the Age of Mankind* (Smithsonian Books, 1988).

25. M. C. Leakey and J. M. Harris, *Laetoli: A Pliocene Site in Northern Tanzania* (Clarendon Press, 1987).

26. T. D. White and G. Suwa, "Hominid Footprints at Laetoli: Facts and Interpretations," *American Journal of Physical Anthropology* 72 (1987):485–514.

27. R. Lewin, *Bones of Contention* (University Chicago Press, 1987).

28. J. T. Stern and R. L. Susman, "The Locomotor Anatomy of *Australopithecus afarensis*," *American Journal of Physical Anthropology* 60 (1983):179–317.

29. R. H. Tuttle, "Evolution of Hominid Bipedalism and Prehensile Capabilities," *Philosophical Transactions of the Royal Society of London* (series B) 292 (1981):89–94.

30. White and Suwa, "Hominid Footprints at Laetoli."

31. R. T. Tuttle et al., "Further Progress on the Laetoli Trials," *Journal of Archaeological Science* 17 (1990):347–362.

32. Scientific convention frowns on proposing new scientific names for specimens in journals that are not peer reviewed, so I stress that this is only a suggestion.

33. R. Lewin, *In the Age of Mankind* (Smithsonian Books, 1988).

34. M. Demas et al., *Laetoli Project: Conservation of the Hominid Trackway Site at Laetoli, Tanzania: Report on the 1995 Field Season* (Getty Conservation Institute, 1995).

35. Ibid.

36. Ibid.

CHAPTER NINE

1. A. Guiterman, in H. Wendt, *Before the Deluge* (Victor Gollancz, 1968).

2. Wendt, *Before the Deluge.*

3. Ibid.

4. J-M. Chauvet et al., *Dawn of Art—The Chauvet Cave: The Oldest Known Paintings in the World* (Harry N. Abrams, 1996).

5. A. Lister and P. Bahn, *Mammoths* (Macmillan, 1994). Brady and P. Seff, "Elephant Hill," *Plateau Journal* 31 (1959):81–82.

6. L. D. Agenbroad, "Hot Springs, South Dakota: Entrapment and Taphonomy of Columbian Mammoth," in P. S. Martin and R. Klein, eds., *Quaternary*

Extinctions: A Prehistoric Revolution (University of Arizona Press, 1984), pp. 113–127.

7. P. J. Scrivner and D. J. Bottjer, "Neogene Avian and Mammalian Tracks from Death Valley National Monument, California: Their Context and Classification," *Palaeogeography, Palaeoclimatology, Palaeoecology* 57 (1986):285–331.

8. Jordan D. Marche, "Extraordinary Petrifactions: The Fossil Footprints at Nevada State Prison," *Terra* (Los Angeles County Museum) 24 (1986):12–18.

9. E. D. Cope, *American Naturalist,* 1883 (cited in Marche, "Extraordinary Petrifactions."

10. O. C. Marsh, "On the Supposed Human Footprints Recently found in Nevada," *American Journal of Science* (1883):159–160.

11. Twain, "The Carson Fossil Footprints," *The San Franciscan* (February 16, 1884).

12. Marche, "Extraordinary Petrifactions."

13. Chauvet et al., *Dawn of Art—The Chauvet Cave.*

14. N. Casteret, *Illustrated London News,* 1948.

15. R. W. Graham et al., "Tracking Ice Age Felids: Identification of Tracks of *Panthera atrox* from a Cave in Southern Missouri, U.S.A.," in K. M. Stewart and K. L. Seymour, eds., *Palaeoecology and Palaeonenvironments of Late Cenozoic Mammals* (University of Toronto Press, 1996), pp. 331–345.

16. W. von Koenigswald et al., "Jungpleistozane Tierfährten aus der Emscher-Niederterrasse von Bottrop-Welheim," *Münchner Geowissenschaftliche Abhandlungen* 27 (1995):5–50.

17. O. C. Vialov, *The Traces of Vital Activity of Organisms and Their Paleontological Significance* (Akad. Nauk. Ukrainian S.S.R., 1966).

18. Agenbroad, "Hot Springs, South Dakota."

19. B. Huevelmans, *On the Track of Unknown Animals* (Hill & Wang, 1965; reprint, MIT Press, 1972).

20. R. H. Tedford, "The Diptrotodons of Lake Callabonna," *Australian Natural History* 17 (1973):349–354.

21. Huevelmans, *On the Track of Unknown Animals.*

22. See W. Ley, *Willy Ley's Exotic Zoology* (Viking Press, 1966), Chapter 3, for an account of Mantell's paper.

23. Ibid.

24. Stephen Jay Gould, "The Dodo in the Caucas Race," *Natural History,* 1996, pp. 22–32.

25. P. V. Rich et al., *Vertebrate Paleontology of Australia* (Monash University Publishing, 1991).

26. B. Brewster, *Te Moa: The Life and Death of New Zealand's Unique Bird* (Nikau Press, 1987).

27. A. McKenzie, *Pioneers of Martin's Bay,* 3d rev. ed., Whitcombe and Tombs (Publishers) cited in B. Brewster.

28. W. L. Williams, "On the Occurrence of Footprints of a Large Bird, Found at Turanganui, Poverty Bay," *Transactions of the New Zealand Institute* 4 (1872):124–127.

29. K. Wilson, "Footprints of the Moa," *Transactions of the New Zealand Institute* 45 (1913):211–212.

30. J. Napier, *Bigfoot: The Yeti and Sasquatch in Myth and Reality* (E. P. Dutton, 1972).

31. R. Erdoes and A. Ortiz, *American Indian Myths and Legends* (Pantheon Books, 1984).

32. Russell Ciochon, John Olsen, and Jamie James, *Other Origins: The Search for the Giant Ape in Human Prehistory* (Bantam Books, 1990).

33. C. C. Swisher et al., "Latest *Homo erectus* of Java: Potential Contemporaneity with *Homo sapiens* in Southeast Asia," *Science* 274 (1996):1870–1874.

34. Ivan T. Sanderson, *Abominable Snowmen: Legend Come to Life* (Chilton Company, 1961).

35. Ibid.; see also M. Shackley, *Still Living* (Thames and Hudson, 1983).

36. Sanderson, *Abominable Snowmen: Legend Come to Life.*

37. Ibid.

38. Robert M. Pyle, *Where Bigfoot Walks: Crossing the Dark Divide* (Houghton Mifflin, 1995).

39. Sanderson, *Abominable Snowmen.*

40. Ciochon, Olsen, and James, *Other Origins.*

41. G. Krantz, *Big Footprints: A Scientific Inquiry into the Reality of Sasquatch* (Johnson Books, 1992).

42. M. R. Dennett, "Bigfoot Evidence: Are These Tracks Real?" *Skeptical Inquirer* 18 (1994):498–508.

43. Ciochon, Olsen, and James, *Other Origins.*

44. G. Krantz, "A Species Named from Footprints," *Northwest Anthropological Research Notes* 19 (1986):93–99.

45. Ciochon, Olsen, and James, *Other Origins.*

46. W. Schad, 1992. The hetorochronic mode of evolution in the vertebrate classes and the hominids. PhD dissertation, University of Witten, Herdecke. 202 p. + 113 Figs.

47. Hermann Poppelbaum, *A New Zoology* (Philosophic Anthroposophic Press [Dornach, Switzerland], 1961).

48. Rupert Sheldrake, *The Presence of the Past* (Fontana, 1988).

CHAPTER TEN

1. William Irwin Thompson, *Time Falling Bodies Take to Light* (St. Martin's Press, 1981).

2. William Bryant Logan, *Dirt: The Ecstatic Skin of the Earth* (Riverhead Books, 1995).

3. Ibid.

4. Ibid.

5. S. Coren, "Are Fingerprints a Genetic Marker for Handedness?" *Behavior Genetics* 24 (1994):141–147.

6. C. F. Johnson and E. Opitz, "Unusual Palm Creases and Unusual Children," *Clinical Pediatrics* 12 (1973):101–112.

7. B. Schaumann and M. Alter, *Dermatoglyphics in Medical Disorders* (Springer Verlag, 1976).

8. F. Gettings, *The Book of Palmistry* (Triune Books, 1974).

9. P. White, ed., *Crime Scene to Court: The Essentials of Forensic Science* (Royal Society of Chemistry, 1998).

10. Ibid.

11. W. J. Bodziak, *Footwear Impression Evidence* (CRC Press, 1995).

12. J. Barbour, *Footprints on the Moon* (American Book/Stratford Press, 1969).

13. J-C. Armen, *Gazelle-boy* (The Bodley Head, 1974).

14. A. Chaikin, *A Man on the Moon: The Voyages of the Apollo Astronauts* (Viking Press, 1994).

15. T. A. Match and K. L. Jones, *The Martian Landscape* (NASA Scientific and technical Information Office, 1978).

16. Roger Rosenblatt, "Visit to a Smaller Planet," *Time* (July 4, 1997): 37.

17. D. S. Mckay, "Search for Past Life on Mars: Possible Relic Biogenic Activity in Martian Meteorite ALH84001," *Science* 273 (1996):924–930.

18. Bruce Holbrook, *The Stone Monkey* (William Morrow, 1981).

19. M. G. Lockley et al., 1996. "Dinosaur Tracks and Other Archosaur Footprints from the Triassic of South Wales," *Ichnos* 5 (1996):23–41.

20. Norman Friedman, *Bridging Science and Spirit* (World Bridge Group, 1997).

21. E. Schrodinger, *What Is Life?* (Cambridge University Press, 1944).

22. R. Bucke, *Cosmic Consciousness* (1901; reprint, E. P. Dutton, 1969); Ken Wilber, *Sex, Ecology and Spirituality: The Spirit of Evolution* (Shambhala, 1995); Anna Lemkow, *The Wholeness Principle* (Quest Books, 1989).

23. G. Murchie, *The Seven Mysteries of Life* (Houghton Mifflin, 1978).

24. Logan, *Dirt: The Ecstatic Skin of the Earth.*

INDEX

Abominable Snowman, 262–267
Aldrin, Edwin "Buzz," 284, 286
Alexander, McNeill, 209–210
Alvarez, Louis, 222
Alvarez, Walter, 222
Ameghino, Carlos, 122
Amphibians
   Carboniferous period, 59–61
   Devonian period, 54–56
Anatomy
   feet, characteristics of, 137–138
   senses, distribution of, 33–35
   See also Biological relationships of
   tracks and organisms
Ankylosaurs, 163–165, 214
Anton, Mauricio, 255
Aquinas, Saint Thomas,
   189
Archosaurs, 79–80
   ankle structure and locomotion, 91–93
   bipedalism, 93–94
   crocodiles, 151–153
   petrified hands, mystery of, 82–87
Aristotle, 183, 252
Armen, Jean-Claude, 235
Armstrong, Neil, 27, 284
Arthropods
   Carboniferous period, 57–58
   Jurassic period, 154–155
   Paleozoic Era, 43–44
   Permian period, 75–76
   Precambrian and Cambrian, 51–55

Auel, Jean, 254
Australia
   conservation of Lark Quarry, 203–204
   dinosaur stampede in, 201–203
   giant wombats in the Pleistocene epoch,
   256–257
Australopithecus, 239–240,
   242–245

Baird, Donald, 30, 97, 102
Bakker, Robert
   brontosaurs, 171, 174
   champion of dynamic dinosaurs, 207,
   209, 211
   dinosaurs footprints as flowerbeds, 151
   foot-soil relationship, 137–138
   megalosaurs in Wyoming, 130
   sauropod birth, 169
   Stegosaurus in Utah, 147
   Triceratops' forelimbs, 216
Barrow, John, 157
Batory, Dana, 160
Bean, Alan, 284
Bennett, Chris, 196
Benton, Michael, 101–103
Big Bang, 1–2, 4–5, 186
Big Blue Banana lake, 111
Bigfoot, 262–267
Biological relationships of tracks and
   organisms, 49–50
   feet, variation in, 137–138
   in hominids, 267–269

hoof-horn correspondence in
ungulates, 38–39
mathematical relationships, 106–108
narrowness/width polarity, 23, 113–115,
119–120, 130–132, 166
sexual dimorphism, 40–42, 87–89
shifting animal forms, 71–74
track to body, 217–219
trackways and footedness, 125–126
Biosphere, 4
emergence of, 46–47
evolution of, 6, 186
holistic view of organisms. *See* Holism
Bipedalism, 25–25, 93–94, 239–240
Bird, Roland
brontosaurs, 148, 170–171, 204
carnivore attack, 177–180
evolution/creation debate, 159–160,
183–184
mystery dinosaur, 208
social behavior of dinosaurs, 174
Birds
early webbed-footed, 204–205
Mesozoic Era, 165–167
moas, 258–262
Presbyornis, 230
Tertiary period, 226
Black, Davison, 267
Bohm, David, 2–3
Bolivia, Cal Orko tracksite, 212–215
Bortoft, Henry, 32
Bourne, Geoffrey, 267
Brady, L. F., 86
Brand, Leonard, 68–71, 185
Brontosaurs
attacked in Texas, 177–181
baby, 169–170
in Bolivia, 214
decline of, 162
long trackways of, 126–127
social behavior of, 145, 149, 174, 176
as swimming sauropod, 170–172
trampling by, 150–151
Broom, Robert, 64
Brown, Barnum, 148, 208–209
Bruce, Robert, 75
Bryan, William Jennings, 182
Bucke, Richard, 296

Buckland, Rev. William, 30, 67–68, 71, 77,
83, 109, 130
Burnshaw, Stanley, 53
Bushmen
animal's nature, reading of, 50
consciousness and spirit of, 3, 279–280
end of the, 95
meaning for, 181–182
rain, reading of, 45–46
as trackers, 9–14

Cambrian explosion, 51
Camels, 233, 235
Canada, ankylosaurs in British Columbia,
163–165
Carboniferous period, 56–57
amphibians in, 60–61
monster millipedes in, 57–58
vertebrate development and reptile
footprints, 59–60
Carenque, Battle of, 199–201
Carnivores, 35–38
and caves, 254–256
purported attack by in Texas, 176–180
Carpenter, Ken, 164
Casamiquela, Rodolfo, 30, 122, 161
Caster, Kenneth, 55
Cave paintings, 52, 250–252, 254, 275
Cenozoic Era, 225. *See also* Tertiary period
Centaur, 237
Cernan, Eugene, 284
Chaikin, Andrew, 286
Chatwin, Bruce, 14
China, lack of fossils in, 156–157
Chirotherium, 82–87, 102–103
Cinderella syndrome, 16, 19
Ciochon, Russell, 263–264, 266–267
Clack, Jenny, 55
Clay, 46–47, 278–279
Clayton Lake State Park, 197
Coelocanth, 54–55
Colorado, Gateway, 99–100
Conan Doyle, Sir Arthur, 11, 156, 159–161,
192
Connolly, Billy, 65
Conrad, Charles "Pete," 284
Consciousness
development of spatial, 277

and evolution, 6–8, 157, 186–188, 275–276, 295–299
    See also Noosphere
Convergence, 268–273
Cook, Captain James, 258, 260
Cope, Edward Drinker, 40, 146, 253
Coren, Stanley, 280
Crawford Prize, 47
Creationism
    and evolution, 182–185, 188–189
    and science, 68–71, 154
Cretaceous period, 159–160, 191
    ankylosaurs, 163–165
    birds in, 165–167
    end of, 221–222
    Iguanodon, 160–162, 191–192
    pterosaurs, 205–207
    Triceratops, 215–217
    Tyrannosaurus rex, 209–212
Crichton, Michael, 11
Crocodiles, 151–153, 196–197
Cruziana, 51–52
Currie, Philip, 164–165

Darrow, Clarence, 182
Darwin, Charles, 12–13, 22, 30, 109, 259
Davies, Paul, 109
Dawson, John William, 59–60
Demathieu, Georges, 30, 93, 101
Devonian period, amphibians in, 54–56
Diictodon, 65–66, 71
Dimetrodon, 61–64, 71–72, 74
Dimorphism, sexual, 40–42, 87–89
Dinosaur National Monument, 98, 128
Dinosauroid, 101
Dinosaur Ridge, 193–195, 197–198
Dinosaurs
    ankylosaurs, 163–165
    brontosaurs. *See* Brontosaurs
    carnivore attack in Texas, 177–181
    death of, 221–224
    diminutive, 120–122
    evolutionary cycle of, 219–220
    freeway of, 192–195, 198
    hopping, 124–125
    Iguanodon, 160–162
    longevity of species, 111
    megalosaurs, 130–132, 135–137

origins of, 101–102, 110
ornithopod, 143–145
Otozoum tracks, 115–118
pterosaurs, 138–141
social behavior of, 143–145, 149, 174, 176
speed of, 135–136, 207–211
stampede in Australia, 201–203
Stegosaurus, 146–147
thunder beings, symbols of, 146–147
trampling by, 150–151
Triceratops, 215–217
Tyrannosaurus rex, 209–212
Diprotodon, 257
Ditfurth, Hoimar von, 189
Dixon, Jeane, 183–184
DNA, 279–281
Dodson, Peter, 217
Duke, Charles, 284, 286–287
Duve, Christian de, 5, 46, 157
Dynamite, as tool of ichnology, 197–198

Earth
    deep structure of, 44–47
    desecration of with land mines, 290–292
    sacredness of, 292–295
Ediacaran biota, 48
Einstein, Albert, 13, 109
Eldredge, Niles, 100
Elite tracks, 19
Ellenberger, Paul, 30–31, 94–98, 100, 115, 117, 124
Ellenberger, Victor, 94–95
England
    Iguanodon in, 160–161
    Purbeck fossil beds, 162–163
Eocene epoch, birds and frogs in, 230–231
Eoraptor, 101
Eubrontes, 113–115
Evolution
    and consciousness, 3–4, 157, 276, 295–299
    convergence, 268–273
    and creationism, 68–71, 182–185, 188–189
    cycles of, 41–42, 219–220, 268–273
    from physiosphere to noosphere, 4–8

teleology and human consciousness,
186–188
tracks and shifting animal forms, 71–74
universe, creation of, 1–3

Farlow, James, 179, 181, 184, 203, 211, 217
Fibonacci, Leonardo, 106
Fibonacci series, 106–108
Fingerprints, 20–21, 23, 280–282
Fish, 44, 71–72
Fleming, Farley, 223
Footprints. *See* Tracks
Fossil Art, 47–52
Fossils
    attempted theft of, 141–143
    dating, 28
    as jigsaw puzzles, 102–103
    matching trackmakers and tracks,
      16–17
    Permian protomammal, 64–67
    Precambrian-to-Cambrian transition,
      47–52
    preservation of and the fossil record,
      90–91
    trace, 50–52
Frazer, Sir James George, 12
Friedman, Norman, 295
Friends of Dinosaur Ridge, 195
Frogs, 230–231
Fulguration, 100

Galopim de Carvalho, Antonio, 200–201
Galton, Peter, 97
Gambaryan, P. P., 124
Geological Society of America,
    193
Geological time scale, standard, 7
Geologists, and time, 27–29
George, Chief Dan, 146
Getty Conservation Institute, 246
Ghost prints, 171–174
Gigantopithecus, 263–265, 267
Gilette, David, 196
God
    and creation, 69–71, 188–189
    and meaning, 3, 181–182
    reverence for among scientists, 68, 109
    See also Creationism

Goethe, Johann Wolfgang von, 32, 35, 48,
72
Goodwin, Brian, 31
Gould, Stephen Jay, 100, 156, 259
Grallator cursorius, 113–115
Green River Formation, 230

Haast, Sir Julius von, 259
Hadrosaurs, 208–209
Hamon, Count Louis, 88–89, 281
Hands, mystery of petrified, 82–87
Hartmann, Hayman, 46, 278
Hatcher, John Bell, 152
Haubold, Hartmut, 30, 77, 87, 93,
173
Heisenberg, Werner Karl, 1
Herodotus, 151
Heston, Charlton, 184
Heterochrony, 40–41, 51
Hill, Andrew, 239, 241, 252
Hillerman, Tony, 142
Hipparon, 237–238
Hitchcock, Rev. Edward
    Anomoepus, 118–120
    bird tracks, interpretation of, 108–109,
      259
    Eubontes, 114, 147
    God, reverence for, 109
    Grallator, 112–113
    Hoplichnus equus, 154
    Otozoum, 97, 115–118
    vertebrate ichnology, founder of, 29–30,
      105
Ho, Mae-Wan, 10, 31–32, 37, 48–49, 70,
    107, 296
Holbrook, Bruce, 2
Holism
    and the earth, 162, 296–299
    movement and locomotion, 44
    organisms, view of, 9–10, 31–39, 48–50,
      280
    wisdom of, 210
    yin and yang, 40–41, 46, 52, 298
    See also Schadian perspective
Hominids
    bigfoot and other ancestral, 262–268
    binary orientation of, 23–25
    contemporary tracks of, 282–283

evolution of, 276–278
in the Pliocene epoch, 239–245
search for origins of, 155–156
tracks and skulls, 267–269
Horgan, John, 3
Horses
dawn horse, 228–229
gait, natural development of, 237–238
humans, close association with,
235–237
locomotion of, 26
Huene, Friedrich von, 85
Huevelmans, Bernard, 257
Humboldt, Alexander von, 83
Huxley, Julian, 4
Huxley, Thomas Henry, 68, 228–229
Hylonomus, 59, 71–72

Ichnologists, defined, 18
Ichnology, vertebrate, founding of, 29–30
Ichnotaxonomy, defined, 18
Igneous rocks, 45
Iguanodon, 160–162, 191–192
Invertebrates, 43–44, 57
Iridium band, 221–224
Irwin, James, 284

Jackson, Phyllis, 21
James, Jamie, 263, 266–267
Jennings, Chester, 99
Johanson, Donald, 239–240, 244
Jung, Carl, 2, 5
Jurassic period, 105, 129
arthropods, 154–155
brontosaurs. *See* Brontosaurs
crocodiles, 151–153
diminutive dinosaurs, 120–122
Eubrontes, 113–115
Grallator cursorius, 113–115
hopping dinosaurs, 124–125
megalosaurs, 130–132, 135–137
middle, rare tracks from in U.S.,
127–128
Noah's Raven, 108–109
ornithopods, 143–145
Otozoum tracks, 115–118
pterosaurs, 138–141

Kaup, Johann Jakob, 83
Kepler, Johannes, 106–107
Khodja Pil Ata, longest dinosaur
trackways, 134–137
Kielan-Jaworowska, Z., 124
King, Michael, 101–103
Knoebber, John, 164
Koestler, Arthur, 109
Korea
baby brontosaurs, 169–170
Jindong Formation, 167–169
pterosaur tracks in, 205–207
web-footed birds, 204–205
Krantz, Grover, 267
K-T boundary/layer, 221–224
Kuban, Glen, 184
Kurten, Bjorn, 225, 254

Land, first animals to walk on, 53–56
Land mines, 290–292
Lark Quarry, 201–204
Lawrence, Jerome, 182
Leakey, Louis, 156
Leakey, Mary, 237, 241–242, 244–245, 252
Lee, Robert, 182
Lemkow, Anna, 296
Leonardi, Dr. Piero, 77
Leonardi, Father Giuseppe, 30–31, 77, 105,
121–122, 192, 213, 227
Lewin, Roger, 242
Life, emergence of, 46–47
Light, tracks of, 4–5
Lingham-Soliar, Theagarten, 254
Link, H. F., 83
Lizards, of the Triassic period, 95–96
Lizzie, 60
Locomotion
and ankle structure, 91–93
beginnings in Precambrian and
Cambrian, 51–52
bipedal, first, 93–94
footedness, 125–126
hopping, 122–125
of horses, 228–229, 237–238
on the moon, 284–285
quadred v. biped, 25–26
speed of dinosaurs, 135–136, 207–211
walking, first animals on land, 53–54

Logan, William Bryant, 46, 59, 278–279
Look, Al, 208
Lorenz, Konrad, 65, 100
Lovelock, James, 57
Lucy, 14, 16, 27, 239–240, 242, 244
Lull, Richard Swan, 117
Lyell, Charles, 30, 83–84, 91, 109, 259

Maasai, 247
MacClary, John Stewart, 148
MacDonald, Jerry, 62–63
Madsen, Jim, 63
Mammals
  definition of, 65
  hopping during the Mesozoic era,
    122–124
  Ice Age, 254–255
  tracks, first known, 99–100
  tracks of early masquerading as
    dinosaurs, 226–227
  See also Protomammals
Mammoths, woolly, 250–252
Mantell, Gideon, 258–259
Mantell, Walter, 258
Marche, Jordan, 253
Margulis, Lynn, 48
Mars, tracks on, 287–290
Marsh, Othniel, 146, 253
Masao, Kato, 282
Mathematical proportions, in anatomy,
    106–108
McAllister, James, 196
McCrea, Richard, 164
McIntosh, Jack, 195
McKenzie, Alice, 260
McNamara, Ken, 40, 56,
    72
Medewar, Peter, 157
Megatracksites
  Dinosaur Ridge, 193–195, 197–198
  Moab, 132–134
Mesozoic Era, 79–80
  Cretaceous period. *See* Cretaceous
    period
  hopping mammals, 122–124
  Jurassic period. *See* Jurassic period
  Triassic period. *See* Triassic period
Meyer, Christian, 97, 135, 201, 212–213

Millipedes, giant, 57–58
Mind. *See* Consciousness
Mines, land, 290–292
Minotaur, 232–233
Mitchell, Edgar, 284
Moas, 258–262
Moody, Pliny, 108
Moon, tracks on the, 284–287
Morphogenesis, 48–52
Morphology
  of dinosaurs, 220
  dynamic of, 270–273
  spectrum of, 36
Morrison Formation, 145–146, 152
Murchie, Guy, 80
Music, predation presented as, 180–181
Muybridge, Eadward, 26
Mylodon, 253

Nakasato Dinosaur Center, 198
National Science Foundation, 195
Newark Basin, 110–111
New Mexico
  Early Permian footprints in, 62–63
  Purgatory trackway, 147–150
Newton, Isaac, 188
New Zealand, moas, 258–262
Nichols, Doug, 223
Nicosia, Umberto, 77
Nixon, Richard, 284
Noah's Raven, 108
Noosphere, 4, 6–8, 156, 186–187, 296–299.
    See also Consciousness
Nopsca, Baron Franz von, 84–85
Nova Scotia, Carboniferous period
    amphibians, 60–61

Olsen, John, 263, 266–267
Olsen, Paul, 96–97, 111, 132, 140
Organisms, growth of, 48–52
Ornithischians, 217–218
Ornithopods, 143–145, 162, 168–169,
    191–193
Ornstein, Robert, 93
Ostrom, John, 174, 195
Ouspensky, Peter, 28–29, 189, 276
Owen, Sir Richard, 59, 68, 83, 109

Padian, Kevin, 140
Paleontology
    conservation of sites, 199–201, 203–204
    Crawford Prize, 47
    expanded dimensions of, 28–29
    limitations of, 19
    and tracking, 29–31, 77–78
Paleozoic Era, 43–44
    animal life in the early, 53–54
    Carboniferous period. *See*
        Carboniferous period
    Devonian period. *See* Devonian period
    Permian period. *See* Permian period
Paley, William, 109
Palichnostratigraphy, 81
Palmistry, 13, 22–23, 88–89, 281–282
Pangaea, 80–81
Paul, Gregory, 203
Peabody, Frank, 85–86, 95
Peking Man, 155
Permian period
    Dimetrodon, 61–64
    protomammals, 64–67
    shape shifting in, 71–74
    spiders, 75–76
Peterson, William, 210
Petrified hands, mystery of, 82–87
Phenomenological approach to science,
    32, 49
Physiosphere, 4–6, 186
Piltdown Man, 155–156
Pleistocene epoch
    cave drawings and mammoths in,
        250–252
    giant wombats in Australia, 256–257
    mammals in, 254–255
    sloth tracks in, 252–253
Pliocene epoch
    hominids, tracking the early in,
        239–245
    horses in, 237–238
Polo, Marco, 129, 152, 196
Poppelbaum, Hermann,
    270
Popper, Karl, 1
Portugal
    Battle of Carenque, 199–201
    Lagosteiros Bay, 153–155

long trackways, 126--127, 200
Precambrian times, 45
Precambrian-to-Cambrian transition,
    47–52
Presbyornis, 230
Prosauropods, 96–97, 117–118
Protomammals, 44
    Dimetrodon, 61–64
    hairy footed?, 94
    Permian, 64–68
    posture and ankle structure, compared
        to archosaurs, 92
Psihoyos, Louis, 149, 164
Pterosaurs, 138–141, 196–197, 205–207
Purbeck fossil beds, 162–163
Purgatory valley, trackway in, 147–150
Pyle, Robert, 266
Pythagoras, 106–107

Quadrupeds, locomotion, 25–26

Raath, M. A., 124
Renders, Elsie, 237
Reptiles, 56–60, 80. *See also* Dinosaurs
Riddenoure, Betty Jo, 148
Rift valleys, 110, 239
Ritts, Bradley, 227
Rocks
    igneous, 45
    sedimentary, 45–47
Rodents, 35–38
Romer, Alfred Sherwood, 61
Rosenblatt, Roger, 290

Sacredness, of the earth, 292–295
Sagan, Carl, 59
Sahni, Ashok, 222
Sanderson, Ivan, 264
Santos, Vanda, 201
Sarjeant, Bill, 18, 30–31, 101–103, 160,
    164
Sasquatch, 262–267
Saurischians, 217
Sauropods, 117–118, 127, 138, 149–150,
    162, 168–172
Schad, Wolfgang
    hominid evolution, 268, 270
    mammals, holistic biology of, 10, 32–33

masculine/feminine bipolarity, 40
phenomenological perspective, 49
senses, 35–38
spiral forms, 66
Schadian perspective, 38, 48–49, 51, 71,
  113, 117, 212, 217. *See also* Holism
Schmidt, Harrison, 284
Schrodinger, Erwin, 296
Schulp, Anne, 212
Science
  and intuition, 3
  origin of the term, xiii
  phenomenological approach and
    objectivity, 32, 49
  and religion, 68–71, 109, 181–189
  and subjective zeal, xiii–xiv
  uncertainty and philosophy, 1–2
  unity v. provinciality, 81
Scopes, John, 182
Scott, David, 284
Seattle, Chief, 146, 292
Sedimentary rocks, 45–47
Seilacher, Dolf, 47–48, 50–51
Senses, from the Schadian perspective,
  33–35
Sexual dimorphism, 40–42, 87–89
Sheldrake, Rupert, 50, 270, 296
Shepard, Alan, 284
Shipton, Eric, 264
Simpson, George Gaylord, 228
Sloths, 252–253
Smith, Richard, 65
Soares, Mário, 200–201
Social behavior of dinosaurs, 143–145, 149
Soergel, Wolfgang, 85
Sophocles, 287
South Africa, Permian protomammal
  fossils, 64–65
Species, number of, 15
Spiders, Permian, 75–76
Spirochete, 48
Stegosaurus, 146–147
Sternberg, Charles, 163–164
Stimpson, George, 231–232
Stokes, William, 140
Strada, Gino, 291
Stratigraphy, 46
Stromatolites, 46–47

Suwa, Gen, 242, 244
Symmetry, recurrence of, 23–25

Takatanka-ohitika, 146
Tanzania, early hominids in the Laetoli
  area, 239–245
Taoists, 46
Tapirs, 227–228
Taxonomy, 18, 81
Tedford, Richard, 257
Teilhard de Chardin, Pierre, 4, 129,
  155–157, 183, 187, 296
Tertiary period, 225–226
  birds in, 226
  cloven-hooved ungulates, 232–235
  early horse tracks, 228–229
  Eocene epoch. *See* Eocene epoch
  Pliocene epoch. *See* Pliocene epoch
  tracks and dating rocks from, 226–227
Tetrapods
  Carboniferous period, 59–60
  defined, 15, 44
  Devonian period, 54–56
  Dimetrodon, 61–64
  and the Mars mission, 287
Texas
  brontosaurus tracks in, 170–171
  carnivore-brontosaurus attack,
    176–180
Theropods, 117–118, 130, 176–180
Thomas, David, 180–181, 196, 203
Thomas, Elizabeth Marshall, 12–13, 45–46
Thompson, d'Arcy, 72, 103
Thompson, William Irwin, 10, 48, 146,
  277, 296
Thulborn, Tony, 124, 202–203, 209
Thunder beings, dinosaurs as symbols of,
  146–147
Time, xi–xii, 2–3, 27–29, 277
Tippler, Frank, 157
Trace fossils, 47–52
Tracking
  modern v. aboriginal, 11–13
  and paleontology, 29–31, 77–78
  Ten Commandments of, 19
Tracks
  attempted theft of, 141–143

relationships with organism. *See*
Biological relationships of tracks and
organisms
changes in the cycle of evolution,
270–273
contemporary (with shoes), 282–283
dating of, 111, 226–227
elite, 19
ghost prints, 171–174
holistic approach to. *See* Holism;
Schadian perspective
hominid, 242–243
as individual signatures, 21–22
as jigsaw puzzles, 102–103
as language, xi–xii
of light and energy, 4–5
matching to trackmaker, 16–17
meaning of, 13–15
naming of, 17–19, 116–117
and nerve-sense animals, 38
number of, 15–16
on Mars, 287–290
on the moon, 284–287
recording of, 19–21
record of and fluctuations of sea level,
132–134
splitting and lumping, 81, 97
See also Trackways
Tracks, named
Ameghinichnus, 122
Anomoepus, 118–120
Aquatilavipes, 165
Archeornithipus, 165
Baropezia, 60
Brasilichnium, 121–122
Caririchnium leonardii, 192
Chelichnus, 67–68
Chirotherium barthi, 87
Chirotherium stortonense, 87
Cynodontipus, 94
Dimetropus, 62–63
Dinehichnus socialis, 143
Diplichnites, 53
Eubrontes giganteus, 113
Grallator, 113–115
Hatcherichnus, 152
Hoplichnus equus, 154
Hwangsanipes, 205

Hylonomus, 59
Iguanodontipus, 161
Jindongornipes, 165
Koreanaornis, 165
Laoporus, 68
Megalosauropus, 130–132, 135
Octopodichnus, 75
Otozoum, 97, 115–118
Paramphibius, 55
Pseudotetrasauropus, 96–97
Pteraichnus, 140–141
Saltasauropus, 124
Tetrapodosaurus borealis, 164
Tetrasauropus, 96–97
Tyrannosauripus petersoni, 210
Uhangrichnus, 205
Tracksites
Bolivia, Cal Orko, 212–215
Clayton Lake State Park, 196–197
conservation of, 199–201, 203–204,
245–247
Dinosaur National Monument, 98, 128
Dinosaur State Park, 114
Gateway, Colorado, 99–100
Khodja Pil Ata, 134–137
Korea, Jindong Formation, 167–169
Laetoli, Tanzania, 239–247, 294
Lark Quarry, 201–204
Newark Basin, New Jersey, 96, 110–111
New Mexico, 62–63, 147–150
Nova Scotia, 60–61
Portugal. *See* Portugal
Purbeck fossil beds, 162–163
See also Megatracksites
Trackways
footedness, 125–126
longest, 126–127, 135–137, 200
Tresise, Geoffrey, 87
Triassic period, 79–80
archosaurs, success in, 91–93
dinosaurs, origins of, 100–102
lizards in, 95–96
mammals in, 99–100
petrified hands, mystery of the, 82–87
prosauropods, giant, 96–97
records of and fossil preservation,
90–91
Triceratops, 215–217

Trilobites, 51–52
Turner, Alan, 255
Tuttle, Russell, 244
Twain, Mark, 88, 249, 253, 281
Tyrannosaurus rex, 209–212

Ungulates, 35–38, 232–235
Universe, evolution of. *See* Evolution
Utah, Dinosaur National Monument, 98, 128

van der Post, Laurens, 11, 16
Vernadsky, Vladimir, 157
Vialov, O. C., 256
Voigt, Friedrich, 82
Voltaire, 155
von Koenigswald, G. H. R., 263

Wade, Mary, 202–203
Walker, Captain Joseph, 266
Webb, David, 233
Wheeler, John, 295
White, Tim, 242, 244
Wilber, Ken, 2, 4, 296
Willard, Bradford, 55
Williams, W. L., 261
Willis, Della, 47
Willruth, Karl, 84
Wills, Leonard, 101

Yeti, 262–267
Young, John, 284, 286